Handbook of Palaeodemography

INED Population Studies

Volume 2

This book series is devoted to publications of international relevance in population studies and demography as promoted by the French National Institute for Demographic Studies (INED, Paris). As one social science among many, demography is interlinked with related disciplines such as sociology, anthropology, history or linguistics, and continuously explores its boundaries with neighbouring disciplines, ranging from epidemiology and biology to economics. The studies published in this series are based on solid empirical research and firm methodological foundations or research findings. Particular attention is paid to long-term and/or collaborative surveys.

The series provides international visibility to works of high academic standard, and promotes an internationally acknowledged style of demographic research, upheld by INED for more than half a century and rekindled in various fields such as the study of demographic situations around the world, the relationship between demographic conditions and development; international comparisons; migration, identities and territories; family studies; gender studies and sexuality; ageing, health and mortality; trajectories, mobility and social networks.

For further volumes:
http://www.springer.com/series/11579

Isabelle Séguy • Luc Buchet

Handbook of Palaeodemography

Contributions by
Daniel Courgeau and Henri Caussinus
Translated by Roger Depledge
Language editing: Catriona Dutreuilh

Springer

Isabelle Séguy
Unité Histoire et Population, INED
Paris, France

Luc Buchet
CEPAM, UMR 7264 (CNRS, UNS)
Nice, France

ISSN 2214-2452 ISSN 2214-2460 (electronic)
ISBN 978-3-319-01552-1 ISBN 978-3-319-01553-8 (eBook)
DOI 10.1007/978-3-319-01553-8
Springer Cham Heidelberg New York Dordrecht London

Library of Congress Control Number: 2013956741

Printed on acid-free paper

Springer is part of Springer Science+Business Media (www.springer.com)

Foreword

This handbook of palaeodemography would not have existed without many years of collaborative work between historian-demographers and anthropo-archaeologists[1] with the aim of comparing all the available data – demographic, anthropological, environmental, archaeological and historical – in order to acquire a clearer picture of the demographic features of historical populations.

Reconstituting the behaviour and demographic mechanisms of societies that have left no relevant written documents means going to places where demographers do not usually venture. Success depends on both the specialist skills of a multidisciplinary team and on sources of sufficient quantity and quality to test the methods on actual data rather than using artificial series created for that purpose.

However, before hoping to correlate data as diverse as archaeological remains, ethnological data and epidemic, climatic and demographic models, it is essential to have a clear understanding of our own sources. This requires designing specific methods that can be applied with tools appropriate to the demography of pre-statistical populations. This sort of approach necessarily involves a degree of controversy and error. It also requires a large measure of modesty. We are as yet far from an "integral palaeodemography" (Integrating Archaeological Demography, Paine, 1997) running from prehistory to the modern period. The reader will realise how much is yet to be done, given the difficulties we shall explain, before we can envisage a "return to discussion of the demographic history of planet Earth" (Bocquet-Appel 2005: 287). We do have the benefit of 25 years' research, but many problems remain to be solved, not least a re-examination of the quality of the sources used, be it for the dating of sites, of material remains or of skeletons.

A number of avenues are possible, but we shall be looking for sociodemographic information in the biological archives, especially the many thousands of preserved human skeletons. Not only do these provide arguments for reconstituting certain demographic behaviours in the past, but they also enrich our knowledge in other

[1] Research conducted as part of a multi-year programme supported by INED and CEPAM (UMR 7264, previously 6130, University of Nice Sophia Antipolis – CNRS).

fields, such as social science, economics and medicine. Bones retain the marks of the health and living conditions of past populations, in the form of bone pathologies, stress indicators and markers of activity, or of cultural practices such as deliberate cranial deformation. They reflect the environment of those populations and certain cultural and demographic behaviours. The skeleton is a sort of biological memory, of interest both to geneticists, who seek to monitor long-distance migration and the effects of population mixing, and to palaeo-epidemiologists looking for bacteria and viruses now dead or still active. This involuntary witness to the actions of everyday life, such as widely differing diets between groups, is sometimes the only remaining evidence of tragic past events – natural disasters, wars and massacres. The study of graves provides specific information about past societies that is not covered by any other historical source; it gives a place to social categories often "forgotten" by written sources, such as women and children. It also provides this information for long stretches of time, since there are human remains practically all over the world, and reveals health at different ages of life and in different socioeconomic contexts (hunter-gatherer, early farming and urbanisation, industrialisation, effects of colonisation and migration). Combined with other sources of information, particularly environmental and archaeological, skeletons are invaluable indicators for the social and human sciences.

After discussing the osteological evidence, the basis of palaeodemographic analysis as we have defined it, and examining how far it may provide demographic information, we will return in greater detail to the estimation of the sex and age of the exhumed skeletons. All palaeodemographic studies are based on the distribution of ages at death as observed in the archaeological population under study. However, despite some published claims, this question is far from settled and two opposing viewpoints still exist, although the controversy is less virulent now than in the 1980s. While some palaeodemographers continue to work from the determination of individual age at death, others have clearly demonstrated the need to first estimate the age-at-death structure of a set of buried skeletons before making any individual determination.[2] This approach, taking all the individuals into account, makes it possible to avoid a certain number of methodological and statistical pitfalls.

The exclusive use of this notion of "collective" age limited the scope of French-speaking palaeodemographers for demographic modelling. Apart from the "estimators" proposed by Jean-Pierre Bocquet-Appel and Claude Masset (see Chap. 5), which provide access to five demographic parameters, none of the tools of contemporary palaeodemography are directly applicable, without major biases, on the basis of a "collective" age indicator. It is to meet this need that new model life tables appropriate to pre-industrial populations have been devised.

[2] This position is apparently less isolated than before, according to the conclusions of the working group on age estimation in palaeodemography at the Max Planck Institute for Demographic Research, Rostock, Germany (Hoppa and Vaupel 2002b), which may help point English-speaking palaeodemographers towards new avenues of research.

This handbook has a simple purpose. It is a practical publication intended for anthropologists and archaeologists who wish to interpret the bone remains they study in demographic terms, without neglecting prospective aspects. Our intention is to remain accessible to a wider public in terms of the understanding and application of methods. We have chosen to present and propose methods that are currently "operational", leaving to one side the many theoretical proposals which are either biologically impossible or technically impractical to apply to archaeological material. In deliberately focusing on the practical aspects of palaeodemographic study, we do not rule out the exploration of other avenues that may lead to the development of other methods.

The field covered is deliberately restricted to the historical periods of Western Europe. The methods we propose are based on comparisons with recent populations, because, while biological drift of a few centuries may be corrected, it would be hazardous to suppose a biological continuum for *Homo sapiens* for more than 100,000 years. Furthermore, the scarcity of material and human remains from the more distant periods calls for a radical review of the problem (see the work of Jean-Pierre Bocquet-Appel), while avoiding extrapolations that might deviate from reality.

This is not the only handbook to propose new approaches in palaeodemography; a number of recent books[3] demonstrate the renewed vitality of the discipline. However, unlike other authors, who often address these problems in a highly theoretical manner, we have had the opportunity to test each of our proposed methods on site data from the Anthropolis database at CEPAM, comprising anthropological and archaeological information on a hundred and more populations from the classical and medieval periods and a total of several thousand skeletons.[4] This has enabled us to review and modify our tools[5] so that they conform more closely to archaeological realities and demographic constraints.

The time has come for new advances, for the discipline has not stood still while this handbook was being written. While the probability vector method has long been central to our work, guiding the adaptations and tools proposed in this handbook, we have also explored a quite different approach with help from colleagues in mathematics and statistics. These new proposals are opening up exciting new prospects; so we have included at the end of the handbook ("Prospective section") the initial findings of these exploratory methods that may ultimately become standard in palaeodemography.

Paris, France

Isabelle Séguy
Luc Buchet

---

[3] Saunders and Katzenberg (2000), Marquez and Hernandez (2001), Hoppa and Vaupel (2002a), and Bocquet-Appel (2008a, b).

[4] The Anthropolis database will shortly be accessible from the CEPAM website.

[5] This explains the discrepancies between our earliest publications and the more recent applications presented in this handbook.

Preface

At the Crossroads of Demography and Archaeology

What can we know of the demographics of early populations for which we have none of the standard sources of historical demography, such as parish records, nominal rolls and censuses? One idea that naturally comes to mind is to analyse the human remains excavated by archaeologists. But these bones must be made to talk and the results interpreted carefully. And that is precisely the purpose of this book which, while quite unusual for demographers, is also familiar. It is unusual in its description of the techniques for estimating age at death from the gradual closure of the cranial bones, the wear and tear on certain joints and the number of cementum rings deposited on tooth roots, all of which are far removed from the world of demography, and yet familiar, because the book establishes an operational framework for these estimations by making full use of the demographer's basic toolkit, starting with life tables. This is why palaeodemography – the demography of early populations – is such a vigorous hybrid discipline. While it boasts some fine discoveries, it is also punctuated by lively controversies that called for a proper treatise to describe recent progress and to outline future developments. That task has now been accomplished in this handbook produced by Isabelle Séguy and Luc Buchet, with help from many colleagues, not least Daniel Courgeau and Henri Caussinus.

Based on a meeting of minds between a demographer and an archaeologist, the book is also the fruit of long-standing cooperation between INED and CEPAM. This joint research unit of the CNRS and the University of Nice Sophia Antipolis is one of France's largest archaeological laboratories, with an unequalled collection of plant, animal and human remains, recorded in digital databases. More than 10 years ago, CEPAM, then headed by Frank Braemer, approached INED with the audacious idea of providing a permanent position for a demographer at CEPAM while seconding one of its own researchers to INED's "History and Populations" unit. With many joint publications to their name, the two scholars have more than fulfilled their contract. The archaeology-struck demographer found herself working alongside the demography-struck archaeologist, to the great benefit of both disciplines. This *Handbook of Palaeodemography* is the culmination of their innovative cooperation.

Outside the narrow circle of specialists, the idea of combining archaeology and demography was not an obvious one. I recall one meeting of INED's Scientific Council at which some members expressed doubts about supporting a major joint palaeodemography project between INED and CEPAM. Were we not moving rather too far from the central concerns of the institute? Fortunately I had attended the 8th *Journées Anthropologiques de Valbonne* at CEPAM headquarters (June 2003) and so could easily give practical examples of the synergies between the two disciplines. The Scientific Council was convinced and gave long-term backing to the joint project.

And yet, is it any wonder that demography and archaeology should work together, given the degree of openness of both disciplines? Archaeologists have a long tradition of calling upon a whole range of specialists, once called "auxiliaries", specialised in epigraphy, numismatics, geological stratigraphy, physical and chemical materials analysis, numerous techniques for dating material objects and animal and plant remains, and, most recently, geographic information systems. Demographers, for their part, focus on their core competence, that of demographic analysis, known as "formal demography" outside France. But they nonetheless engage in mutual exchange with a wide range of disciplines, including sociology, economics, geography, history, politics, philosophy and epidemiology. All of these fields are represented at INED, faithful to an interdisciplinary approach that is making its way in equivalent research centres elsewhere. Palaeodemography (which might just as well have been called "archaeodemography") is no more than demographic analysis for population historians who wish to use the resources of archaeology. Specifically, palaeodemography enables historians interested in population dynamics to exploit archaeological data in cases where no written archives exist. This is what defines palaeodemography, more than any specific time period: it covers all periods from prehistory to the twentieth century, including antiquity, the Middle Ages and the early modern period.

Palaeodemographers examine buried skeletons; so they naturally use some of the techniques of the forensic scientist. Our authors did consult the specialised literature of the forensic sciences, but make a clear distinction in terms of objectives: the forensic scientist seeks to identify the age of an individual, whereas the demographer is more interested in the age-sex structure of a population, in order to describe its general dynamics and specifically the probabilities of survival at given ages. There is a basic reason for this: repeated attempts to find a reliable biological indicator for age at death have failed. Individual variations severely limit any correlation between age estimated in this manner and actual age (these ages can be compared in the few cases where we have both human remains and their names). Coefficients are often below 0.5, even for the closure of the cranial sutures, one of the most frequently used indicators at present.

As a result, the authors opt for a collective, probabilistic strategy for estimating the distribution of ages at death. The question discussed at length in this handbook is a subject of much debate: How can we reconstruct a plausible distribution of ages at death for a buried population? A "buried" population is never directly representative of the "burying" population, i.e. the individuals normally destined to be interred in the burial ground or cemetery under study, who, in turn, are quite likely to differ significantly from the surrounding society, because of demographic biases (caused by migration) or social ones (such as the aristocratic membership of certain religious communities of pre-Revolutionary France or selective burial practices). This handbook alerts readers to the biases in representativeness that may arise at every stage, but does not fall into the converse trap of standardising the buried population at all costs: its specific features must not be averaged out, but recognised and properly accounted for when choosing and specifying the best-fitting model.

The authors thus advance by stages.

After determining the scope and limitations of the various bone-based estimation methods, they describe the range of models that can be used to reconstruct the demographic dynamics underlying the available data, for all their flaws and biases. One instrument well known to demographers working with poorly recorded societies is the set of model life tables published by Sully Ledermann from 1956 on: these tables are constructed from logistic relationships that empirically link the observed probabilities of dying at various ages. At a higher level, we have the "logit system" devised by William Brass in 1969, which consists of taking the life table of a known population as a reference standard in order to adjust the unknown table of the study population, using the variations in median age at birth and the relationship between child and adult mortality. Brass first established a standard for the mortality regime of Western Europe and then an "African Standard". Since then, bodies such as the OECD, United Nations and the INDEPTH network of demographic surveillance sites have produced a large number of regional standards. The authors of this handbook prefer to use as references the life tables they have constructed themselves, based on data from pre-industrial countries that have not completed their demographic transition. This enables them to extrapolate mortality in the earliest years of life, seldom accessible from the archaeological data because of the poor conservation of child skeletons. In short, the standardising or modelling method involves applying to ancient societies where data are lacking the mortality distribution of societies for which reliable records are available.

This method, in turn, must be based on estimates of the age distribution at death. The handbook moves on to address this point. The authors opt for the "probability vector" method proposed by Claude Masset in 1973. Although originally designed to estimate the age distribution of populations recorded in historical demography, the method is extended here to buried populations: the authors closely follow the path traced by Jean-Pierre Bocquet-Appel while introducing various corrections and variants.

This part of the handbook actually raises a disturbing question: How can we fill the gaps in our knowledge without unduly substituting our model for reality? Faced with obscure or incomprehensible evidence, we are tempted to force the unknown into the mould of the known, which is the very mechanism of bias itself, unless we exercise considerable caution – as our authors certainly do – to ensure that the model remains no more than a set of hypotheses that are both perfectible and refutable. In the social sciences as in the others, any model is reductionist in the sense that it simplifies and arranges reality, but this reductionism is productive if the truths that emerge from it exceed in both quantity and quality the details that have to be sacrificed. Rather than just filling in gaps by means of isolated estimates, the demographic models in this handbook use a coherent network of assumptions, a structure in the strictest sense of the word, to plausibly and verifiably link together the various parameters that define the dynamics of a population. Therein lies their strength.

But the transfer still calls for considerable caution and a constant awareness of the nature of these assumptions and their mode of construction. The aim is not to replace poorly supported archaeology or an inadequate timeline with a preconceived pattern. The authors are careful to specify that the demographic modelling will be all the more satisfactory where the osteological remains are properly dated and soundly supported by archaeological evidence. A further pre-caution is mentioned recurrently throughout the handbook: a model must never be applied mechanically but must take account, as far as possible, of the social and historical context, which may be influenced by migration, the presence of a garrison or religious institution, the selective use of cremation, the effects of epidemics, wars and so on; the chapter on archaeological examples provides a very judicious reminder of this point.

There is thus a tension in the handbook between two necessities: to construct a demographic model enabling the archaeologist-demographer to capture demo-graphic dynamics despite highly fragmentary evidence and yet to contain that model within the bounds of probability. For example, there are thresholds of life expectancy or mean length of life below which no population is viable, just as the model may produce disparities between mortality risks at different ages that lie outside the range of possible values. As always in demography, the work of indirect estimation is both modest and risky, both tentative and daring; it must be bold while keeping within certain bounds to avoid missing the target of the "plausible" and "demographically possible" – keywords in this handbook. One might say that estimates of age at death must remain "demo-compatible", to coin a word. Model-ling has thus a double effect: it releases researchers from their impotence by giving them access to unseen structures and it places reasonable limits on their ambitions. This handbook steers a course between the two necessities. In scientific terms, what could be healthier than this intellectual tug-of-war?

In the last part of this handbook, Isabelle Séguy and Luc Buchet hand over to two eminent statisticians, Daniel Courgeau and Henri Caussinus, who have generously contributed their skills to this endeavour. They both return to the "probability vector" method of age estimation illustrated by Masset and Bocquet-Appel. They place this method within the wider set of statistical proportional fitting procedures, which start from the margins of a table (row and column totals) to reconstruct each of its cells (for our purposes, the number of deceased individuals in each age group and at each historical period or "stage").

The aim here, once again, is to start from the known and fill in the unknown, except that Courgeau and Caussinus adopt a truly Bayesian approach to do so: the gaps are not filled in arbitrarily by a prior model but rather the posterior estimation of probabilities is improved by making the best use of pre-existing elements of certainty or likelihood, so that the reconstructed table of data differs as little as possible from the initial table. This discussion is necessarily a technical one, but it has already produced tangible results: the demographic estimations applied to certain illustrative populations throughout the handbook are substantially revised as a consequence.

These methodological innovations by Daniel Courgeau and Henri Caussinus will fulfil their promise once they have been converted into reference tools accessible to researchers with less exceptional statistical skills. Whatever the future application of these ideas, this final section has the great advantage of opening up new lines of inquiry. For Isabelle Séguy and Luc Buchet, their *Handbook of Palaeodemography* is not some definitive treatise designed to hand down canonical truths that are cast in stone. The subject is by its nature a complex and evolving one. The handbook is, as it were, a guide to a site in an advanced stage of excavation, where the first structures have been unearthed by such talented forerunners as Claude Masset and Jean-Pierre Bocquet-Appel, while the authors, with their sound knowledge of the literature and their own rich experience, have dug down further and classified the discoveries made so far. As for every excavation, it is clear that more exploration is needed. The authors and their collaborators may be justly proud of their work.

François Héran

Acknowledgements

Our thanks go first to François Héran, who during his 10 years as Director of INED provided enduring support for our research into the demography of archaeological populations. Since its foundation in 1945, INED has attached particular importance to historical research, alongside more "formal" demography. With his appetite for intellectual adventure, François Héran secured a place for this discipline within the institute by promoting long-term observation and multidisciplinary approaches. It is thanks to his unfailing confidence that the INED key project that would ultimately lead to this handbook first saw the light of day. We trust that the result, the fruit of joint work among anthropologists, archaeologists, demographers, historians, mathematicians and statisticians, meets his expectations for open-minded and inventive demography.

We also enjoyed the dynamic support of Frank Braemer, then Director of CEPAM, who saw in this cooperation with INED an opportunity to develop a research field previously explored without the help of demographic techniques. With his support for a discipline crucial to any archaeological approach, Frank Braemer placed humankind once more at the core of our research. We are grateful to him for hosting an INED researcher at CEPAM and for providing the programme with the necessary funding.

Our thanks also go to Chantal Cases, current Director of INED, and Martine Regert, current Director of CEPAM, for maintaining their support and confidence in our project.

This book would not be what it is today, after more than 10 years' research, without the invaluable help of statisticians on temporary contracts at INED, who, far from shrinking from the unusual nature of the data to be processed, dug deeply and with enthusiasm into the somewhat stony ground of palaeodemography. Magali Belaigues-Rossard deserves our special gratitude for designing the first applications. Her pioneering work laid the foundations for this book in the year 2000 and opened the way to subsequent studies.

We are also grateful to Marième Bouali, Eve-Line Boulle, Monica Ginnaio, Efi Markou and Stéphane Renard for handling data entry and setting up the bibliographic database and to Brahim Ahmedou, Paul Beurnier, Bertrand Buffière, Annie

Carré, Alain Confesson, Nadège Couvert, Luc di Benedetto, Benoît Haudidier, Nicolas Lannoy, Carole Perraut and Magali Suchecki for their contributions at various stages, including statistical analysis.

Armelle Alduc-Le Bagousse, Véronique Gallien, Claude Masset and Didier Paillard gave us generous access to their archaeological and anthropological data. Also, many colleagues and friends offered countless helpful comments and words of encouragement. We are eternally thankful to Arnaud Bringé for his availability, patience and experience. Not least, Daniel Courgeau and Henri Caussinus exceeded our expectations by concluding the book with an innovative approach which prevents us from writing the word END on the last page of this book!

We are extremely grateful to our translator Roger Depledge who, with erudition and plentiful good humour – of a truly British variety! – took on the Herculean task of translating these pages from the original French. Our thanks also go to Catriona Dutreuilh for helping to bring our work to an international audience, under a project initiated and encouraged by the INED publishing department, and for her editing of the manuscript. Last but not least, we wish to thank Nicole Berthoux for her meticulous work on the figures.

Isabelle Séguy
Luc Buchet

Contents

Part II Reconstructing Demographic Parameters

Chapter 1
Introduction

So, What Is "Palaeodemography"?

1.1 Scope and Current Research Areas

Palaeodemography is the study of past populations that have left no or few written documents containing evidence of their demographic behaviour. Its objectives are similar to those of historical demography, the difference being more the nature of sources used than the periods of study. Whereas historical demography is based on written sources, palaeodemography[1] relies on material sources.

Palaeodemographic research has developed in various directions, all based on archaeological data. Key research strands include evaluating the number of human beings in a given territory from material remains (Trément 1999b) or urban remains (Storey 1997); searching for evidence of a demographic transition in the Neolithic on the basis of the spatial-temporal distribution of archaeological sites (Bocquet-Appel and Demars 2000; Bocquet-Appel et al. 2005); and estimating demographic parameters from anthropological data (estimated sex and age of human skeletons) (Acsádi and Nemeskéri 1970; Buchet 1978; Simon 1982; Blondiaux 1989; Gallien 1992; Guillon 1998; Theureau 1998; Buchet 1998: Paillard et al. 2006, to name but a few). These approaches use widely differing sources, each with their own specific biases. They are outlined briefly below.

1.1.1 *Spatial Distribution of Settlements and Population Estimates Based on Archaeological Remains*

Methods for estimating the size of a given population may use a variety of material remains. Some attempt to correlate the area of the excavated living space with population density according to technical-cultural stages defined by analogy with

[1] Also called "prehistoric demography" in the 1970s and 1980s, because most studies concerned pre- or proto-historical populations.

I. Séguy and L. Buchet, *Handbook of Palaeodemography*, INED Population Studies 2,
DOI 10.1007/978-3-319-01553-8_1,

current situations (e.g., Hassan's work (1981) on Palaeolithic hunter-gatherers); others have attempted to estimate the population of a given territory from food remains, assumed to be proportional to the size of population and length of occupation of the site (Zvelebil 1981), or from the resources available, under a Malthusian-type hypothesis that the ecosystem has a fundamental influence on settlement density (Hassan 1975; Lo Cascio 1999). All these studies are based on ethnological comparisons, models and simulations; their conclusions are to be treated with considerable caution. Little research is conducted in these areas today.

Other methods are based on archaeological data collected from systematic prospection at micro-regional scale (Leveau et al. 1999; Bintliff and Sbonias 1999; Van der Leeuw et al. 2003). This information is analysed by chronological subdivision in order to identify regional variations in settlement or, more exactly, the variation in the number of persons per square kilometre (Goudineau 1980; Leveau et al. 1993). Given the uneven state of conservation of remains, taphonomic problems (erosion, subsidence, etc.), the scarcity and difficulty of systematic prospection and the variety of occupation densities, the causes of error are many and varied.

Quite properly, researchers currently exhibit a certain prudence and prefer to restrict themselves to artefacts and their quantitative variations, without seeking to extrapolate numbers or migration patterns of human beings. Some archaeologists monitor developments in the density of sites per square kilometre (Bertoncello 1999; Trément 1999a, b; Verdin 1999); others use the appearance, development, continuation or decline of various types of habitat (Van Ossel 1992; Durand-Daste et al. 1998; Ouzoulias et al. 2001; Bats et al. 2003). In recent years, archaeologists have used spatial analysis methods taken from quantitative geography to move from mere observation of the number and distribution of settlements towards a dynamic approach to sites in their quantitative, qualitative and structural aspects. At present, topics for study are rather the interactions between societies and their environments (Berger et al. 2005), with analysis of the anthropic processes (socio-economic, cultural, political) or environmental factors (soil, hydrology, climate, etc.) that determine patterns of settlement and their changes over long periods, moving away somewhat from the "human entity".

Archaeological information restricted to the localization and radiocarbon dating of archaeological sites is the basis for the current research of Jean-Pierre Bocquet-Appel (Bocquet-Appel and Demars 2000; Bocquet-Appel et al. 2005; Bocquet-Appel and Bar-Yosef 2008). The series of published Palaeolithic sites form a spatial-temporal sample that has been processed using geostatistical techniques to reveal areas of presence and areas of absence. These findings are used to reconstitute the process of expansion and contraction of modern humans and Neanderthals in Palaeolithic Europe, a period marked by major shifts between ice ages and interglacial periods. The same data, used in ethno-archaeological research to determine population density on the basis of technical and climatic conditions, allow for a dynamic approach to the hunter-gatherer populations of the Upper Palaeolithic.

1.1.2 *Testing the Hypothesis of a Neolithic Demographic Transition*

A new avenue for research has been explored in recent years by palaeodemographers focusing on the possible change in demographic regime brought about by the shift from a subsistence economy (hunting-gathering-fishing) to an agro-pastoral economy. Starting from Ester Boserup's theory (1965) that population increase is a factor favouring the intensification of agricultural practices, palaeodemographers have suggested that the appearance of agriculture in the Neolithic was preceded or followed (opinions diverge on this point), by a sudden increase in population. They have sought to measure the demographic and health impact of this change (Cohen 1977; Jackes 1988; Cohen and Armelagos 1984; Armelagos 1990; Armelagos et al. 1991; Meiklejohn and Zvelebil 1991; Pennington 1996) and even its genetic consequences (Ammermann and Cavalli-Sforza 1984).

More recently, Jean-Pierre Bocquet-Appel and his fellow researchers (Bocquet-Appel 2002; Bocquet-Appel and Dubouloz 2003, 2004; Bocquet-Appel and Naji 2006; Bocquet-Appel and Bar-Yosef 2008) have looked for evidence of a Neolithic demographic transition in necropolis data, but distinguishing solely between adult and immature exhumed skeletons. According to Bocquet-Appel, the development of agriculture in the early Holocene contributed to population increase via higher fertility associated with a reduction in intervals between successive births, as evidenced by the presence of a larger number of immature skeletons in cemeteries dating from the beginnings of agriculture. This Neolithic demographic transition is, as it were, the mirror image of our current demographic transition characterised by a reduction in mortality and a subsequent fertility decline. Here, the sequence of events is reversed, with increased fertility followed by a rise in mortality, due no doubt to the appearance of new pathogens (because of close cohabitation with domestic animals), before a return to equilibrium at a very slow rate of population increase.

1.1.3 *The Sociodemographic Approach Based on Bone Remains*

The third major area of current palaeodemographic research is based on anthropological analysis of population samples represented by skeletons, and the use of new methods to estimate certain demographic parameters.

Like the historian-demographer who reconstitutes trends in sociodemographic behaviour from hearth taxes, censuses, church records of baptisms, weddings and funerals and many other sources that originally had nothing to do with demography, so the palaeodemographer seeks to reveal the demographic characteristics of past populations from their bone remains alone. The anthropological data taken from exhumed skeletons are of interest not only to particular fields of demography (such as age-sex composition, mortality, migration, variations in rate of natural increase) but also to other biological and sociocultural disciplines. It is the cross-referencing of these various sources of information that makes it possible to retrace the living conditions of past populations, according to their socioeconomic status, technical and cultural standards and health.

The bone material is taken from burial sites of widely varying size and period, from the earliest collective graves to modern cemeteries. Palaeodemographers are thus faced with a large number of biases – from the collection of skeletons to the estimation of sex and age at death – that vary according to archaeological context. But the quality of the sample and the accuracy of the estimated age at death[2] are crucial to the reliability of results. The margin of error associated with an age estimated on the basis of biological criteria which vary greatly from one individual to another, may have considerable repercussions for palaeodemographic analysis. Despite the pioneering work of Claude Masset, published from 1971, and Jean-Pierre Bocquet-Appel in 1977 (1977b), these repercussions have not always received the attention they deserve from some members of the palaeodemographic community.

While not ignoring other sources of archaeological information, our work focuses on anthropological data, where analysis is highly informative but requires rigorous methods and appropriate tools. In this handbook, palaeodemography refers to the study of demographic structures on the basis of human bone remains.

While this line of study remains a difficult endeavour, 40-odd years of critical research have firmly established its credentials.

1.2 Major Milestones in Palaeodemographic Research[3]

1.2.1 First Attempts

The first palaeodemographic analyses of archaeological skeletons go back to 1685, when Robert Le Prévost de Cocherel, with the aid of a master surgeon, attempted to discover the age and sex of twenty skeletons from a tumulus on his estate (Masset 1994). But this research remained anecdotal and was not really resumed until the early twentieth century. Using tables taken from the work of Thomas W. Todd and D. Lyon (1924, 1925), Earnest A. Hooton, in 1930, studied an archaeological population of indigenous people in New Mexico, and Henri Vallois in France, in 1937, began studies with palaeodemographic intentions on prehistoric and

[2] Errors in determining sex are of lesser impact than those concerning age because sex determination methods, if properly applied, give relatively certain results. Moreover, sex is often ignored in analyses of populations because of the difficulty of determining the sex of children.

[3] A more comprehensive study was published in French in *Annales de démographie historique* (Buchet and Séguy 2002).

historical populations in Europe. The results obtained were highly inaccurate, however, due to the very approximate estimates of the age and sex of the exhumed bodies and the inadequate representativeness of the samples observed.

1.2.2 Fascination for Palaeodemography

It was not until the 1960s that serious research in palaeodemography really began. A Hungarian team (Janos Nemeskéri, Gyula Acsádi and Eric Fugedi) devoted themselves exclusively to experimental research and attempted to reconstitute the demographic structure of a historic population on the basis of age-sex estimates of skeletons taken from the major medieval necropolises excavated in Hungary. To that end they paid particular attention to sex and age determination and established precise rules to eliminate operator subjectivity. Their work led in 1970 to the publication of the first "handbook" of palaeodemography: *History of Human Life Span and Mortality* (Acsádi and Nemeskéri 1970) in which the authors propose a method for reconstituting the size of a population group and its variations and for determining the processes of conquering or abandoning land. After its publication, the anthropological community became fascinated by palaeodemography, though two competing schools of research rapidly emerged.

North American palaeodemographers, using the existing demographic tools (UN 1956; Coale and Demeny 1966; Coale et al. 1983) or recently developed ones (Brass 1971, 1975), concentrated their efforts on developing life tables from cemetery data and using data from anthropological demography (Angel 1972; Weiss 1973; Swedlung and Armelagos 1976; Buikstra and Konigsberg 1985). However, they failed to see the weaknesses of the data on which they were building their models, notably regarding the accuracy of individual ages at death. Despite the warnings of a historian-demographer (Peterson 1975) and of anthropologists (Masset 1971, 1973b, 1976b, 1982; Bocquet-Appel 1977a), these "anthropo-palaeodemographical" models were used in numerous site studies. Combined with the new possibilities of computer processing, they were used as the basis for demographic simulations (Howell 1976; Howell and Lehotay 1978), with results that must be viewed with caution due to the unreliable nature of the input parameters (population structure, mortality distribution and age-specific fertility rate). The publication of the results from the Libben site (Lovejoy et al. 1977), the pride and joy of American palaeodemography, cruelly exposed their serious discrepancies with respect to the accumulated experience of anthropological and historical demography (Box 1.1).

In France, meanwhile, the vitality of palaeodemographic research was expressed in the work of Jean-Noël Biraben (1969, 1971), Claude Masset (from 1971) and Jean-Pierre Bocquet-Appel (from 1977b). The great milestone was Claude Masset's demonstration, beginning in 1971, that a palaeodemographic approach based directly on individual ages at death estimated from skeletons is inapplicable because these ages cannot be accurately calculated. Research then turned towards

Box 1.1 Bias Caused by Incomplete Demographic Observations

Starting from the individual estimated age at death for each of the 1,300 skeletons exhumed at the Libben site (Ohio, U.S., AD 800–1100), C. Owen Lovejoy et al. (1977) established a life table and then calculated life expectancy at birth, life expectancy at age 15, birth rate, fertility rate and average number of children per woman. They concluded that the low mortality observed among children and the high mortality among young adults were specific to ancient populations with little exposure to risks of infection. In 1982, a simulation run by Nancy Howell on the basis of these results clearly revealed anomalies in the mortality curve, which she attributed to biological behaviour and social patterns very different from those of the present day, or to errors in age determination.

Figure 1.1 shows that the mortality curves of a modern population (here Sweden, Wargentin's table, 1755; Dupâquier 1977) is close to that of present-day populations of hunter-gatherers (here the Bande Fulani of Senegal; Pison 1989). Despite the under-recording of child deaths and the biases caused by self-reported adult ages, the two series of probabilities of dying are undeniably similar. However, the profile of the mortality curve for the Libben site, based on osteological data, diverges so clearly from them that it is difficult to see this as a feature specific to mortality in earlier periods. This divergence reflects errors in the determination of the skeletons' sex and age at death that were recognised and corrected by the authors using another indicator (Meindl et al. 2008, p. 264).

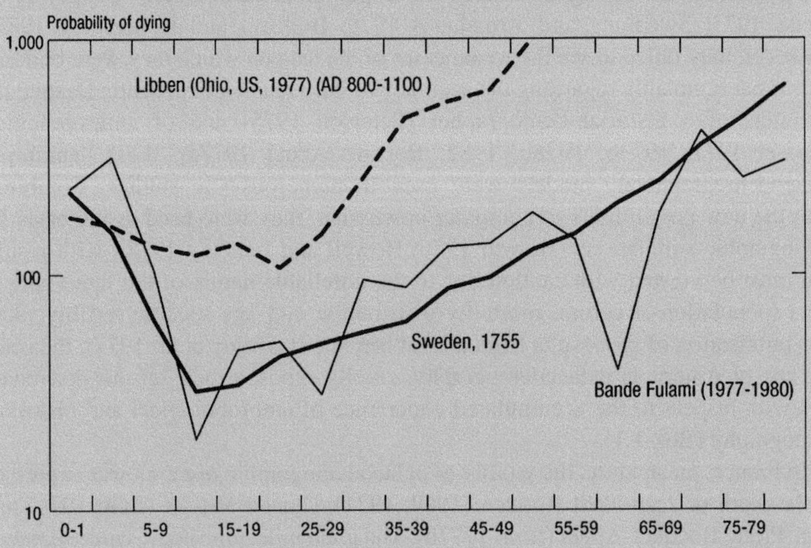

Fig. 1.1 Comparison of mortality curves taken from archaeological, ethnological and statistical data

other areas, such as defining certain demographic characteristics of a population from "palaeodemographic estimators" (Bocquet-Appel and Masset 1977, 1985, 1996), and developing the "probability vector" method (or ALK method) to estimate age distribution in a buried adult population (Masset 1982).

A rift was opening between French palaeodemographers, who attacked the age determination errors and the biases in archaeological data (Masset 1982; Bocquet-Appel 1985, 1986; Masset and Parzysz 1985), and North American palaeodemographers, who maintained that their atypical results were evidence of ancient mortality, the great unknown (Buikstra and Konigsberg 1985). There followed 10 years of violent controversy until Lyle W. Konigsberg and Susan R. Frankenburg acknowledged, in 1982, the two main causes of error[4] that had been pointed out by Jean-Pierre Bocquet-Appel and Claude Masset.

1.2.3 The French School and Research in the French-Speaking World Since 1992

The critical work of Jean-Pierre Bocquet-Appel and Claude Masset, comprising ingenious palliative solutions, gave French palaeodemography a new lease of life. The main strength of the French school was its remarkable similarity of methods, both for determining age and sex (same osteological indicator, same reference collection) and for getting the sources to "speak" (juvenility index, estimators). We will return to these methods in subsequent chapters. The total effect was positive: palaeodemographic studies gained in rigour and the extensive set of sites analysed using the same methods – and which can therefore be compared – now constitute a valuable palaeodemographic sample to underpin future research.

More recently, other research themes have emerged. The discovery of cemeteries directly linked to epidemics of plague encouraged anthropologists (Signoli et al. 1995; Signoli 1998; Bizot et al. 2005; Tzortzis 2009b) to turn their attention to "catastrophe graves", funeral groups characterised by simultaneous burial in a single structure of several individuals who had died from a single cause. These graves testify to the urgent burial of people who died in a sudden, unexpected event, such as a massacre, epidemic, natural disaster or severe famine. Anthropologists are particularly interested in those containing a large number of persons who died in a very short lapse of time (Bocquet-Appel and Arsuaga 1999). This research involves both palaeodemography, which some authors address with questionable methods (see Chap. 3) and palaeo-epidemiology, because advances in molecular biology now

[4] First, the influence of the reference population's age structure on the age distribution at death estimated for burial populations (using the ALK method), and second, the impossibility, whatever the indicator used, of obtaining an accurate estimate of age at death, given the statistically weak correlation between biological age and chronological age.

make it possible to identify certain pathogens responsible for major epidemics (Drancourt et al. 1998, 2007; Raoult et al. 2000; Raoult and Drancourt 2002).[5]

1.2.4 Research Outside the French-Speaking World Since 1992

By the early 1990s, although the criticisms expressed by Jean-Pierre Bocquet-Appel and Claude Masset (1982) had been taken on board by some palaeodemographers (Jackes 1992; Konigsberg and Frankenberg 1994; Konigsberg and Herrmann 2002), they were disregarded by others and many monographs continued to present life tables constructed on the basis of individual age estimates (Mensforth 1990; Benedictow 1996; Macchiarelli and Salvadi 1994; Coppa et al. 1995; Piontek et al. 1996; Alekseeva and Buzhilova 1997; Alesan et al. 1999). Other researchers, relying in some cases on data whose validity was beginning to be questioned, devoted themselves to wide-ranging overviews of the development of human longevity since the Pleistocene (Smith 1993; Helmuth 1999; Bogin and Smith 1996; Bogin 2001; Boldsen and Paine 1995), or to the demographic consequences of Neolithization (Armelagos 1990; Armelagos et al. 1991; Cohen and Armelagos 1984; Eshed et al. 2004; Jackes and Meiklejohn 2008).

More significant advances have been made in developing comparative techniques and using mathematical and statistical methods, both to determine the age at death of burial populations (Saunders and Katzenberg 1992; Konigsberg and Frankenberg 1992,[6] 1994; Konigsberg et al. 1997; Meindl and Russell 1998; Aykroyd et al. 1999; Konigsberg and Holman 1999; Jackes 2000; Holman et al. 2002; Boldsen et al. 2002; Gowland and Chamberlain 2002, 2005) and to test the validity of archaeological samples (Saunders and Hoppa 1993; Rogers and Saunders 1994; Lucy et al. 1995; Hoppa 1996; Paine and Harpending 1996; Freter 1997). Valuable theoretical contributions were developed by James W. Wood and his colleagues on this point (Wood et al. 1992b), opening up a wider discussion (Jackes 1993; Cohen 1994; Wood et al. 1994), while other researchers, using historical data, were able to directly compare the palaeodemographic results with those obtained from vital records for the same site (Molleson 1995; Scheuer and Bowman 1995; Ortega Muñoz 2004).

[5] The results published by the Rickettsia Unit of the Marseille medical school are controversial (Wood and DeWitte-Aviña 2003.

[6] Although the method recommended by Lovejoy et al. (1985a) was tending to replace that of Lyle W. Konigsberg and Susan R. Frankenburg (1992) for the estimation of average age at death and age distribution at death with minimum bias, many palaeodemographic studies (e.g., Storey and Hirth 1997; Nagaoka et al. 2006, who abandoned it in 2007) continued to use the method of Lovejoy et al. (1985a) despite its evident biases. In the prospective section, Chap. 12, the reader will find a critical presentation of Lyle W. Konigsberg and Susan R. Frankenburg's work which contributed in 1992 to a reorientation of palaeodemographic research.

Work begun earlier on life tables (Siven 1991a, b; Konigsberg and Frankenberg 2002) to remedy the biases identified in anthropological collections was pursued, while other researchers explored the potential of mathematical models using event history analysis or the maximum likelihood method (Gage 1990; Paine 1989; Milner et al. 2000, 2008; Konigsberg and Herrmann 2002; Wood et al. 2002; Muller et al. 2002). The early years of the new century saw a major shift in methodology, with a number of palaeodemographers clearly recommending that mortality by age be estimated directly "from the whole skeletal sample not classified by age". The conclusions of workshops held in 1999 and 2000 under the auspices of the Max Planck Institute for Demographic Research (Rostock, Germany) were published in 2002 in what amounted to a methodological protocol: *The Rostock Manifesto for Palaeodemography* (Hoppa and Vaupel 2002b, pp. 1–8).

This work has undoubtedly helped to revitalise palaeodemographic research. The publication of a number of methodological books testifies to the discipline's new-found vigour (Saunders and Katzenberg 2000; Marquez and Hernandez 2001; Hoppa and Vaupel 2002a; Buchet et al. 2006a; Bocquet-Appel 2008a, b; Chamberlain 2006).

Part I
The Data of Interest

Chapter 2
Epistemology of the Discipline

2.1 Representativeness of the Anthropological Sample

2.1.1 Biological Archives: Contribution and Limitations

The palaeodemographic approach to ancient populations as defined above is based on a single source: human bone remains,[1] which are a precious biological archive. It can only be used in those cases where the remains have survived. To adopt an effective demographic approach, the anthropologist attempts to determine the age at death and the sex of skeletons, while the archaeologist uses stratigraphic and typological observations to provide topo-chronological data for the site.

Palaeodemography depends, in part, on the excavation of burial sites that may be very different in size and nature, studied in the context of research excavations or rescue archaeology. In the best case, the site is exhaustively excavated, but usually the dig is only partial. The issue of representativeness arises in all anthropological or palaeodemographic studies. It must be clearly stated, with well-defined limits to the validity of the conclusions proposed. As early as 1957, Gyula Acsádi and János Nemeskéri defined the three preconditions they considered necessary for any demographic study based on osteological remains: the cemetery should be exhaustively excavated and an accurate topo-chronology established, the exhumed skeletons should be well conserved, and there should be biological or social links between the individuals. Since that date, there has been relatively little discussion of the issue[2]

[1] Consequently, the source is only available where funerary practices involve conservation, such as burial. Cremated remains are hard to use and some funerary practices do not preserve the bones (corpses abandoned, exposed, or immersed). In some cases the bones cannot be studied simply because of their state of conservation.

[2] Many studies compare the results obtained for a single site from cemetery data and historical demography data, with the occasional wide chronological discrepancy (Piontek and Weber 1990; Molleson and Cox 1993), without checking the respective representativeness of the sets of data. Some studies take care to compare results from the two sources for a single place and time, but

I. Séguy and L. Buchet, *Handbook of Palaeodemography*, INED Population Studies 2,
DOI 10.1007/978-3-319-01553-8_2,

(Walker et al. 1988; Wood et al. 1992b; Hoppa 1999), although in the real world of archaeology and palaeoanthropology all three of Acsádi and Nemeskéri's conditions are never met. The reason is that, from the very moment of burial, social factors (selection by age or social status) and biochemical factors (differential preservation of bones) distort the image of the population one may obtain. In the field, excavation of funeral sites is more often partial than exhaustive and the preservation of the skeletons varies in quality (Fig. 2.1).

In the laboratory, the inaccuracy inherent to the methods for estimating sex and age always leaves room for unwelcome uncertainties and margins of error. Between the buried population and the palaeodemographic sample, defined as the entire set of exhumed skeletons whose age and sex have been anthropologically determined, the loss of information is considerable. This loss must be identified and, if possible, quantified (see Antibes, Fig. 2.2 and Sect. 11.2.2).

2.1.2 From the World of the Dead to the World of the Living: The Question of Representativeness

What, therefore, does the exhumed sample represent? Does it show us the world of the living or does it only reflect the world of the dead? The question is now, as it has always been, a critical one.

The first stage is to determine whether the sample is representative of the buried population: do we have all the skeletons that there should be in the burial ground? Was there selection before burial, or were cemetery areas divided by age, sex or social status? Was the excavation partial or exhaustive? If the sample faithfully reflects the sequence of deaths at that time and place, it is possible, with the help of certain hypotheses, to estimate a number of demographic parameters characterising the population that used the site.

The second stage is to determine what population used the cemetery, i.e. the "burying" population. Was it a special group defined by marked social or migratory characteristics, or did it include all the residents in that place? Where there is more than one burial ground for a single community, was the population using the cemetery in question a representative subset of the whole population, or did it have specific characteristics (e.g. socioeconomic or religious)? This concept of "burying" population is an important one: it is what establishes the link between the deceased and the survivors, the world of the dead and the world of the living.

The third stage is to attempt to trace the path back to the living population. This hazardous exercise requires the adoption of preliminary hypotheses about fertility, mortality and migration and the use of demographic models. While our English-speaking colleagues are happy with archaeological and palaeodemographic

---

with no consideration of the problems of estimating the age at death of individual skeletons (Saunders et al. 1995; Saunders and Herring 1995; Ortega Muñoz 2003).

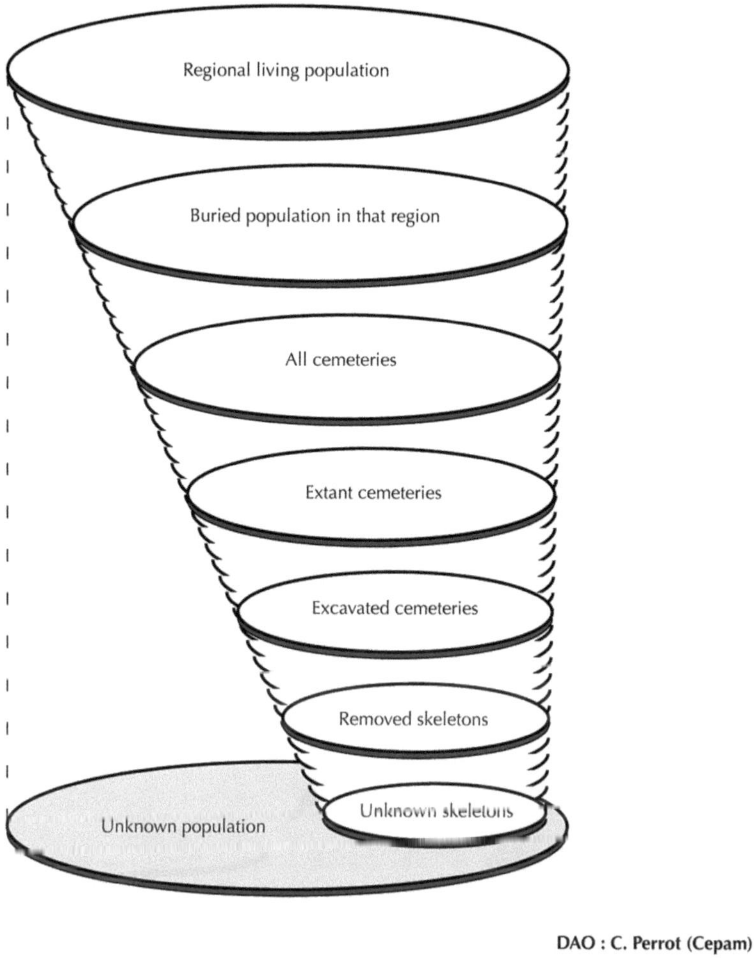

Fig. 2.1 Constituting a skeletal sample. *Interpretation*: Obtaining an idea of the population living around the site, assuming that burials were local, calls for knowledge of all the contemporary cemeteries likely to have been used by the residents of the region involved, whose settlement sites are not always archaeologically identified. This would require the cemeteries to be (a) preserved, (b) located by archaeologists, (c) excavated, and the skeletons to be (d) removed and (e) examined. Experience shows that these conditions are hardly ever all met, making it difficult for the anthropologist-palaeodemographer to determine the representativeness of the exhumed sample

modelling, the French-speaking community has remained somewhat reticent towards that approach, preferring to observe and compare skeletal samples.

It is no easy matter to determine the representativeness of an osteological sample. Some anthropologists have focused on the biases due to funerary recruitment and have attempted to estimate the proportion of the buried population that thus escapes the anthropologist's eye. However prudent this approach may be, it

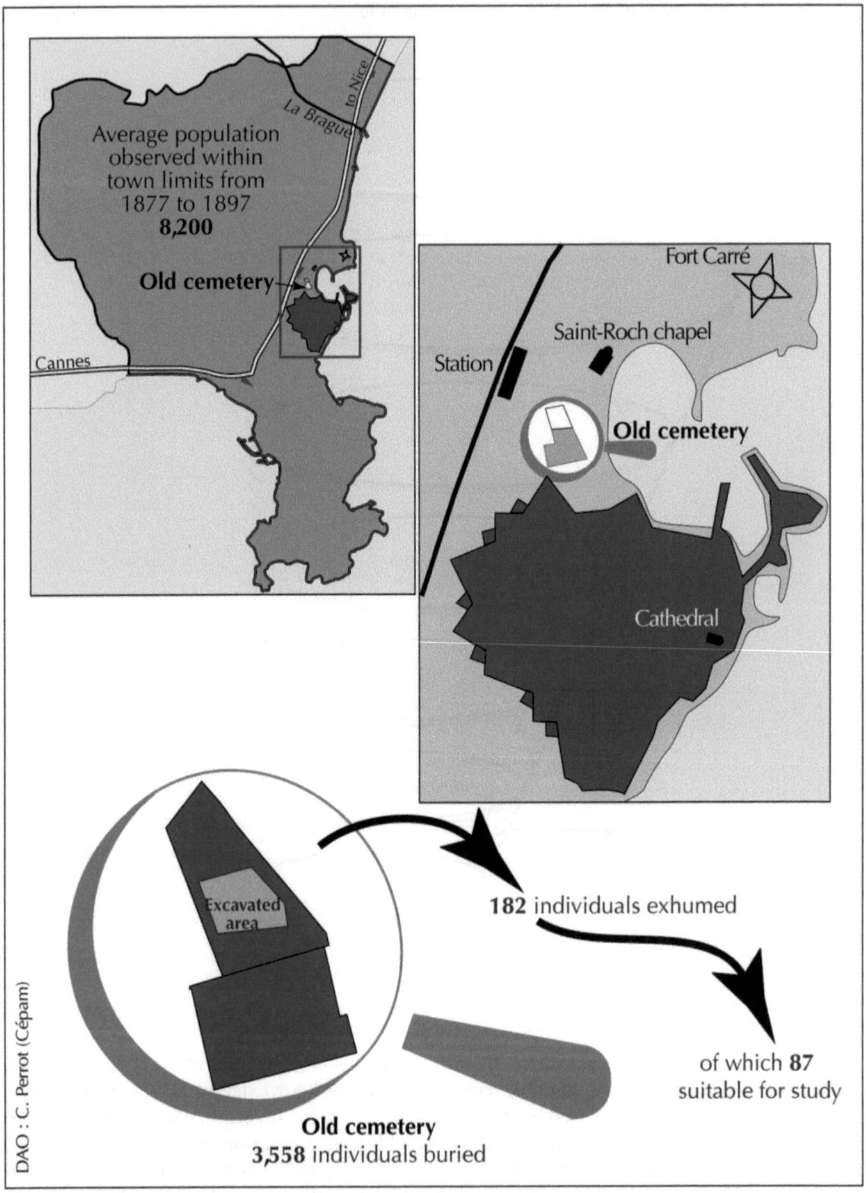

Fig. 2.2 Representativeness of a skeletal sample (Antibes site). *Interpretation*: Each stage in the process leading from the living population to the skeletal sample under study is marked by a loss of quantitative and qualitative information:

- From the "living population" (i.e. all the persons present on the territory under study during the period considered) to the "burying population", who may only be a fraction, not necessarily representative, of the population occupying the site;
- From the "burying population" (those using the cemetery under study) to the "deceased population", which is the result of selective mortality (by age, sex, socioeconomic status, or individual pathology) applied to the population exposed to the risk of death;
- From the "deceased population" to the "buried population" in the cemetery under study, which may only be part of the deceased population (particularly where the deceased are buried in their

sometimes leads to discovering biases where in fact there are none. For example, with a cemetery reserved for a religious community, it would be pointless to reconstruct from the buried population the children and adults of the opposite sex who are missing or very few within that community. With a military cemetery, do the exhumed skeletons provide an image of the villagers living nearby or rather of the sub-population using the cemetery, namely the garrison stationed there? Depending on the approach chosen, one may conclude that the palaeodemographic sample is representative or non-representative.

The approach is a difficult one, requiring the researcher to go beyond the usual standard frameworks (parish cemetery, "natural population") and compare, wherever possible, the "ground archives" with the historical archives in order to seek out the features specific to each burial site and, consequently, its recruitment. For example, there is a high risk that a change in the recruitment of the cemetery population (e.g. a burial ground originally reserved for a small group of socially privileged persons which was opened up to the entire community) may be wrongly interpreted as a change in health conditions.

We cannot, therefore, stress too strongly the precautions that must be taken when attempting to assess the living population from the osteological remains of a few of its members. Even in optimal archaeological conditions it is never certain that the proportion exhumed is significantly representative of all the components of the buried population. For example, a brief calculation shows that hardly 50 % of the expected population has been found in the rural medieval cemetery at Frénouville (Normandy, sixth–seventh centuries AD), even though the burial area has been fully excavated (Buchet 1978; Pilet 1980), because most of the children are missing. The palaeodemographic sample taken from the urban cemetery in Antibes (Alpes-Maritimes,

Fig. 2.2 (continued)
place of origin, or where there is more than one burial ground, or where funerary practices vary according to the deceased person's age or social category);
- From the "buried population" to the "exhumed population", which is a potentially biased sub-sample because the archaeological excavation was not extensive or the skeletons were preserved differentially by site or age (with fewer of the youngest and oldest) or sex (with more men), although the subject is still disputed;
- From the "exhumed population" to the "analysable population" (the palaeodemographic sample) from which the anthropologist attempts to determine sex and estimate age, and which comprises only the well-preserved skeletons (whose proportion depends on the quality of the excavation and storage conditions).

In this case (Antibes from 1877–1897, see the study in Sect. 11.2.2), the observed living population totals around 8,200 individuals, of whom three-quarters lived in the city centre (blue area), and one-eighth in the broad expanse contained within the city limits (grey zone). The remaining eighth is the population counted separately, corresponding to people temporarily residing in the city but domiciled elsewhere.

The burying population is more or less the living population, and a similar hypothesis is made for the overlap between the deceased population and the buried population.
As the cemetery was only partially excavated, only a small proportion of the individuals recorded as buried there were collected (5 %), from whom fewer than half made up the palaeodemographic sample.

nineteenth century), although it includes only 2.5 % of the population that was buried there, has turned out to be statistically representative of the age distribution of buried adults (for more detail, see the site studies in Chap. 11). It is consequently not so much the sampling ratio that matters as the sampling method. A random sample is more likely to represent all the components of a population than a sample taken from a burial selection (which is nonetheless representative of the selected population).

2.1.3 Problems with Small Samples

The Antibes case (see Chap. 11) exemplifies another recurring problem in palaeodemography: the small number of skeletons observed. Few sites are able to provide more than a hundred or so analysable skeletons, which then need to be classified by chronological phases of varying duration.

The many studies that seek to circumvent the difficulties of the discipline, and merely compare samples with each other, either chronologically or geographically, inevitably run into this problem, without necessarily realising its full extent. Any conclusions drawn from these comparisons lack a degree of statistical significance. The small number of individuals generally analysed may provide a deceptive picture and random variations are like mirages in the field of palaeodemography.

It is difficult, therefore, to identify a general trend from an insufficient number of skeletons, because it is masked by the dispersion of their individual characteristics. As a result, findings that may appear to reveal major changes in mortality from one site to another, or one period to another, in fact merely reflect random variations affecting the small number of skeletons observed.

The reliability of palaeodemographic findings depends closely, therefore, not only on the thoroughness of the method used but also on the sample size. This is why we believe it is important to specify the statistical limitations of the results obtained by applying validity tests and calculating confidence intervals.

For small samples, non-parametric tests[3] are to be preferred, because they do not assume that distributions are normal. They are more sensitive to the medium than the mean (ranking). Note, however, that these tests have less statistical power than parametric tests (for equal sample sizes).

2.1.4 The Migration Question

The survival of DNA in archaeological bone remains gives us a direct access to the genomes of past populations and makes it possible, among other things, to

[3] We may cite Mann–Whitney and Wilcoxon's tests, which can be used to compare two independent samples, and Krukskal-Wallis's test, which determines whether k populations are identical and whether at least one of them tends to be different.

reconstitute migration patterns. This approach is as yet limited owing to problems of contamination, conservation and recovery of DNA from the bones, and also to the cost of the analyses. Pending the information on large series promised by current research on DNA, the only way of identifying migrants in a skeletal sample is to observe each individual's phenotype by macro- and microscopic examination, supplemented by physico-chemical analysis of the bones collected. However, it must always be borne in mind that although the phenotype does reflect the genotype, any resemblances between individuals found in this way do not guarantee relatedness, just as relatedness does not always translate into resemblance.

The phenotypical differences between populations originate from the history of settlement under the combined effect of biological, social and environmental factors.[4] The weight of each factor is hard to discern: some people migrated, others transmitted ideas and material cultures, and others adopted them. It is this alternation between change and equilibrium, this permanent biological and cultural cross-breeding, that causes populations to be born, die or transform themselves and that underlies their homogeneity or heterogeneity at a point in time t of their existence. By analysing the morphological effects of these changes on skeletons, the anthropologist may hope to reveal migrations, but in order to conclude that a biological phenomenon, such as morphological heterogeneity, corresponds to a historical phenomenon, such as migration, anthropological data cannot be used alone; they must be cross-referenced against all the data available, both archaeological and historical.

This exercise is not risk-free and to base historical conclusions on osteological data is not a straightforward task (witness the misuse of anthropological analysis by certain ideologies). For that reason few anthropologists go there, preferring to assume that their populations are closed and to conclude that the observation of bone remains is pointless for the study of settlement patterns.

2.1.5 Is It Futile to Study the Demography of Archaeological Populations?

The pitfalls encountered by palaeodemographers when they attempt to describe the population to which an archaeological sample belongs have been clearly signposted. Yet the techniques for overcoming these obstacles with a satisfactory margin of confidence remain to be established. The picture of the living population as provided by a set of deceased persons supposed to represent them is distorted by a range of factors that are hard to identify and, therefore, difficult to correct.

Nevertheless, the palaeodemographic sample does, to some extent, reflect the mother-population. For some of that population's demographic features it does

[4] The influence of migration, and the effects due to small samples, are well described in Langaney 1988.

provide some meagre information, albeit obscured by funeral practices, and it translates what the living thought of themselves via their presentation of their dead and their ideas about death. We believe, therefore, that it is possible to cautiously move towards creating population models from archaeological and anthropological indicators.

2.2 Ideas About Age

2.2.1 Age, a Word with Many Meanings

Age as a quantitative, continuous, generally reliable fact has been one of the foundations of demography since the first "political arithmeticians" established the link between mortality and age more than three centuries ago.[5] The use of this variable in exploring and revealing socio-demographic phenomena, with respect to the age of the individuals who experience or initiate them, gives credence to the idea that age is the determinant of most demographic behaviour (Véron 1994). However, a concept as "ordinary" (Pressat 1979) as age, marking the time elapsed from an individual's birth until the demographic phenomenon analysed, should not conceal the various realities it covers for different speakers in different times and places.

In palaeodemography, age measures the time elapsed between birth and death (discerning events during the buried person's lifetime remains extremely difficult). It is not calculated from historical civil records or declarations of age (in, say, a funerary inscription), but rather estimated from biological indicators of growth for juveniles and of ageing for adults. It is therefore not a civil or chronological age measuring events occurring between two calendar dates but a biological age expressed variably in each individual within a well-defined developmental pattern. In other words, an individual's biological age necessarily falls within a range of estimates reflecting the biological diversity of the whole population. Use of biological age leads us to consider the idea of age with a certain "distance", to look beyond an *exact* age to a *probable* age, ascribing to this variable an unusual degree of relativity.

Neither is the age we work with the exact measurement of time elapsed between birth and death, but rather the reflection of the individual's place at a given time within their social environment. Hierarchy and the social segmentation of ages vary from one time and place to another. In truth, analysing demographic phenomena on the basis of such a relative variable involves a number of paradigm changes, which we venture to outline below.

[5] This section is based on a paper given to the AIDELF conference in Dakar, 2002 (Séguy and Buchet 2006b).

Fig. 2.3 The ages of man. Illumination from Jean Corbichon's translation (fourteenth century) of *De proprietatibus rerum,* Bartholomeus Anglicus (thirteenth century), © BnF

2.2.2 Civil Age, Biological Age and Social Age: An individual's Place in Society

Age is not only the measure of the number of years lived but also, and perhaps above all in the periods of interest to us, a state. For both adults and children, every age in life is characterised, in social, political, economic and legal terms, by the powers it confers or denies, which distinguish it from other ages. The breakdown of life into various stages is a universal phenomenon, but the stages are not the same from one culture to another. They correspond to individuals' perceptions of the continuities and discontinuities in their own lives (Haraven 2000).

The social structure of medieval France, for example, used a division of ages inherited from classical tradition and identified by Gregory of Tours, who distinguished seven[6] ages (Fig. 2.3). Three ages in childhood: *infantia* up to 7 years,

[6] The number 7 had a great symbolic value in Antiquity and throughout the Middle Ages.

pueritia from 7 to 14, and *adolescentia* from 14 to 21; and four ages in adulthood: *pueritia adolescens* up to 25 or 30 years, *juventus* up to 40 or 45, *senectus* up to 60 or so, and *senium* beyond 60 (Lett 1997).

This division has the merit of roughly corresponding to physiological observations: appearance of first permanent teeth, puberty, menopause, and old-age dependency. Educators, doctors and lawyers agreed that the stages of biological development marked the child's progress within society: from birth to the first deciduous teeth at about 2 or 3 years, which corresponded to the age of weaning[7]; from the age of reason, at about 7 or 8 years, and the first permanent teeth, to puberty, at about 14 or 15 years,[8] which marked the end of childhood and entry into the world of adults. In mediaeval France, the legal age of majority was 14 or 15 years for boys and 12 years for girls (Alduc-Le Bagousse 1994, p. 32). At that point adolescents had rights, such as that of pleading or testifying, and duties. This was also the average age at which boys entered apprenticeship or, in the aristocracy, began to bear arms; whereas for women, the age of majority and the age of marriage often coincided.[9]

In the eyes of their contemporaries, adolescents were adults; in the eyes of the anthropologist, they were still immature. And what of the demographer? By including them as "children", are we not overlooking the "adult" risks to which they were exposed, such as pregnancy and accidents of warfare?

Other ages raise problems for the palaeodemographer, particularly the youngest, since infants remained outside society until they were enrolled into the Christian community by the sacrament of baptism[10] (Lett 1995). The many newborns who died before being christened were often banished to the margins, literally, of the Christian cemetery (Treffort 1997; Séguy 1997; Séguy and Signoli 2008; Tzortzis and Séguy 2008; Séguy 2010).

2.2.3 *Towards Socially Significant Age Groups*

Clearly therefore, social age categories, like biological ages, do not strictly align with civil age groups. While for demographers age is regularly divided in a linear but rather artificial fashion, for anthropologists (in the broadest sense), historians and palaeodemographers, age corresponds to uneven segmentations (Fig. 2.4). Although it can be convenient to use predefined age groups, this may also mask serious breaks and discontinuities.

---

[7] Late weaning is attested by both written sources and anthropological analyses (Gallien 1992; Herrscher 2003).

[8] Age at which it is agreed that the girls of that period had their first periods (menarche) (Post 1973; Lett 1997).

[9] Many mediaeval literary sources illustrate the early age of marriage for girls throughout the Middle Ages, at least in aristocratic society.

[10] Infant baptism, just after birth, was far from general practice before the twelfth century, even in aristocratic families, who were the most Christianised (Alexandre-Bidon and Lett 1997).

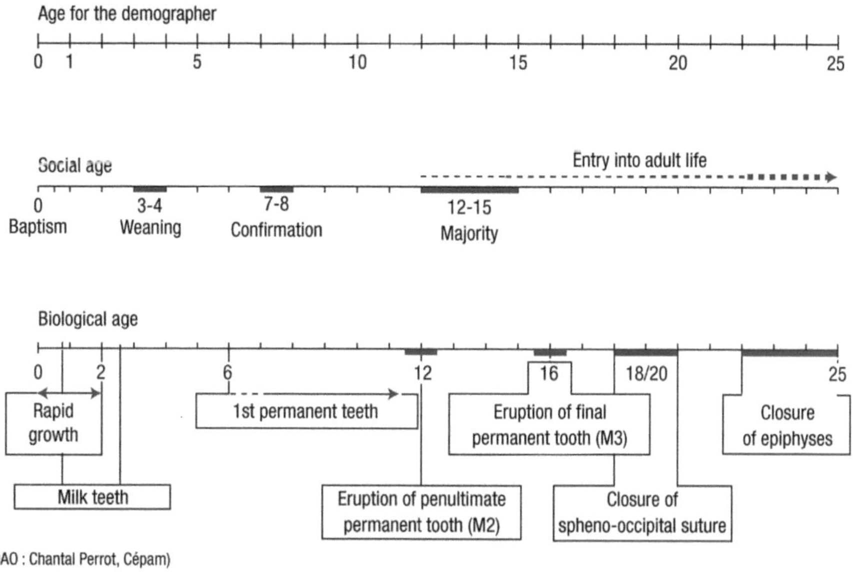

Fig. 2.4 Biological stages and social time

Demographic analysis must be able to detect any inflection points in the measurement of risk; and mortality, fertility and indeed migration,[11] must be measured with respect to the individuals concerned. Just as the scale chosen to observe demographic phenomena – short, medium or long term – determines the vision we may have of them (Véron 1994, p. 378), so age grouping by socio-biological criteria may well provide a new viewpoint for certain demographic phenomena, particularly for children (Séguy and Buchet 2006; Buchet and Séguy 2008).

2.3 Considerations Concerning Reference Populations

2.3.1 The Hypothesis of Biological Uniformity

All currently available methods for estimating age were developed on sets of *recent* skeletons whose sex and age at death are known; these sets are called "reference populations". The morphological criteria recognised to be discriminating

[11] Far from being sedentary, children in the Middle Ages often left their homes, or even their towns or villages, for events as varied as family recomposition after a separation or death of one of their parents; oblation (gift of a child to a monastery); marriage; starting work or apprenticeship; entering domestic service; fosterage (the aristocratic practice of entrusting an adolescent's education to another lord). The practice of entrusting infants to wet-nurses appears to have been only marginal at that time; this was not true of child slavery, which involved much larger-scale migration.

parameters in the reference population are then used on ancient series. These methods assume, therefore, that the biological parameters used are constant, or vary little, over time.

This is the hypothesis of biological uniformity (Howell 1976), invoking the non-variability of biological phenomena over time, that underpins all palaeodemographic research. It posits that (a) the biological processes relating to human mortality and fertility in the past are similar to those observed at the present day by demographer-anthropologists and that (b) biological development takes place within the same timeframes for all populations, irrespective of time and place.

These last two postulates have been the subject of lively discussion within the palaeodemographic community for the last 15 years or so.

Although we cannot accurately measure them or be sure of their linearity, growth processes do appear to vary over time and space. Over a short time period (a few generations), changes have been measured in a number of biological processes, such as in the order of eruption of teeth, in the mean age of puberty (Biraben 1982; De La Rochebrochard 2000) and in growth processes related to recent improvements in living conditions (Hoppa 2000b; Piontek et al. 2001). The use of standard correlation tables constructed from contemporary observations should therefore, in theory, be restricted to populations whose dietary, health and economic conditions are close to those used as models. In practice, until such day as more ancient reference materials become available, ages at death of all children and adults buried since the first Upper Palaeolithic sites are still determined on the basis of contemporary reference populations. For these reasons, in addition to the individual variations reflected in a mean age, there is a non-measurable margin of uncertainty between actual age at death and estimated age.

Palaeodemographers are aware of the fragility of the biological uniformity hypothesis underpinning their work and have pondered the possible consequences of a drift of biological markers over the centuries. If the growth or ageing processes do not occur at the same speed in the archaeological population as in the reference population, major discrepancies between estimated age and actual age may occur. However, although the possibility of a centuries-long drift of biological age indicators cannot be dismissed,[12] anthropologists have chosen to neglect it, for lack of any means to measure it,[13] while hoping that any discrepancies are not too great.

---

[12] For example, with respect to the drift in the closure of cranial sutures: (Masset 1982; Bocquet-Appel and Masset 1995; Simon 1983, 1987; Molleson and Cox 1993). More recently, Hoppa (2000b) has revealed morphological changes in the pubic symphysis between two chronologically distinct samples.

[13] According to Masset and Castro e Almeida, "This is as yet only a statistical link, which we are largely unable to interpret. . . To settle this point, we lack too many data that lie inaccessible in the cemeteries", (1990, p. 130).

2.3.2 *Towards a Pre-industrial Biological Standard?*

One solution is to come as close as possible to the standards of pre-industrial populations, albeit not ancient or medieval, who are largely non-urbanised, have little or no access to modern medicine and have barely, if at all, begun their demographic transition. This is how Bocquet-Appel (1977b) and Masset (1982) proceeded, by choosing collections of skulls from the late nineteenth century. This is also the solution adopted in this handbook.

However, although well-documented nineteenth-century bone remains are available, establishing an acceptable reference population remains a delicate task. Each usable series has its own characteristics and it is often hard to interpret the differences observed between series. When the reference collection intended for studying cranial sutures was being established (see Sect. 4.1), major differences in the age distribution by cranial closure were found between the Portuguese collections used by Claude Masset[14] and the Simon collection[15] in Geneva. Is this the effect of geographical distance (ethnic features), local health conditions (more deaths from tuberculosis and epidemics in Lisbon), difficulties in interpreting the closures (now impossible to verify since the Portuguese collection was destroyed by fire in 1978) or small sample size?[16]

For the purposes of estimating children's age at death, establishing a reference collection is even more problematic. Since there is no properly accessible osteological collection, the solution is to establish a reference collection on the basis of dental indicators. This task too is a delicate one because the only data available for the nineteenth century concern tooth emergence; for example, the major series published by Eduard Mühlreiter in 1920 gives the mean ages for tooth eruption among children in Vienna between 1870 and 1890. However, a comparative study of two populations, one from the Roman period and the other from the nineteenth century, concludes that there is no significant drift in the tooth mineralisation process (Saunders et al. 2000), so we can accept the hypothesis that observations of contemporary teeth may serve to construct a reference population. Care must be taken, however, to measure the degree of mineralisation and not emergence, which cannot be observed on dry bone.

2.3.3 *Influence of the Sex and Age Structure of the Reference Population*

Without an appropriate method of correction, the profile of the sex and age structure of the reference population determines that of the population under archaeological study, as Masset has shown. There are basically two causes for this, biological and

[14] See Chap. 4, Box 4.1, concerning the three Portuguese reference collections.

[15] Concerning the creation of this collection, Gemmerich Pfister 1999.

[16] Differences in age distributions by cranial closure may be observed between populations. They do not contradict the theory of biological uniformity, which holds that age distributions of stages of closure remain invariable.

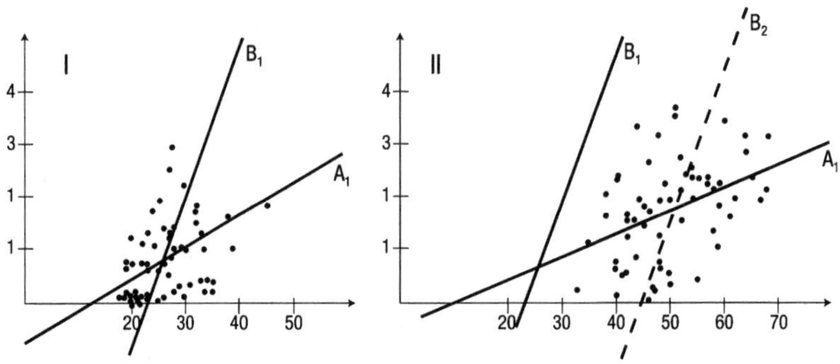

Fig. 2.5 Influence of structure of reference population on structure of study population. *Interpretation*: The regression lines A_1 and A_2 of sutures by age vary little between populations I and II (biological uniformity). The regression lines B_1 and B_2 of ages by suture, on the other hand, depend on the distribution by age at death of the selected reference population. Populations I and II have quite different age distributions. If regression B_1, based on the suture characteristics of population I, was used to estimate the age of the individuals in population II, the results obtained would be quite unrealistic. *Source*: Based on Masset 1982, p. 23

statistical. Whereas the biological indicator of age is a linear function of the subject's age within limits that vary little from one population to another, an individual's civil age is only partly a function of the biological indicator, because of considerable individual variability. In other words, while a given biological age corresponds to a stage of suture closure, that stage of suture closure does not correspond to a precise chronological age.

Furthermore, Fig. 2.5 clearly illustrates the difference in approach between the probability vector method (based on the B regressions) and those proposed by Jean-Pierre Bocquet-Appel and Jean-Noël Bacro (2008) and Henri Caussinus and Daniel Courgeau in the prospective part of this book, Chap. 13, which are based on the A regressions.

Choice of reference population alone may explain alleged differences between two archaeological sites. In Fig. 2.5, depending on whether the individuals included in the comparison collection are mostly young (reference population I) or older (reference population II), the estimated distribution by age at death in the archaeological series will reflect these differences.

Mathematically speaking, the matrix approach used to estimate a "collective" age at death takes no account of the structure of the reference population, since the frequencies are observed by age group.[17] However, in practice (Fig. 2.6), the

[17] This observation holds whatever the mode of calculation: frequencies of biological stages by age group (see Chap. 5) or frequencies of age groups for a given stage (prospective part, Chap. 13).

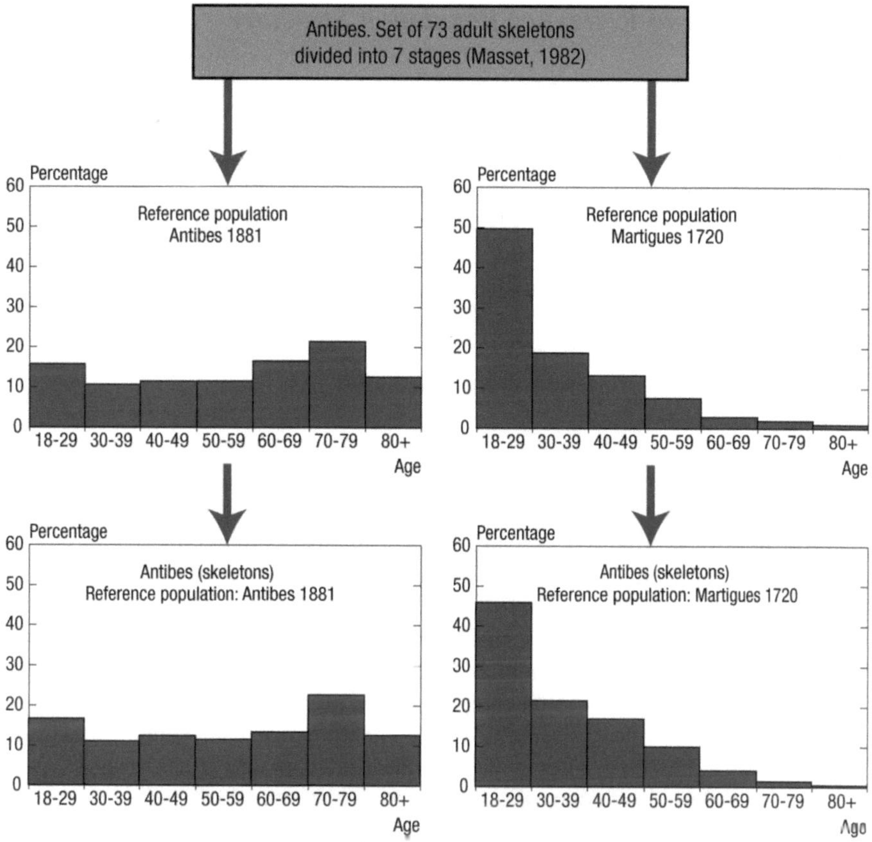

Fig. 2.6 Influence of reference population structure. *Interpretation*: The archaeological population taken from the dig at the old cemetery in Antibes (late nineteenth century) was adjusted to the age structure of deaths in the town of Antibes in 1881 (Buchet and Séguy 1999, 2003) and to that observed during the plague epidemic in the town of Martigues in 1720 (Signoli et al. 2005; Séguy et al. 2006b)

statistical distribution comes up against the dual problem of the small number of individuals in the comparison collections (see the inventory proposed by Usher 2002) and the even smaller number of archaeological skeletons to which these calculations are applied.

Similarly, the sex distribution in the reference population may have an effect when the osteological indicator of ageing evolves differently for men and women (such as the closure of cranial sutures, see below). In fact, palaeodemographers must remain alert to the nature of the reference collections that underpin their results.

2.3.4 *Choice of Reference Population Structure*

One solution for reducing this bias[18] would be to establish the largest possible comparison collection, with each age group having roughly the same number of individuals, and, if possible, a balanced sex ratio. This constraint can, to some extent, be integrated when establishing a comparison sample based on contemporary individuals (see Chap. 4), but cannot be introduced in dealing with older comparison collections, without running the risk of moving away from the biological model of pre-industrial populations. Alternative solutions have been considered.[19]

2.3.4.1 The Idea of a Standardised Population

In order to control the influence of the reference population and prevent it from being reproduced in the study population, Claude Masset (1982) thought of "standardising" his collection so as to produce a neutral reference population. To that end, he allocated to each age class the same number of individuals, proportional to the number of years included (the first had 12, the last 4, and all the others 10).[20] Where there were insufficient subjects in an age class, he randomly created fictional individuals with the same sutural characteristics as those in that age group. In their attempts to estimate the distribution of ages at death in their study populations, this "standardised" comparison collection provided anthropologists with a reference population that was both common to all (enabling inter-site comparisons) and assumed to be bias-free since it had no intrinsic structure of its own (the age distribution is linear, reflecting the underlying histogram, which is flat).

This proposal was adopted and improved by Christian Theureau in 1996 and 1998. Noting that the palaeodemographic results were influenced not only by the age distribution of the reference population but also by the stage distribution of the biological indicators, he established a new reference population with the same numbers in each age class and each stage. For that purpose, he collated the data from two comparison collections, in Hungary (Nemeskéri and Harsányi 1958; Acsádi and Nemeskéri 1970) and Portugal (Bocquet-Appel et al. 1978; Masset 1982). But only the individuals in the Portuguese collection are of known sex and

[18] The influence of the reference collection would not be eliminated, but would be reduced to a flattening of the estimated distribution.

[19] All the proposed solutions are based on the probability vector method, the only one used by French palaeodemographers until recent work by Jean-Pierre Bocquet-Appel (2008b) and Henri Caussinus and Daniel Courgeau in this handbook. Starting from the stage distribution observed in a given age group, the new proposed solution avoids the problems of a possible centuries-long drift and the influence of the structure of the reference population, since it depends only on the assumption of biological uniformity.

[20] Some authors have criticised this choice of methodology on the grounds that natural populations never present an equal probability of dying in all age classes (Buikstra and Konigsberg 1985; Konigsberg and Frankenberg 1992; cited by Schmitt 2002).

age. In so doing, the author falls into other statistical traps: first, ages and stages cannot be standardised simultaneously; one has to be chosen over the other (stages, say); and second, he bases his estimates on observed data (age at death for the Portuguese collection) and *estimated data* (individual age by osteological indicators, with no margin of error, for the Hungarian collection).

2.3.4.2 A Pre-industrial Population Structure

Use of a standardised reference population is not satisfactory because a population never has a uniform age structure. The distribution of ages at death is not flat and the proportion of old people is generally much higher than that of the young (except for the special case of military cemeteries). Consequently, the distribution of deaths obtained from a standardised reference population considerably overestimates the proportion of young adults, at the risk of leading to erroneous explanations linked to social conditions or health.

However, it is useful to maintain this unity of method and measurement that is the strength of anthropological research in the French-speaking world, and propose a population structure that is acceptable as a reference population for historical periods. We believe that it should come as close as possible to the characteristics of pre-industrial populations (see Chap. 4).

2.3.5 Other Insidious Biases in Current Reference Collections

In addition to the previous two problems (over-contemporaneous assumptions and the influence of age and sex structure), the currently available comparison collections suffer from other failings, no doubt considered to be minor, that nonetheless influence estimates by sex and age of buried populations.

The fact that these collections generally only contain a few tens or hundreds of individuals rules out any representative statistical coverage of the entire spectrum of possible realities. Collections with thousands of individuals may suffer from other biases, every bit as serious, such as attribution of age by biological criteria rather than civil records (unclaimed corpses and violent deaths, for example), serious illnesses unrepresentative of the general health of the population (in the case of anatomical pathology collections).

Most collections are made up of the unclaimed corpses of patients who died in hospital.[21] If health status and socio-economic category affect the chosen biological age criteria, then this type of recruitment may well impact the estimate of age at death.

[21] They also include the corpses of people who donated their bodies to science, and those taken from morgues, autopsy rooms and prisons (see Yann Ardagna's 2004).

2.3.6 Towards an Ideal Comparison Collection

Even if the ideal comparison collection remains utopian, attempting to define one is a way of identifying the failings and advantages of existing collections.

In absolute terms, an ideal collection would comprise several thousand individuals, with equal proportions of males and females of all ages (including extreme old age), properly registered in civil records. These individuals would be fully preserved (skeleton, skull and *postcranium*) to allow for a multi-criteria approach to age estimation, while maintaining a relative population homogeneity. Not least, to ensure a certain uniformity in growth and ageing processes, the population would be precisely situated in time and space as close as possible to the pre-industrial populations for which it was to act as yardstick.

Chapter 3
The Osteological Data

With the exception of DNA analysis (see below) – although sex identification is not a research priority in molecular biology[1] – each method uses a specific reference population made up of recent skeletons of known sex and age at death.

At present, bearing in mind what has been said about reference populations, it is not possible to combine more than one method in a single estimation. To estimate age using a multi-criteria approach, all selected indicators must be observed for each individual. Yet few osteological collections comprise statistically large enough numbers of skeletons for which both the entire set of bones and accurate indications of sex and age are available.[2] The collection proposed by Claude Masset (see below) would have met these criteria in all respects, and had the further advantage of representing a pre-industrial population. Unfortunately, it was destroyed by fire and most of it is no longer accessible to further measurement. There are other collections (Ardagna 2004) that might meet these requirements, and this point deserves further research.

In general, anthropologists must therefore establish as many comparison collections as the number of age indicators they have selected. More than one collection may sometimes be used for a single age indicator. One example is the various collections of contemporary teeth used by Ursula Wittwer-Bachoffen (Germany), Joël Blondiaux (northern France) and Claude Rücker (southern France), to estimate age from the growth rings in the dental cementum.

It must be remembered that all these methods, whatever the biological age indicators used, assume that the indicators remain constant, or very little, over time, under the principle of the biological uniformity of *Homo sapiens sapiens*. Estimation consequently entails a risk of error that cannot be measured so long as there is no properly documented ancient reference collection available.

---

[1] For a brief overview of research on ancient DNA, see Christine Keyser-Tracqui et al. 2002.

[2] Most published multivariate approaches are in practice based on large numbers of comparison collections of highly variable quality.

I. Séguy and L. Buchet, *Handbook of Palaeodemography*, INED Population Studies 2, DOI 10.1007/978-3-319-01553-8_3,

3.1 Identifying Sex

The methods used to determine sex vary both in the bone element studied and the examination procedure (morphoscopic, morphometric or biochemical).

3.1.1 Adults

Analysis of well-defined osteological characteristics – usually of the pelvis and skull – determines the sex of an adult skeleton with a satisfactory success rate (over 90 %). The quality of results depends upon both the indicator and the state of preservation of the skeleton. Experienced anthropologists can often determine the sex of a skeleton they are examining with a risk of error of less than 5 % if they can use more than one characteristic and the skeleton is well preserved. The fewer the indicators, the higher the risk of error.

The method developed by Jaroslav Bruzek (1991, 1992, 2002; Bruzek and Ferembach 1992; Bruzek et al. 1996, 2005), adapted by Pascal Murail (Murail et al. 1999, 2005; Bruzek and Murail 2006), using scopic or metric criteria of the hipbone, is usually recommended at present. For "extra-coxal" features, an initial overview was published by Denise Ferembach et al. in 1979 and a more recent one by Dominique Castex et al. in 1993. The effectiveness of certain methods (hipbone, skull and post-cranium) has been tested[3] on the Simon collection in Geneva and the collection at the Institut de Médecine Légale in Nice.[4] Although the hipbone is usually presented as the only indicator to be used, the results obtained from skull and femoral features are just as good.

In 1963–1964, Eugene Giles and Orville Elliot proposed discriminant functions to determine sex from human skulls. This method was not as successful as hoped because it can only be used on exceptionally well-preserved skulls, which are rare in archaeology.

Where the osteological remains are suitable, palaeogenetic techniques can be used.[5] Once the fossil DNA has been extracted, the sex of an adult can be determined from the size polymorphism between the alleles of the gene for amelogenin[6] on the X and Y chromosomes (Mannucci et al. 1994; Sullivan et al. 1993; Faerman et al. 1995).[7] However, in addition to the problems of DNA preservation in soil, this technique is difficult and costly, so cannot be applied on a routine basis (Hänni 1994; Orlando 2005).

---

[3] By Isabelle Aymard, in a dissertation in 2004. Contrary to a commonly expressed opinion, Aymard did not note any influence of age on the results (see also, Aymard et al. 2005).
See also Phillip L. Walker's work (2008).

[4] See Hilmi 2005.

[5] Fattorini et al. 1993; Larsen et al. 1996, 2000; Stone et al. 1996.

[6] Protein found in developing tooth enamel.

[7] This was the method used to identify the remains of Tsar Nicholas II's family (Gill et al. 1994; Akane et al. 1992).

3.1.2 *Juveniles*

In practice, it is impossible to determine a child's sex from their bones. Some features of the skull and hipbone give approximate results, but for all the attempts that have been made,[8] there is as yet no satisfactory method. Consequently, children are studied considering both sexes together. As in the case of adults, and with the same reservations, genetic markers may be analysed.

3.2 Estimating Age

In the absence of civil records, estimating the age of individuals, alive or dead, is not an easy task, and where research data are incomplete, demographers have to use biological indicators of age.[9] Growth and ageing are revealed by a succession of morphological changes occurring at precise times within a relatively circumscribed period. From observation it is possible to define what point an individual has reached in their growth or ageing and thereby deduce their age.

Where the subject of study is bone remains, palaeo-anthropologists and forensic scientists use the same methods to determine biological age at the time of death, with certain differences which are not without importance in palaeodemography. Forensic scientists work in almost all cases with isolated corpses whose age they attempt to estimate as accurately as possible, but with no concern for the demographic implications. They may combine a number of techniques and trust their judgement (past experience) without being required, as are palaeodemographers, to standardise their observations in order to guarantee the neutrality of the observer and the reproducibility of the estimate. Not least, in forensic science, the margin of error inherent in the individual approach is of little importance, whereas it has considerable effects in palaeodemography, although anthropologists with forensic training often tend to "forget" the margins of error that apply to their estimates.

Age at death estimations of buried individuals are not based on the same criteria for children, i.e., individuals who has not completed their growth, and adults, whose skeletons bear the marks of more or less advanced biological ageing. Within a given social context, biometric development differs little from one child to another. However, it can vary where health conditions affect growth (malnutrition, childhood diseases, epidemics, etc.). Conversely, adult age indicators are based on criteria of biological ageing such as changes in bone tissue and joint surfaces,

---

[8] Boucher 1957; Vito (de) and Saunders 1990; Fazekas and Kósa 1978; Hunt 1990; Majó 1992, 1997; Majó et al. 1993; Mittler and Sheridan 1992; Schutkowski 1986, 1987, 1989, 1993; Sundick 1978; Weaver 1980.

[9] For example, the work of Nicholas Townsend and Eugene A. Hammel (1990), who have also shown that in some conditions children's biological age could be more reliable than their declared "chronological" age.

Table 3.1 Chronological age and biological age

Indicator	Men	Women	Both sexes	References
Ectocranial sutures	0.59	0.34	0.56	Meindl and Lovejoy (1985)
	0.57	0.53	–	Buchet, Séguy, in the present book
Endocranial sutures	0.59	0.35	–	Bocquet-Appel and Masset (1982)
	0.51	0.35	–	Acsádi and Nemeskéri (1970)
Neck of humerus	0.44	0.34	–	Bocquet-Appel and Masset (1982)
Neck of femur	0.56	0.58	–	Bocquet-Appel and Masset (1982)
Pubic symphysis	0.37	0.68	0.36	McKern and Stewart (1957), Gilbert and McKern (1973)
	0.84	0.69	0.78	Meindl et al. (1985a)
Auricular surface	–	–	0.60	Bedford et al. (1993)
	0.55	0.63	–	Falys et al. (2006)
	–	–	0.72	Meindl et al. (1985a)
Dental cementum	0.96	0.93	–	Wittwer-Backofen and Buba (2002)
		–	0.88	Gabard et al. (2007)

Disclaimer. The reported correlations reported between chronological age and the age indicators used are the sole responsibility of the authors

development of teeth and periodontium, and the closure[10] of the neurocranial bones.[11] Their great variability from one individual to another makes the choice of indicator a difficult one and considerably increases the margin of error in an estimate. None of the currently known age indicators has a high statistical correlation with chronological age, and a 100 years of scholarly debate have not brought agreement among palaeodemographers on the choice of the best age indicator (Table 3.1).

3.2.1 *Juveniles*

For children, estimating age is relatively easy. Human growth involves a succession of morphological changes in bones and teeth that occur at precise times within a short period. It is therefore possible to determine at what point in the growth process the child died and thus deduce their age (always bearing in mind any variations due to health conditions).[12]

[10] Fusion of two bones.

[11] Part of the skull housing the brain and sense organs.

[12] Many methods are proposed. They are based on length of diaphyses (bone shafts) (Bass 1971; Fazekas and Kósa 1978; Johnston 1962; Stloukal and Hanáková 1978; Sundick 1978), stages of tooth eruption and mineralisation (Schour and Massler 1940, 1941; Ubelaker 1978, 1989; Moorrees et al. 1963a, b; Chaillet 2003; Buchet et al. 2005, 2006b), closure of epiphyses of long bones (methods collated by D. Ferembach et al. 1979) and the maturation of hand bones (Lalys 2002).

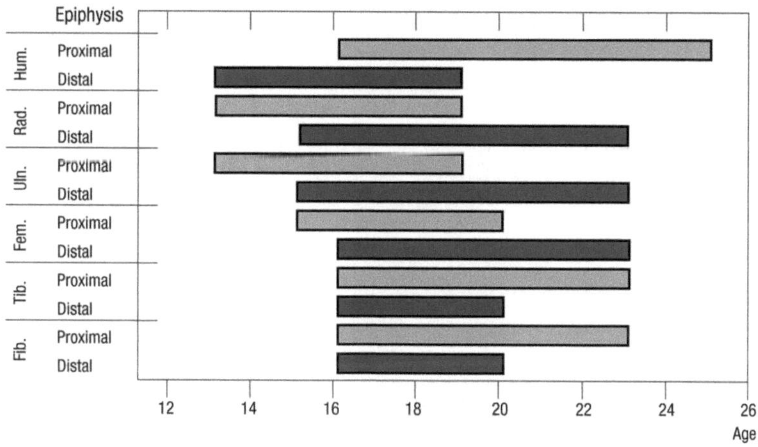

Fig. 3.1 Variation in epiphyseal fusion in major long bones. *Source*: Luc Buchet, after Brothwell 1972

The most frequently used biological age indicators are ossification of the skeleton, length of the long bones and tooth growth.

The state of ossification of the skeleton is determined by the appearance of points of ossification, and then the closure of the epiphyses[13] during the growth process. This method is appropriate for the first months of life. After the first 2 years, children's morphological development is subject to individual variation and the margin of error for each estimate is fairly wide (methods collated by Denise Ferembach et al. 1979; Johnston and Zimmer 1989). This variability in skeleton maturation is due to genetic factors (Tanner 1962; Garn and Rohmann 1966, Scheuer and Black 2000), diet, diseases and living conditions. Furthermore, according to Arthur. B. Lewis and Stanley M. Garn (1960) and Wilton M. Krogman (1962), emphyseal fusion may occur earlier in girls.

In anthropological terms, the ossification of the spheno-occipital synchondrosis (union between the sphenoid and occipital bones at the base of the skull), which occurs around age 19, marks the end of childhood and the entry into mature adult life.

As shown in Fig. 3.1, the length of the long bones correlates with children's age (Stloukal and Hanáková 1978; Sundick 1978; Alduc-Le Bagousse 1988; Adalian et al. 2002, 2006; Lalys 2002; Lalys et al. 2006). However, estimates based on this indicator are also affected by individual variability and the child's sex (a datum at present virtually inaccessible from the skeleton).

Tooth growth was first measured by observing eruption stages according to the tables of Isaac Schour and Maury Massler (1940) and then Schour and Massler (1941), which were long accepted as authoritative. The wider use of x-ray examinations,

[13] Ends of long bones, not fused in immature subjects and fused in adults.

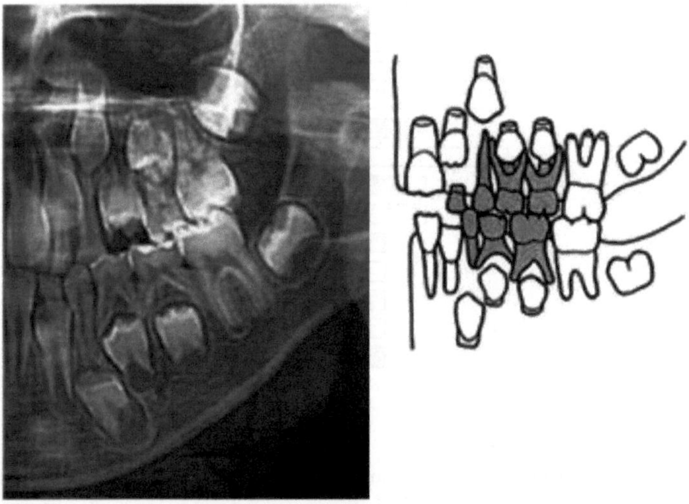

Fig. 3.2 Estimating a child's age from dental mineralisation observed on a skeleton and an x-ray. *Source*: drawing Luc Buchet after Ubelaker 1978; x-ray plate, CEPAM

particularly panoramic plates, made it possible to study successive stages in dental mineralisation without damaging the alveolar tissue. Anthropologists were provided with a large number of tables for estimating the age of juveniles from the degree of mineralisation of their deciduous and permanent teeth. Those most often used come from Ubelaker (1978, 1989) (Fig. 3.2) and Moorrees et al. (1963a, b) (see Box 3.1).

Box 3.1 Brief History of Methods for Estimating Children's Dental Age Used Until 2007

Anthropologists have long used tooth development as an indicator of children's age at death. The first method of quantifying tooth growth was to define a stage of tooth eruption. However, many factors of variability distort the correlation between age of eruption, osteological age and chronological age.[1] Among these factors, one may note that children from disadvantaged backgrounds present later development and dates of eruption (Heuzé 2004); that warm climates appear to bring forward tooth eruption; that the eruption of the permanent teeth of boys and last-born siblings occurs later than for girls; and that the latest eruption is observed in the European population.[2]

[1]Based on Chaillet 2003, p. 116: "The study of dental emergence shows a high percentage of individuals wrongly classified in terms of dental maturation".

[2]Analysis of causes of variation at www.dentaire.ups-tlse.fr/pedagogie/cours/eruption/texte.html.

(continued)

Box 3.1 (continued)

The first charts for attributing age from the stage of tooth development were drawn up by Schour and Massler in 1940. The sample used apparently comprised a mere 30 or so subjects suffering from chronic diseases (Halcrow et al. 2007). In 1949, Veikko O. Hurme calculated the mean age in years for the eruption of each tooth, with the standard deviation, for each sex and dental arch. However, his work has only rarely been used by anthropologists. In 1963, Coenraad F. A. Moorrees, Elizabeth A. Fanning and Edward E. Hunt Jr proposed a different approach, based on close observation of x-rays, noting the degree of mineralisation of each tooth. Because of the technical observation difficulties they encountered, their method could only be used for three deciduous teeth[3] and 10 permanent teeth (the maxillary incisors and eight mandibular teeth). The results obtained were better than those from the degree of eruption,[4] but the method still has certain disadvantages.[5] Anthropologists have the further difficulty of determining the tooth's stage of mineralisation (14 stages per tooth, prone to observer subjectivity). Furthermore, since the diagrams are not available for all teeth, the study of an ancient population can be complicated, due to the differential preservation of human remains in an archaeological setting.

The charts published by Douglas H. Ubelaker (1978, 1987, 1989) are based on those of Schour and Massler, modified to facilitate the study of Amerindian populations. Although he did not originally intend to propose a universal model, his work was used by European anthropologists, probably because it was published in "Recommandations pour déterminer l'âge et le sexe sur le squelette" (Ferembach et al. 1979) and approved by the Workshop of European Anthropologists (WEA 1980).

In Ubelaker's method, all the teeth, deciduous and permanent, are taken into account as a whole for the 21 growth stages described. The margins of error are set wider than those of Schour and Massler, but the confidence interval is still not known.

Having to take all the teeth into account raises problems in archaeology (for taphonomic[6] reasons, teeth are often missing) and the remains under study rarely coincide with one of the charts, which obliges researchers to "juggle" with more than one estimate (this problems arises with any sort of typology). In the early 1970s, Arto Demirjian (Demirjian et al. 1973) published dental maturity tables using eight criteria of calcification as bases

[3]Milk teeth.

[4]The method was tested before publication on 874 X-rays with a 90 % success rate for the lower jaw and 75 % for the upper jaw (with no error exceeding one stage) (Grøn 1962).

[5]The authors themselves list a number of causes of variation (Moorrees et al. 1963b, p. 1500).

[6]Taphonomy is the study of processes occurring after death (decomposition, diagenesis, etc.).

(continued)

Box 3.1 (continued)

for calculating a maturity index that can be transformed mathematically into a dental age.[7] The estimate of dental age was no longer based on a "snapshot" of the final phase of dental eruption, but included the entire mineralisation process for each tooth. It used a point score given to each tooth according to its stage of development. Dental age in the table corresponded to the sum of these scores for all teeth. This was an individual age and, allowing for the margin of error, the problem still arises of allocating that individual to a given age class where the estimate covers more than one.

In his thesis, defended in 2003, Nils Chaillet preferred a Bayesian approach for estimating the age of juveniles rather than linear regression techniques or Demirjian scores.

Chaillet also demonstrated the value of using multifactorial evidence,[8] under a Bayesian approach, from which he claims to derive age estimates from 4 to 18 years with a precision of ±12 months.[9] The main weakness of his proposal is its absence of discussion of the composition and role of the reference population (three populations in this case), which may considerably distort the estimates.[10] A second criticism concerns the difficulty with this method of attributing certain individuals to a precise age class. He merely observes that he correctly attributes 60 % of individuals to 5-year classes in a contemporary population sample (Chaillet 2003, p. 103).[11]

The advantage of scoring systems based on observing individual dental development, as proposed by Demirjian and Moorrees et al., is that the operator is not required to "stick" at all costs to a rigid typology. However, the reproducibility claimed by the authors is not fully guaranteed, particularly for the Moorrees et al. method, which is so complex that it introduces a dose of subjectivity (how can one be sure that one-half or one-quarter of a tooth has been reached if it is not complete?). For that reason our proposal is based on a simpler interpretation (see Sect. 4.2). The problems with the Demirjian method are of a different order: no proposals for a collective age, and a reference population (established in Quebec) inappropriate to other populations (Frucht et al. 2000; Chaillet 2003; Heuzé 2004). Practical reasons, too, may explain why these methods have not caught on with anthropologists, who, rather than a long examination, prefer to use Ubelaker's table.

---

[7]Demirjian used two selections of four teeth: M2, M1, PM2, PM1 and M2, PM2, PM1, I1.

[8]His databases are taken from Southern France: panoramic dental X-rays (Chaillet 2003), fetuses (Adalian et al. 2001) and hand bones (Lalys 2002).

[9]The margin of error he gives is close to that we obtain in our study for the same region, using the sole indicator of dental mineralisation.

[10]See Chap. 4 concerning the influence of the reference population.

[11]He does initiate a short but instructive discussion of the relevance of the division into 5-year age classes (Chaillet 2003, pp. 103–105)

3.2.2 Adults

The problems of estimating age at death for adults are far from fully resolved and are still the subject of scholarly dispute. As in the case of sex determination, a large number of methods have been devised. Mary Jackes (2000) and Aurore Schmitt (2002) propose good comparative approaches for those most commonly used.

While growth of the bones of the postcranial skeleton is completed by around age 25, closure of the skull bones continues beyond age 70, which explains the early interest taken in this feature by anthropologists seeking to ascribe an age to the skeletons they were studying. The first to propose a table of age estimates by cranial suture closure was Hermann Welcker in 1866, followed by Paul Broca in 1875 and, not least, in 1906 by Jakob Frédéric, whose work was used by Rudolf Martin in the first edition of his *Lehrbuch der Anthropologie*, dated 1928. In 1924 in America, T. Wingate Todd and David W. Lyon Jr also took an interest in suture closure as a biological phenomenon. Although their results are only poorly applicable to age estimation,[14] they were used in 1930 by Earnest Albert Hooton and then in 1937 by Henri-Victor Vallois, who thus succeeded in setting French palaeodemography on the wrong track until the 1970s.

With the renewal of interest in palaeodemographic research, other approaches were examined that used various biological age indicators. While making no claim to an exhaustive inventory, below are some of the methods that most attracted the attention of palaeodemographers.

3.2.2.1 Modifications of the Enamel Organ

The first precise method for defining age from a tooth was proposed by Gösta Gustafson in 1947, using six criteria. This method was adapted by Emanuel Vlcek and Lubor Mrklas in 1975, William R. Maples and Prudence M. Rice in 1979 and Dave Lucy et al. in 1995. It was popular for a time, mainly with forensic scientists, but was seldom used by anthropologists because of technical difficulties in applying it and not least because it was a destructive test (initially it involved using a thin layer taken from the tooth). Henri Lamendin (1988; Lamendin et al. 1990, 1992) proposed a simplified method using two of Gustafson's criteria (root dentin transparency and periodontal recession), but the results were not satisfactory for palaeodemographers (Foti et al. 2001; Megyesi et al. 2006). Each of Gustafson's

[14] Their aim was not to determine age from the sutures. They were examining whether among mammals there was a typically human pattern in the order and speed of suture closure, and how these might vary by sex and race. They concluded that there was a single pattern independent of sex and race. An age estimation table was constructed from their work, taking no account of the biases in the sample that reduced the age of the subjects studied, especially the women (it has since been shown that the closure pattern is not the same for the two sexes, see Masset 1971, 1982). One result was a massive over-estimation of deaths of young women in those palaeodemographic studies that used this method.

six criteria has continued to be studied separately but only one, cementum apposition, looks likely to become an acceptable age marker (see below).

Dental abrasion, for example, is easy to observe (Scott 1979; Walker et al. 1991) but a poor indicator of age at death because of the important role of diet. Similarly, although *secondary dentin* is laid down within the pulp cavity throughout life, it may be disturbed by factors such as periodontal recession.[15] *Root dentin transparency* (due to occlusion of the root's secondary tubules[16]), one of the two criteria adopted by Henri Lamendin, is of interest but hard to estimate without preparing the tooth (Pedinielli et al. 2004). The extent of *periodontal recession,* Lamendin's other criterion (Lamendin et al. 1990) is hard to assess on an archaeological skeleton, particularly when the teeth are discovered outside their sockets.

Furthermore, since the degree of periodontal recession is closely linked to dental hygiene (Jousset et al. 2006), the constitution of a reference population relevant to historical populations is a controversial task. Conservation in an archaeological environment also raises problems for observing *root resorption,* which in any case correlates poorly with age (Johanson 1971; Solheim and Kvaal 1993).

Whereas the methods proposed using these five criteria are poorly suited to providing a satisfactory age at death for palaeodemographers, there has been a growing interest in recent years in an age estimation method based on microscopic observation of *dental cementum.*

Cementum is a crucial part of the support tissue for the tooth and is laid down annually in concentric rings around the root. Examination of cross-sections of the roots with an optical microscope reveals the successive layers of cementum and can be used to deduce the subject's age. After initial work (Lipsinic et al. 1986) that was virtually ignored, it was mainly the research done by Peter Kagerer and Gisela Grupe in 2001, and Ursula Wittwer-Backofen, Jutta Gampe and James W. Vaupel in 2004, that attracted anthropologists' attention. Their observations tend to demonstrate a satisfactory correlation between this indicator and actual age.

In addition, Vitamin D intake, bone disorders and infectious diseases such as bone tuberculosis may cause anomalies in the successive accumulation of cementum layers.

The rings in dental cementum are therefore an unexplored marker for bone anomalies attributable to crises in health and subsistence[17] (Blondiaux et al. 2006). Some of these crises, if they affect an entire buried population, may be interpreted as demographic crises.

---

[15] Resorption of the tooth socket, especially with age.

[16] Dentinal tubules are small channels perpendicular to the tooth's main channel.

[17] Studies have shown a correlation between certain pathologies causing a deficiency in calcium metabolism and anomalies in the regularity of the cementum rings. Other factors may also be involved in laying down wide layers, mainly large hormonal variations, such as during pregnancy.

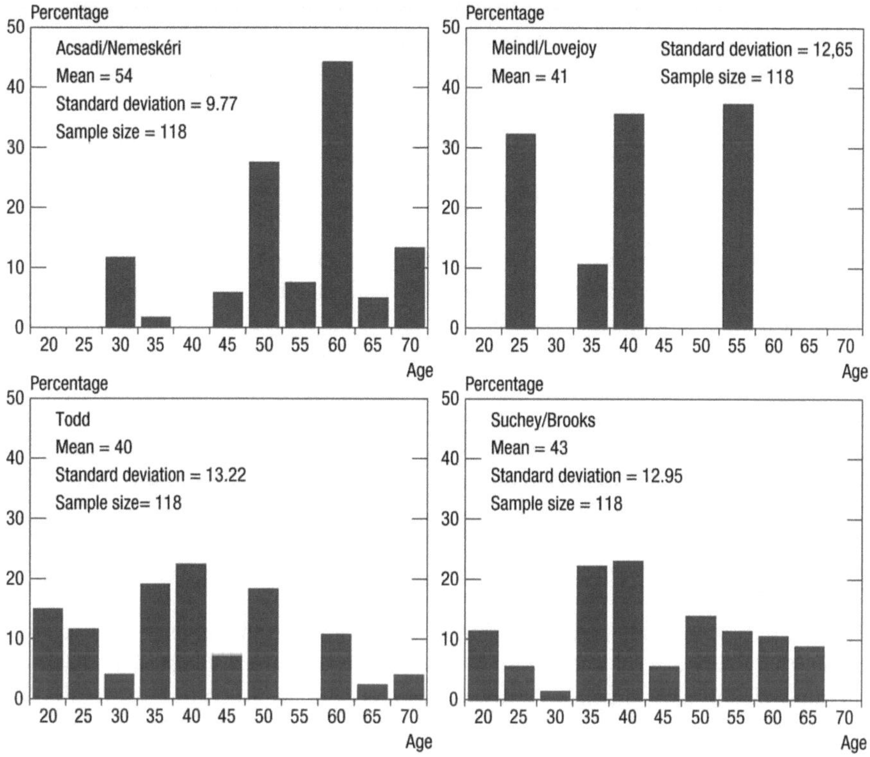

Fig. 3.3 Age distribution of a known sample by pubic symphysis. *Key*: The four methods used (Acsádi and Nemeskéri 1970; Meindl and Lovejoy 1989; Todd 1920; Brooks and Suchey 1990) produce four different distributions of age at death. *Source*: Luc Buchet, based on Jackes 2000

3.2.2.2 Degenerative Joint Disorders

Alterations of the pubic symphysis[18] drew the attention of early researchers such as Todd in 1920. His work was taken up by Thomas W. McKern and T. Dale Stewart in 1957 and regularly improved upon, up to and including publications by Darryl Katz and Judy M. Suchey in 1986 and Sheilagh T. Brooks and Judy M. Suchey in 1990.

In Europe, Hungarian anthropologists (Nemeskéri et al. 1960) proposed a similar approach. This indicator, particularly popular among American researchers (Meindl et al. 1985a) is only approximate, particularly for women over 40 (Fig. 3.3), perhaps because of a poor choice of reference population. It also raises practical problems because the pubic symphysis is a particularly fragile area and seldom preserved on archaeological skeletons.

The sacroiliac joint is often affected by early degenerative disorders (at ages 30–40) because the layer of cartilage covering the auricular surface is very thin

[18] Pubic symphysis: joint located between the two pubic bones.

(Paquin et al. 1983; Brunner et al. 1991), so has attracted interest for estimating age at death (Lovejoy et al. 1985b; Meindl et al. 1985a; Schmitt 2000, 2005). However, Tracy L. Rogers (1990) considers that the criteria defined by this method are only applicable to young adults.[19]

3.2.2.3 Alterations of Bone Tissue

Alteration of the bone trabeculae of the neck of the femur and neck of the humerus, often progressing to osteoporosis, can be observed by radiography; it was proposed as an age indicator by János Nemeskéri et al. in 1960 and is often associated with other criteria (Acsádi and Nemeskéri 1970; Bocquet-Appel et al. 1978; Walker and Lovejoy 1985). This bone loss affects everyone but varies from one individual to another (Robling and Stout 2000a, b) because it depends on the bone mass acquired by that individual during their growth period. This bone indicator, too, is poorly suited to archaeological material because of the alteration of bone tissue, especially cancellous bone, in soil.

3.2.2.4 Changes in Bone Structure

The number of osteons and the diameter of the Haversian canals[20] are two data points that change with age, and microscopic examination of histological cross-sections of long bone diaphyses provides details of bone structure. First to use this method was Ellis R. Kerley in 1965, followed by Marianne Bouvier and Douglas H. Ubelaker in 1977, Samuel D. Stout and Sarah J. Gehlert in 1980, Jean-Pierre Bocquet-Appel et al. in 1980, Ubelaker in 1987 and many others. These authors' results rarely agree and implementation of the method again raises the problem of bone destruction, not to mention the frequent post mortem deterioration of the bone structure by micro-organisms (Garland 1987). The method has only rarely been adopted in palaeodemographic studies.

3.2.2.5 Combining Indicators

Some authors have attempted to improve the results by simultaneously using more than one indicator. Gyula Acsádi and János Nemeskéri in 1970 proposed combining four indicators: closure of cranial sutures – observed on the endocranial surface –, alterations of the symphysis and the degree of osteoporosis observed in the neck of the humerus and femur. This method contains two errors. It assumes that each indicator correlates with age in an identical manner, which is false (see Table 3.1): the higher the indicator's correlation coefficient, the greater its relative weighting. And in their calculation the authors omitted to include the specific margins of error for each age indicator used.

---

[19] In 1975, Herbert Ullrich indeed recommended restricting its use to men only.

[20] Osteon (or Haversian system): basic structure of compact bone, consisting of concentric layers of bone tissue around the Haversian canal containing the bone's nerve and blood supplies.

Since each indicator correlates differently with age, Christian Theureau proposed (1996, 1998, 2006–2007) weighting coefficients.[21] In so doing, he corrected the first error, but the second remains. Furthermore, the poor state of conservation of skeletons taken from archaeological digs often deprives the observer of one or more indicators, so that the age may well be estimated from a different number of indicators from one individual to another.

3.2.2.6 Qualitative Estimation of Age

To avoid these biases, some authors prefer to stick with a qualitative estimation of age, using a number of indicators to divide adult skeletons into *adultus, maturus* and *senium* (Liptak 1980; Alesan et al. 1999; White and Folkens 1991, 2012) or "young adult", "young mature", "mature", "old mature" and "old" (Signoli 1998, 2006; Signoli et al. 2002; Ardagna 2004; Rigeade et al. 2006). Similarly, children may be divided into "babies", "infans I", "infans II" and "juvenile" (Alesan et al. 1999), with no agreement on age categories from one author to another.

This sort of approach is too often subjective, particularly when it uses pathological criteria such as arthritis, which correlate only very approximately with age. The assignation of a skeleton to a category may vary between anthropologists and tends to lead to age rounding. By increasing the number of biological age indicators, this method does not escape the problem of their poor correlation with actual age nor that of attraction to the mean. Nor do the categories proposed coincide with the age groups habitually used in palaeodemography, making comparison between sites impossible.

3.2.2.7 Closure of Cranial Sutures

None of the results obtained to date with these methods are convincing and, pending a good indicator – with its own good reference population – examination of the cranial sutures remains of interest.

Until the 1970s, anthropologists who wanted to attribute an age to a skeleton on the basis of the degree of cranial suture closure referred to the table which was drawn up by Paul Broca in 1875 and later revised in treatises of anthropology by authors such as Henri Vallois (1948) and Georges Olivier (1960). Estimation was highly subjective, given the margins of each segment (Fig. 3.4). It was Claude Masset (1982) who first attempted to rationalise observation so as to make suture closure a practical criterion for palaeodemography. Ten sectors are selected along the sagittal, coronal and lambdoid sutures and the degree of closure is assessed for each segment on a scale from 0 to 4 (Fig. 3.5). The mean of these figures is the "closure coefficient", a basic data point for the calculations that will be explained in later chapters.

[21] Theureau's proposed weighting for the age indicators (1998, p. 43): 0.34 for endocranial sutures; 0.25 for pubic symphysis; 0.18 for humerus neck and 0.23 for femur neck.

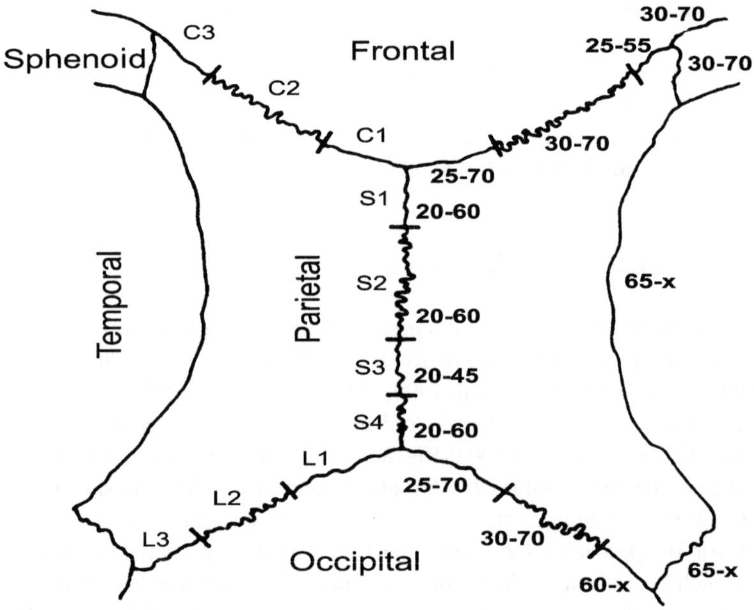

Fig. 3.4 Estimating the age of adults from the degree of cranial suture closure. N.B. Until the 1980s, the age estimate was directly calculated from macroscopic examination of the cranial sutures (men and women) using the guidelines shown. *Source*: diagram after Broca, revised by Olivier 1960

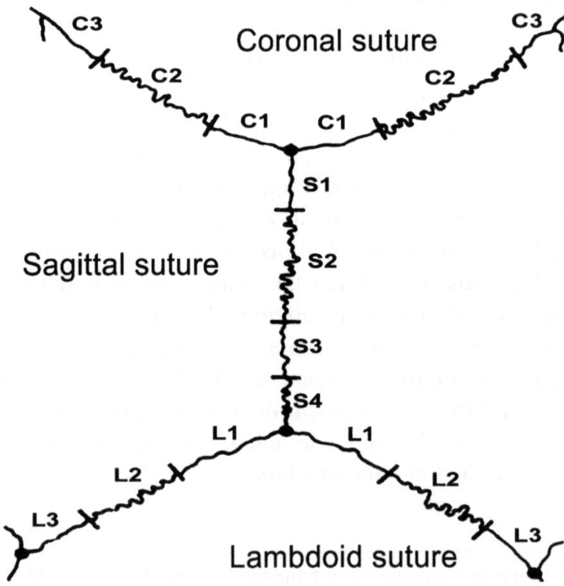

Fig. 3.5 New method for estimating the age of adults from the degree of cranial suture closure
N.B. Following Masset (1982), 10 sutural sectors are selected for calculating a cranial closure coefficient (for paired segments, the mean is taken for each pair). *Source*: drawing by Luc Buchet after Olivier 1960

3.3 Problems with the Methods for Determining Age and Sex

Critical analysis has focused primarily on the method based on cranial suture closure. The drawbacks and advantages of this method are therefore well known, making it a more dependable tool.

Six main causes of systematic error in estimating age were identified by Masset (1971, 1973a, b, 1982, 1995). They are responsible, in particular, for the inaccurate image of excess female mortality between ages 18 and 29 and the virtual absence of older people in palaeodemographic reconstructions. Although these criticisms concern sutures, they could equally well be directed at other indicators. The most frequent errors are summarised in Table 3.2 along with their effects and initial solutions proposed (Bocquet-Appel and Masset 1977; Masset 1982).

Alongside the biological errors, such as sexual dimorphism[22] and the age-sex composition of the reference population, there are the statistical errors, primarily the attraction of the mean[23] (Figs. 3.6 and 3.7). The regression that links biological condition with a subject's "real" age provides an estimate of individual age at death, but again with a margin of error. To ignore this margin, in the belief that any over-or under-estimates cancel out, leads to over-estimating individuals of mean age compared with those who are older and younger; this explains, for example, the absence of old people in earlier palaeodemographic studies based on individual ages (see Fig. 3.6: based on the mean age, the adult population will be distributed between ages 30 and 60, rather than ages 18 and 75. Similarly, if the margin of error associated with children's ages is not taken into account – Fig. 3.7 – they will be distributed between ages 6 and 14 rather than ages 0 and 18) (see also Box 3.2).

For the study of a single grave, by an archaeologist or forensic scientist, the estimation of age at death,[24] as usually proposed, is satisfactory. In palaeodemography, we deal with a set of estimates and our objective is to reconstitute the demographic structure. So how should the margin of error associated with each estimate be taken into account?

To summarise, because of the poor correlation between age estimated from bone indicators and real age, no method determines an individual age accurately enough to be used in a demographic approach to historical populations. To solve this problem, some researchers, in particular Masset and Bocquet-Appel in France, have taken another path: determining the age at death of a buried population via a probabilistic approach.

---

[22] The obliteration of cranial sutures is a slower process in women than in men, requiring the use of separate closure tables for each sex, otherwise the results of the female 18–29 age classes will be exaggerated.

[23] An individual age is always estimated within a range but the range is often ignored in palaeodemographic studies using a sample made up of a set of individual ages.

[24] Henceforth "individual age", as opposed to "collective age" as estimated for a set of individuals.

Table 3.2 Main causes of error in determining age and sex: consequences and responses provided by Bocquet-Appel and Masset

Type of pitfall	Description	Consequences	Responses provided by Bocquet-Appel and Masset
A. Taphonomic pitfall			
Differential conservation of bones in earth	The structure of the adult bone resists biochemical degradation better than the bones of children and old people[a]	Slight over-estimation of young adults and under-representation of deceased infants and old people	The estimator method avoids the errors both in adult age determination and the under-representation of infants
B. Biological pitfalls			
1. Sexual dimorphism	The obliteration of cranial sutures is a slower process in women than in men, especially in young adults (a phenomenon long unknown to anthropologists)	When a cranial suture table drawn up on the basis of a male sample is used for a female skeleton, this causes a major under-estimation of age at death and exaggerates the size of the 18–29 age class[b]	Masset drew up separate endocranial and ectocranial suture tables for each sex[c]
2. Long-term drift in suture closure	The closure of cranial sutures in historic subjects may have been more rapid than in the reference population	Under-estimation of age at death of buried individuals	No solution at present[d]
3. Age-sex composition of reference population	The mean age of skulls at the same suture closure stage is highly dependent on the age-sex structure of the reference population	Cemetery populations cannot be compared if the estimates of age and sex are not based on the same reference population (a precaution omitted by many anthropologists)	Use of a "standardised" reference population, in which all the age classes are of the same size (proportionally to their duration)
C. Statistical pitfalls			
1. Regression errors	The connection between age at death and age indicator is biological (ageing). The statistical connection from a set of skulls of a given age to the corresponding mean suture closure stage can be calculated,[e] but not the reverse. Age does not depend on any age indicator	Any statement based on this type of regression is necessarily false	To go from a given suture closure stage to a mean age, the proper regression must be used, i.e. it must be based on a correct model

(continued)

Table 3.2 (continued)

Type of pitfall	Description	Consequences	Responses provided by Bocquet-Appel and Masset
2. Attraction to the mean	Individual age is estimated within a range of error, which tends to be ignored when the individuals are aggregated (it was long supposed that the errors cancelled out)	If this range of error is ignored, mean ages are over-estimated in comparison with the oldest and youngest subjects (hence the absence of old subjects in the earliest palaeodemographic studies)	Use of a frequency matrix allocating the reference population to biological stages and age groups[f]

Source: Buchet and Séguy 2002

Key

[a]Bone resistance to post mortem physical-chemical degradation is minimal for infant skeletons, is maximal for adult skeletons and declines for older individuals. This can significantly distort apparent mortality curves (Masset 1994, p. 382). The relationship between age at death and the actual state of conservation of bones in earth is less clear (Masset 1973a; Baud and Gossi 1980; Guy and Masset 1997; Guy et al. 1997; Walker et al. 1988); poor conservation makes it particularly difficult to use the most fragile age indicators, such as the pubic ramus of the hipbone

[b]The image of excess female mortality persists in the most recent studies (Simon and Leemans 1991; Cocquerelle 1993; Guy 1995), although it is not clear whether this is due to errors in determining individuals' age and sex (caused by a possible long-term shift in the biological parameters – e.g., sutures, pubic symphysis), selective burial practices (Henry 1959), or a population structure affected by an imbalance between the sexes. Historically, there have indeed been cases of excess female mortality, corresponding to women's deteriorating living conditions (Perrenoud 1975). Although this issue is not as acute as has been claimed in palaeodemographic studies, it is still worthy of attention, if only as a trace of socio-cultural habits associated with women's status in society

[c]This method has been adopted by most French anthropologists, but the use of "unisex" tables of correlation between suture closure stage and age is still frequent, e.g. Todd and Lyon (1924, 1925) in English-speaking countries and Nemeskéri in Eastern Europe (Masset 1982)

[d]The question of whether such a drift exists remains unresolved

[e]Except, of course, if it is a representative sub-sample of the reference population

[f]Masset 1973b; Henry 1954, p. 273

Other authors have attempted to combine the results from various biological criteria, hoping to achieve greater accuracy. But the estimate obtained is no better. The age estimated by a given indicator does not strictly correspond to the estimate that may be provided by another indicator (Fig. 3.8). This is due partly to the fact that the organism's response to growth stress (such as malnutrition or epidemic) affects bone growth and tooth development differently. It is also due to the use of reference populations that differ from one indicator to another, or even for a single indicator.

Furthermore, not all age classes from birth to old age are always properly represented in the reference populations. These biases distort the results (see Box 1.1 in Chap. 1). Use of such reference populations is not recommended.

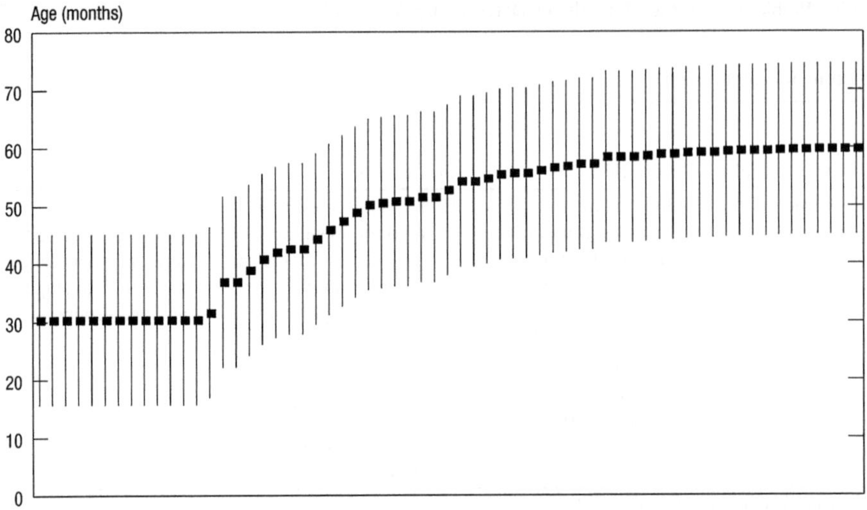

Fig. 3.6 Distribution of individual ages at death (mean and standard deviation) for a male population of the Merovingian period: Cutry, Meurthe-et-Moselle, N.E. France. *Key*. The bone age indicator used is the closure of the ectocranial sutures (Masset 1982), which explains the size and regularity of the margin of error (± 14.8 years). Each square dot represents one individual. *Source*: Luc Buchet in Legoux et al. 2005

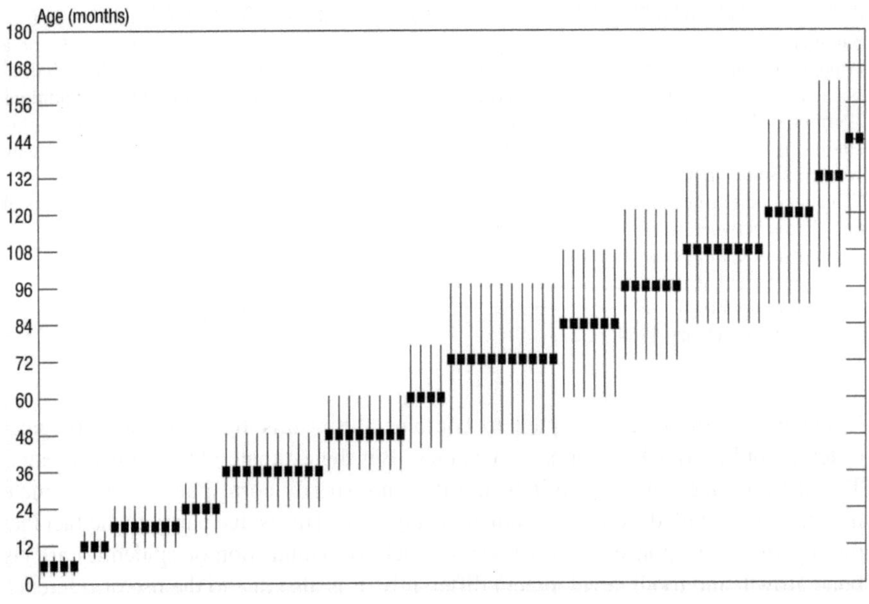

Fig. 3.7 Distribution of individual ages at death (mean and margin of error) for a child population: Rouen, Seine-Maritime, N.W. France. *Key*. Ubelaker's method (1978) is used to estimate the ages of the 81 children of both sexes in the Saint-Jean church cemetery in Rouen. This dental indicator's margin of error increases with subject age. Each square dot represents one individual. *Source*: Luc Buchet from data compiled by Véronique Gallien (INRAP/CEPAM)

Box 3.2 Why Not Adopt for a Palaeodemographic Study the Age Estimation Method Based on the Sacroiliac Joint?

(Lovejoy et al.'s *method* 1985b; *revised by Schmitt, 2000–2005)*

We have repeatedly described the mediocre quality of bone age indicators and have also explained why, pending a better solution, we have opted for the closure of the cranial sutures, despite the criticisms that have been levelled at it. Meanwhile, considering it too risky to use as an indicator of age at death "a poorly understood biological phenomenon" (Schmitt 2002), some anthropologists have rejected cranial closure and sought new indicators, or have proposed a re-interpretation of known indicators such as age-related changes to the sacroiliac joint.

Upon completing their research, both the inventors (Lovejoy et al. 1985b) and the "renovators" (Schmitt 2001, 2002, 2004, 2005; Schmitt and Broqua 2000), stated that this was the long-awaited solution to the problem of estimating adults' age at death: "This approach is at present the best response for estimating the chronological age of adult subjects from the skeleton" (Schmitt 2005, p. 98).

Unfortunately, although it may satisfy forensic scientists or archaeologists wishing to know the age at death of an individual, neither this indicator nor this method are appropriate for a palaeodemographic study.

Indicator

Although the degenerative features do correlate with age, the correlation is not perfect: in some cases of hyperostosis (mainly in males), there are differences between the left and right sides of the joint; but more importantly there is observer subjectivity, even once the method has been simplified ("the researcher's experience is a major factor of bias" Schmitt 2005, p. 93).

Reference Population

Even after the most recent adjustments (Schmitt 2005), problems concerning the structure of the reference population remain. Since the structure of the anthropological reference series does not apply to the study population, Aurore Schmitt attributed to her reference population the structure of a population taken from Ledermann's contemporary standard tables, with a life expectancy at birth of 30 years. In so doing, although she intended otherwise, she applied to archaeological series a structure that is not appropriate for pre-industrial populations.

(continued)

Box 3.2 (continued)

A further problem with the reference sample is that it appears to be applicable to a European population only. That is the conclusion Schmitt came to after obtaining poor results with her method on an Asian population: "The use of methods elaborated on western samples is not appropriate for people from different geographical origins" (Schmitt 2004, p. 4). Consequently one may suspect that the reference used is also inappropriate for studying historical populations of uncertain origin (e.g., those from the Migration Period).

Opting for Individual Ages

The choice of a probabilistic approach is commendable but the error consists in applying it to each individual separately rather than to a total population, because the responses this method furnishes for determining age at death cannot be used for a palaeodemographic study: how are we to allocate to age classes a set of individuals whose individual estimated ages may be 20–29 years, 20–39 years, 20–49 years, 30–59 years, etc.?

To sum up, although the age estimation method based on the sacroiliac joint is as good as any other for a pathology report, it cannot be used for the purposes of palaeodemography.

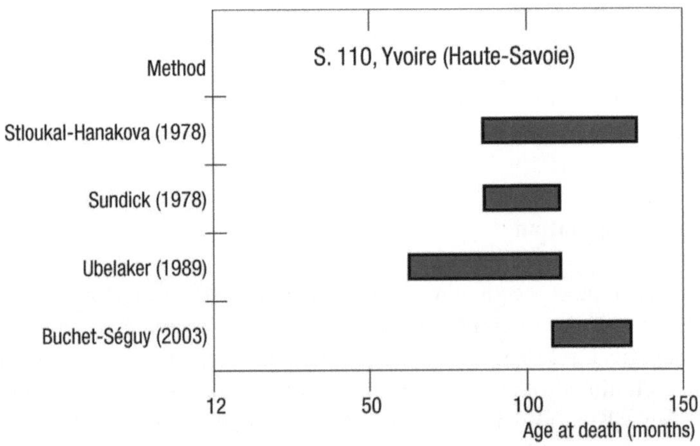

Fig. 3.8 Estimates of the age at death of a child by four different methods. *Key*. According to the method and indicator used – dental mineralisation stage (Ubelaker; Buchet, Séguy); diaphyseal length (Stloukal-Hanáková; Sundick) – the age estimate of the same child varies over a large range (Skeleton 110, Yvoire site, Haute-Savoie, S.E. France. Dig: Joël Serralongue, Haute-Savoie departmental council; anthropological study: Marième Bouali, Luc Buchet, CEPAM, 2006)

Methods for determining age from a skeleton differ for adults and juveniles. The anthropological analyses must be made separately, but the general principles must be common to the two approaches. For a set of buried subjects, the aim is to estimate either the probable age distribution or mean age at death by reference to one, and only one, comparison collection whose main biases have been "neutralised" in advance. The limitations of these methods are shown by the use of a reference population that is not necessarily contemporary with the study population, which amounts to assuming that human biological features have not varied over time. One solution envisaged is to use a "pre-industrial" reference population.

Chapter 4
Establishing a Reference Population

The grounding of our approach must be broad and stable, so that future research can build upon commonly recognized components. To that end, our work must comply with a few essential rules that are outlined below.

Distinguish children from adults in a palaeodemographic approach
Since the processes of growth, for children, and of ageing, for adults, are significantly different, two approaches – based on a unity of method – need to be used according to the age group under consideration, i.e. adult or juvenile. This division is also consistent with a number of demographic observations, particularly the fact that variations in child and adult mortality are totally independent of each other.

Establish the comparison collections on which to base the estimation of sex and age at death
This is a crucial point; indeed it must precede any study. No comparison collection is completely neutral, and using an existing collection does not preclude the task of examining its components and identifying its strengths and weaknesses.

The option of assembling a specific collection for the purposes of a study also requires certain precautions in order to avoid bias.

Prefer an age estimation approach based on a single, clearly defined criterion
Given the data now available, we prefer to put aside multi-criterion approaches, which admittedly appear to correlate better statistically with age, but which use a wide range of collections whose intrinsic qualities have rarely been verified (particularly, individual ages, sometimes established indirectly from biological indicators). Our work will consequently be based on a single criterion, osteological or dental, taken from a comparison collection whose ages are clearly established by civil registration data.

I. Séguy and L. Buchet, *Handbook of Palaeodemography*, INED Population Studies 2, DOI 10.1007/978-3-319-01553-8_4, © Springer International Publishing Switzerland 2013

4.1 Establishing a New Reference Population for Adults

4.1.1 Claude Masset's Reference Collection

One of the largest collections of skulls, that of Ferraz de Macedo Lisbon has been studied by Claude Masset. After some adaptation and the addition of skeletons of older subjects from another Portuguese collection kept at Coimbra, he established a reference collection that for 20 years was the basis of most anthropological and palaeodemographic studies in the French-speaking research community (Box 4.1).

By convention, we shall use the term "Masset reference *collection*" for the set of individuals formed by the combination of these two collections. These are raw data observed by Maria Emília de Castro e Almeida and Antonio Vicente, cited by Claude Masset.[1] We shall use "Masset reference *population*" for the modified reference collection he uses, particularly for attributing age at death for a set of buried adults (probability vector method; see Sect. 5.2).

4.1.1.1 Opting for Ectocranial Sutures

Cranial sutures are probably not the best indicator for determining an individual age at death, but they are no worse than any other for a "comprehensive" approach. Furthermore, the cranial vault is an anatomical feature that is generally well preserved in an archaeological context, unlike the proximal end of the femur or the pubic symphysis, for example. The need to have a larger sample for anthropological study is an initial justification for the choice of sutures as an age indicator.

Masset's research (1982) involved observation of the external and internal cranial sutures, of which he clearly prefers the internal ones. However, we have decided *to use the ectocranial sutures only* as an osteological indicator of age at death, for two reasons. First, a technical one: excellent bone conservation can cause problems for examining the endocranial sutures, because the only access is the foramen magnum and observation is difficult, even with an endoscope, and accuracy may be affected. The other, and perhaps most important reason is linked to the age structure of the Masset reference collection: the ectocranial sutures were measured on individuals aged between 18 and 95, whereas the endocranial sutures, harder to reach, were only measured on the Ferraz de Macedo collection (ages from 18 to 70). Using the endocranial sutures amounts to ignoring the oldest subjects, so that life stops, as it were, at 70.

The choice of 10 segments for observation has not been challenged, because correlation is only improved if fewer segments are used. In 1993, Danielle Soulier and Raoul Perrot proposed using a single segment of the coronal suture (C3).

[1] Our warmest thanks to Claude Masset for making all the data available to us.

Taking Masset's series, for example, a correlation of 0.44 is obtained for women (N = 521) from 10 segments, and 0.31 from C3 alone.

Box 4.1 The Three Portuguese Reference Collections

The Ferraz de Macedo Collection in Lisbon

Dr. Francisco Ferraz de Macedo originally collected some 1,000 skulls at the end of the nineteenth century, but only 849 remained in 1978, when the collection was destroyed by a fire at the Lisbon science faculty. Only a few are left, kept in the same place as the Bocage collection (see below).

To produce his "probability vectors", Masset used the work of Maria Emília de Castro e Almeida and Antonio Vicente, who had assessed suture closure on the skulls in the collection in 1973 and 1976.

The skeletons were those of people buried in two Lisbon cemeteries in 1876 and exhumed 5 years later to be placed in a mass grave; their years of birth ranged from 1806 to 1858. In addition to their age and sex, Ferraz de Macedo noted their occupation, marital status and cause of death (as given at the time).

Most of these people had modest living standards, and were therefore representative of the largest social categories of that age, so the entire set was reasonably consistent in this respect. However, the sex ratio is imbalanced, since the series comprises more female skulls (481) than male ones (368), and the age distribution exhibits two anomalies: a total absence of subjects aged over 70 and, with respect to their recorded ages, statistically significant age heaping.

Until the preparation of this handbook, Masset's thesis (1982) was the most extensive study of this collection.

The Coimbra Collection

Approximately 2,300 skulls of individuals who died between 1910 and 1936 are conserved at the Department of Anthropology of the University of Coimbra. As in Lisbon, each skull is accompanied by documentary evidence of sex, age, place and year of birth, occupation, family status, and probable cause of death. The social origins of the individuals concerned are modest, as in Lisbon. Seventy skulls belong to individuals aged over 70; they were selected by Claude Masset to compensate for the absence of subjects of that age in the Ferraz de Macedo collection.

Publications: Bocquet 1977b; Bocquet et al. 1978; Bocquet-Appel and Xavier de Morais 1987; Masset 1982.

(continued)

Box 4.1 (continued)

The Bocage Museum Collection in Lisbon

The major skeleton collection conserved in the Bocage Museum (National Museum of Natural History, Lisbon) is not widely known, because its collecting protocol was only initiated in the 1980s. It comprises 1,692 skeletons from Lisbon cemeteries and covers the nineteenth and twentieth centuries (1805–1975).

Each skeleton is accompanied by basic documentary evidence (age at death, place of birth, occupation, domicile, and date and cause of death). At present this information is accessible for nearly half the individuals, and the remaining data are still being processed. The two sexes are equally represented, and age at death ranges from birth to 98 years.

A short publication has been devoted to this collection: Cardoso 2006.

4.1.1.2 Exact Age and Declared Age

Rapid analysis of the distribution by sex and age of the reference collection reveals a bias noted but not corrected by Masset (1982). It is a common bias in historical demography, due to approximate declarations of age, leading to a preference for even numbers, particularly for women, and multiples of 5, mainly for men (Fig. 4.1).

Even when individuals are put in 5-or 10-year age classes, this bias remains a problem. It is important, therefore, to correct the declared ages to come as close as possible to a distribution by "exact" age. Historians and anthropologists, regularly faced with this bias, have proposed various methods for correcting it. We have chosen the moving average, in this case over a 5-year period,[2] to allow for age discrepancies due to the preference for multiples of 5. Starting with the observed distribution for each sex separately, we calculated the theoretical number of individuals that each annual age class should include if there were no age heaping. The corrected age distribution is shown in Fig. 4.2.

To "correct" the reference collection accordingly, "real" rather than theoretical individuals had to be placed in each age class. To that end, we randomly drew n individuals[3] aged $x \pm 2$ years and replaced them in the target age class. All in all, the total number of men and women remains constant, their biological characteristics (ectocranial suture closure coefficients) are untouched, and only the distribution by 5-year age classes is slightly modified (Fig. 4.3).

---

[2] A 3-year period was also tried, but this still exhibited artificial irregularities (in annual age classes). However, after placing the individuals in 5-year age classes, the results obtained were close to those given above.

[3] 16 women and 42 men.

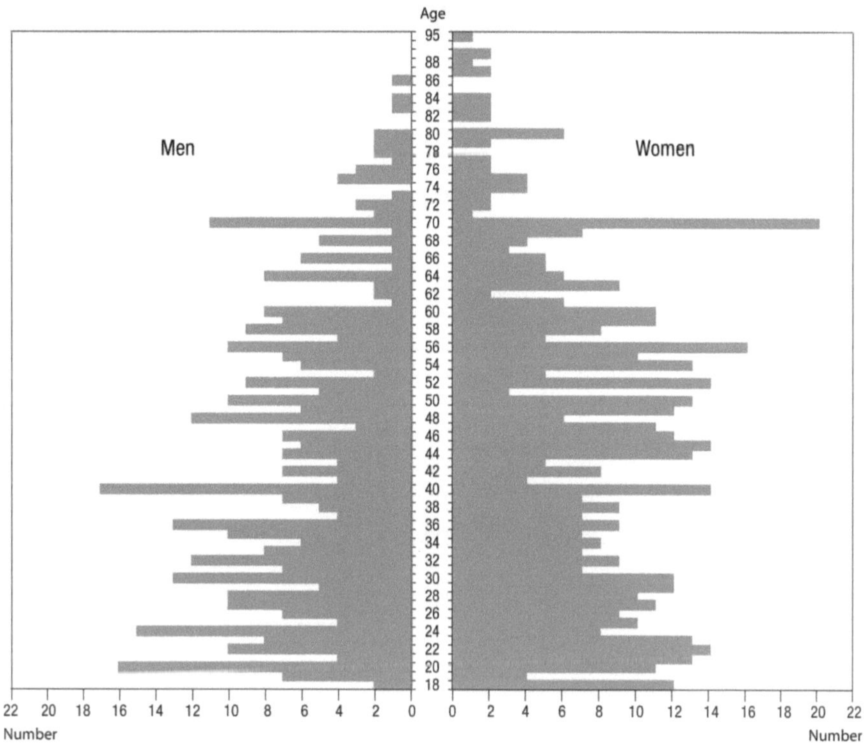

Fig. 4.1 Pyramid of declared ages in the Masset reference collection: age heaping

4.1.1.3 Observation of Individuals Aged Over 60 at Death

Even corrected, the Masset reference collection is affected by another major bias, common to most osteological collections: the low number of older people for observation. Claude Masset faced this problem with the Ferraz de Macedo collection, which contained no individual over 70, and partially evaded it by adding skeletons from the Coimbra collection. However, the sample size remains small and the biological variability observed among the over-60s is still not wide enough, in our view, to obtain a reliable model for estimating age at death based on cranial suture closure. Although we do not claim to have fully solved the problem, we have attempted to supplement the over-60 sample with individuals from other reference collections.

Of the various osteological collections available,[4] we selected that of the National Museum of Natural History in Lisbon (Cardoso 2006), for obvious reasons

[4] Yann Ardagna presents a comprehensive inventory in his thesis (defended 2004, Marseille): "The conservation of biological archives and associated documents in biological anthropology. Applications for French and Hungarian anthropological collections".

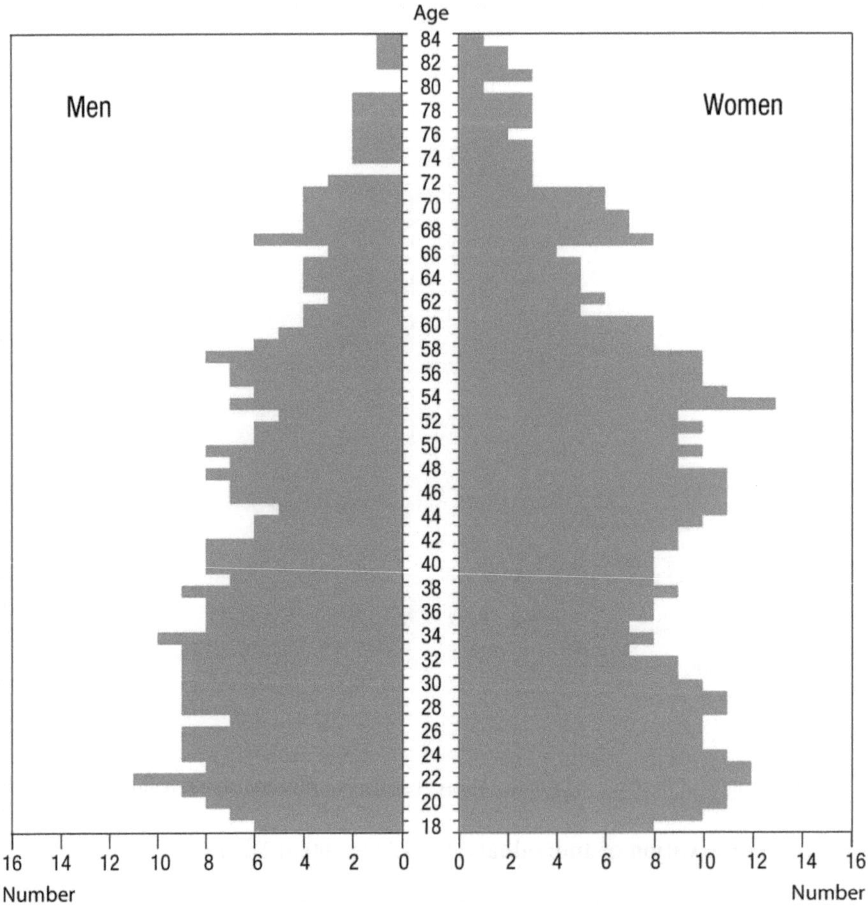

Fig. 4.2 Age distribution of men and women in the Masset reference collection, after smoothing

of geographical and genetic proximity to the population used by Masset. Particular attention was paid to the observation of men and women aged over 60. A total of more than 200 individuals (91 women and 115 men) were studied[5] and added to the 521 women and 394 men in the initial collection.

This new reference population, taken from three Portuguese collections and henceforth denoted $P_{Reference}$, presents a more balanced distribution by age at death, with 612 women and 509 men.

[5] Anthropological study: Luc Buchet and Marième Bouali (October 2006). This series will be called the "Museum collection" when it is distinguished from those of Ferraz de Macedo and the University of Coimbra.

We wish to express our warmest gratitude to Maria da Graça Ramalhinho, director of the Museu Nacional de História Natural (MNHN) and Hugo Cardoso, anthropologist at the Museu Bocage, MNHN, for their help and hospitality during our stay in Lisbon.

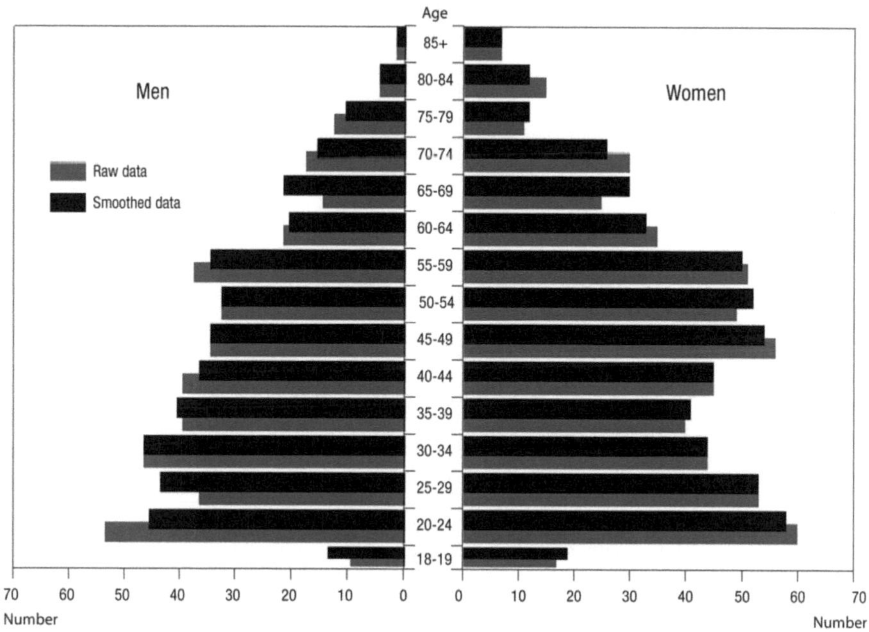

Fig. 4.3 Age distribution by 5-year age class of men and women in the Masset reference collection: before and after correction of declared age

4.1.1.4 Distribution by Sex and Coefficient of Suture Closure: Different Ageing Processes for Men and Women?

Although this addition of some 200 further individuals modified the age structure of the reference collection, it had little effect on the distribution by coefficient of suture closure (Fig. 4.4), despite selection by age. This is evidence of the weak link between age and closure of cranial sutures: a subject may be old according to civil records and yet present juvenile features. Masset's observation that there is a highly marked sexual dimorphism in the rate of closure of cranial sutures[6] still holds for the supplemented sample.

The larger sample does not, however, correct certain anomalies observed in the distribution of men by coefficient of suture closure, such as the low number of men at coefficient 3 or the prevalence of certain other coefficients (see Fig. 4.4). We have corrected the most blatant bias, considering it to be due to interpretation error[7]

[6] This later obliteration of female sutures was noted by Mihály Lenhossék in 1917 (cited by Abdelhamid Grait in his biological and medical science dissertation: "Determination of age at death from the postcranial skeleton", University of Lyon I, 2006, http://anthropologie-et-palaeopathologie.univ-lyon1.fr).

[7] Since Ferraz de Macedo's collection has been totally destroyed, the sutures can no longer be re-interpreted to determine with certainty whether a specific male suture closure pattern exists.

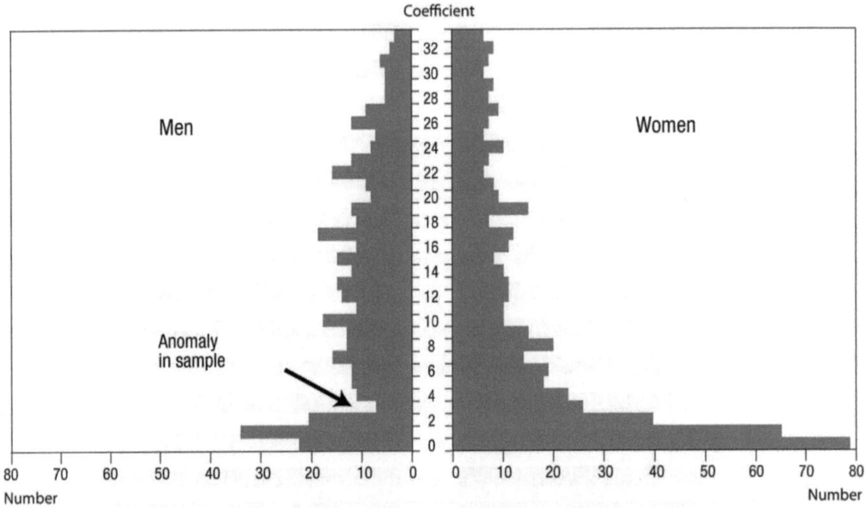

Fig. 4.4 Distribution by coefficient of suture closure and sex in the Masset collection

and not an ageing process specific to men, by applying to the male coefficients 2, 3 and 4 the pattern observed in the female sample.[8] We excluded from the adjustment the coefficients 0 and 1, which are easy to interpret, like those of complete closure, unlike the intermediate values which are more prone to observer subjectivity. Clearly, this correction to the distribution by coefficient of suture closure makes no difference to the age distribution observed in the male sample.

This pattern of closure of the cranial sutures, occurring visibly earlier and faster for men than women, explains why we have chosen a model estimating age at death according to sex.

Tables 3.1, 3.2, and 4.1 (available at Springer Extra) give the distribution of coefficients of cranial suture closure by age groups and sex, i.e., the biological characteristics of the reference population ($P_{Reference}$) we have established.

4.1.1.5 Representativeness of the Sample with Respect to the Demographic Context (Late Nineteenth-Century Lisbon)

The establishment of osteological collections follows certain rules (legal and practical, such as the geographical and social catchment area of the hospital in question) which may affect the representativeness of the sub-population formed in this way. We set out to check whether or not our reference population – $P_{Reference}$ – was representative in demographic terms (age-sex distribution) of the population of Lisbon in the late nineteenth century.

[8] We looked for a distribution presenting a ratio between successive coefficients very similar to that observed for women (i.e. ≈ 0.62).

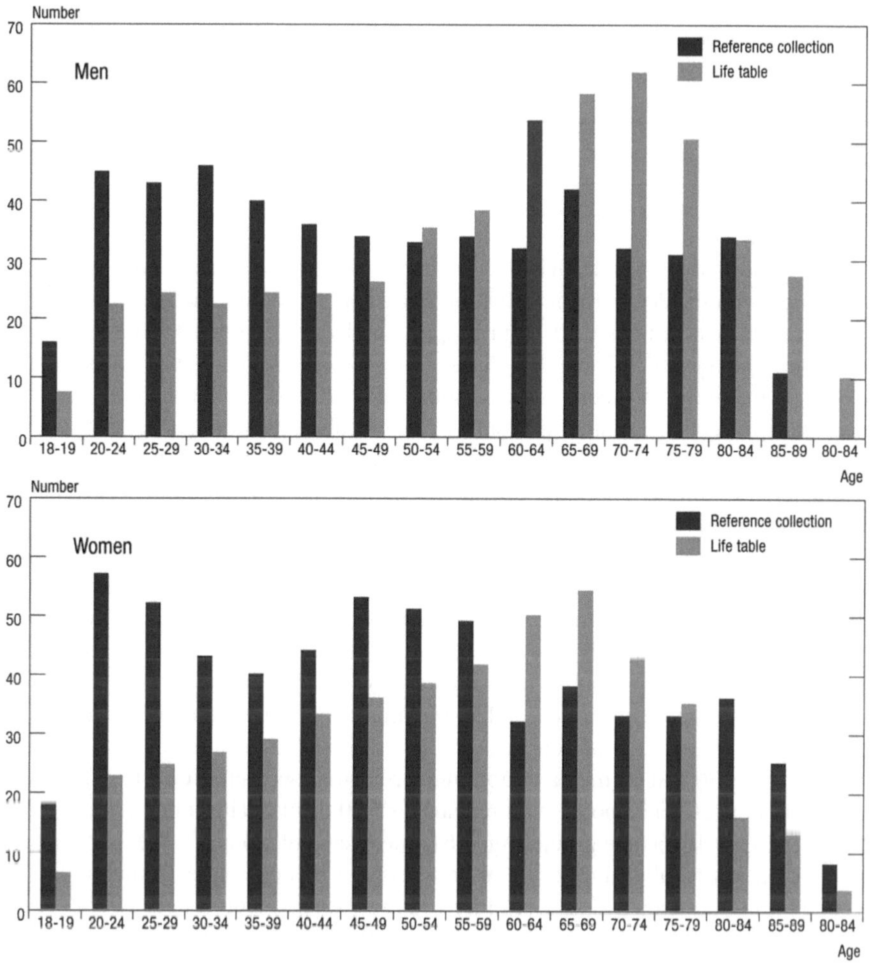

Fig. 4.5 Distribution of deaths by age in the reference collection ($P_{Reference}$) compared with that of Lisbon in 1889. *Key*: *Top* graph men, *bottom* graph women

Using statistical data from the city of Lisbon[9] for the age-sex distribution of the living population in 1890 and of deaths in 1889, we calculated the age distribution at death and the associated death distribution (Fig. 4.5).

---

[9] Source for the 1890 census: *Censo da População do Reino de Portugal no 1 de Dezembro de 1890*. Volume II. Lisboa, Direcção da Estatística Geral e Comércio, 1896 (INED shelfmark: S2Q 1890/2). The age pyramid, established for 1 December 1890, was extrapolated back to 1 July 1889, the nearest date to the death of the individuals under study.

Source for the 1889 death records: *Movimento da População. Terceiro ano 1889–1890*. Lisboa, Ministério das Finanças, Direcção Geral da Estatística, 1892 (INED shelfmark: S3 Q 1889–1890).

Table 4.1 Deaths before and after age 50, for men and women (%): expected (Lisbon mortality in 1889) and observed ($P_{Reference}$) distributions

Age	Expected death at age *x*, women	Observed death at age *x*, women	Expected death at age *x*, men	Observed death at age *x*, men
Age 18–49	37.7	50.2	29.1	51.1
50 and over	62.3	49.8	70.8	48.9

The age distribution at death in the reference collection presents an inverted profile compared with the demographic situation in Lisbon at that time: whereas two-thirds of deaths occurred after age 50,[10] our $P_{Reference}$ collection comprises more than 50 % of subjects under 50[11] (Table 4.1). This skew, which may be due to young adults dying prematurely of tuberculosis or some other disease in the Lisbon hospital, necessarily affects the sampling of suture patterns observable in this population.

Consequently, the reference collection presents a greater diversity of biological characteristics among the under-50s than the over-50s, who are under-represented in our new sample.

As it is, the reference population reflects the sutural patterns of a young population, not often observed in historical demography, except in very special cases of "disaster" mortality (massacre, war, violent plague-like epidemic). A reference population of this sort will tend to under-estimate the ages of any archaeological population to which it is applied as a yardstick. It therefore needs to be adjusted slightly so as to more accurately reflect the mortality of pre-industrial populations.

Analysis of sex distribution at death in the population of Lisbon shows that between ages 30 and 60 men are more numerous than women. Conversely, the sex ratio[12] in our reference population reveals a shortfall of deaths among young men and men in the 40–60 age group (Fig. 4.6). The excess of women in the osteological population is probably due to social conditions in which single women (such as young immigrants, aged 20–24, or older widows with no income, aged 40–59) often had to fall back on the public hospital in case of illness or extreme poverty.

Taking this approach, our reference population ($P_{Reference}$), taken mostly from a hospital, is not representative of the population of Lisbon, either in life or in death. The shortfall of men is particularly significant between ages 40 and 60. This skew, which it would be very hard to correct, may well explain the divergent suture closure patterns by sex (the small size of the male sample would not cover the full range of biological diversity). It is hard to correct because of the risk of over-weighting female biological processes while simultaneously masking males oness. The alternative is to leave the male sample as it is, pending further additions, on the understanding that the age at death estimates for buried men are, as things stand, slightly biased.

[10] This is true for late nineteenth-century Lisbon, but is also more generally applicable to pre-industrial populations. For example, the proportions of deaths before and after age 50 in France in 1770–1779 were 35 % and 65 % (based on Blayo 1975).

[11] In the original Masset collection, 59 % of all women and 65 % of men were aged under 50.

[12] Number of men per 100 women.

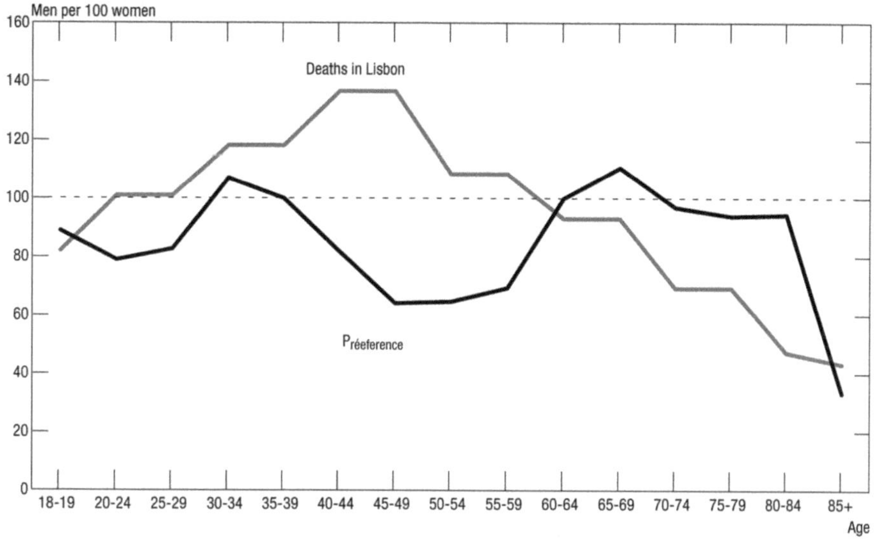

Fig. 4.6 Sex ratios at death in Lisbon, 1889, and in the $P_{Reference}$ population

4.1.2 Adjusting the Reference Population

4.1.2.1 Extreme Ages Observed and Division into Age Classes

The minimum age in $P_{Reference}$ is 18 years, which corresponds to the ossification of the spheno-occipital synchondrosis,[13] rather than to any strict biological transition between juvenile and adult skeletons. This borderline, based on a single observation, may well conceal a biological change that varies between men and women, which remains to be proven (see above). The oldest age observed is 95 years for women and 89 years for men. Consequently, the "90 years and above" age group cannot, on the basis of our evidence, be observed in the same manner for the two sexes, so the last age group used for the "both sexes" sample will be "85 years and over". The size of the sample, and the biases previously described in declared ages, argue against using annual figures. On the other hand, the division into 10-year classes, proposed by Bocquet-Appel and Masset, significantly reduces the precision of any demographic calculations because it assumes that deaths are uniformly distributed across this age interval. The wider the interval, the more it masks any inflection points in the risk of death between two given ages.

We have therefore opted for 5-year classes, which are closer to demographers' standards (providing convenient comparison with demographic indicators that use this interval). Since it is easier to combine groups than subdivide them, observation by 10-year classes of course remains possible.

[13] The joint between the sphenoid and occipital bones in the skull base.

Two groups do not fit into this pattern: the 18–19-year-olds, whom we have chosen to separate off[14] so as to begin observation at age 20, which facilitates calculation of mean age at death (one data point in our mortality models) and comparison of various age groups of the same amplitude. Similarly, the oldest subjects, the "85 and above" age group, do not strictly speaking belong to a single 5-year interval: life does not end at 89 precisely. We do, however, assume a certain equivalence in order to include the oldest ages in our comparative analyses.

4.1.2.2 Expected Distribution by Age at Death in a Pre-industrial Population

To avoid applying the age structure of the reference population to the archaeological populations under study, Masset proposed "standardising" his reference population, i.e., allocating the same number of individuals to each age group in proportion to its amplitude in years. Unfortunately, this age at death distribution does not correspond to any known demographic reality. Even in extreme cases, where the risk of death is the same at all ages, the death distribution is affected by the age pyramid of the living population (Séguy et al. 2006a).

Rather than measure the characteristics of various cemeteries against the yardstick of an unlikely death distribution by age, we prefer a reference population whose death distribution by age reflects the "normal" mortality of pre-industrial populations. Logically, therefore, $\mathrm{P_{Reference}}$ should comply with the death distribution by age observed in late nineteenth-century Lisbon.

It might be thought preferable to fit the reference population to the average mortality pattern of pre-industrial populations (the mean values of the tables in our corpus, see Chaps. 7 and 8), but since the demography of late nineteenth-century Lisbon is statistically representative[15] of mortality by age in pre-industrial populations (Fig. 4.7), the choice of Lisbon does not introduce any bias.

To obtain the chosen distribution, the demographic characteristics of $P_{Reference}$ had to be modified while maintaining its biological ones. For this reason we modified the age distribution at death, not by randomly redistributing individuals from one age group (as in Masset 1982) but on the principle of matrix redistribution. Tables 4.2 and 4.3 (at Springer Extra) describe the breakdown by age and

---

[14] They can always be brought back in later, either with the 20–29 age group, as anthropologists do, or with younger groups, as demographers do (15–19 age group).

[15] The chi-squared test comparing the two distributions is non-significant at 2.5 % level for both sexes combined, at 1 % for men and at 1.2 % for women. The observed mortality distribution for Lisbon does not significantly differ, therefore, from that defined by the pre-industrial mortality standard. Female mortality in Lisbon between ages 25 and 45 is lower than that in the Pre-industrial Standard. In the light of the remarks we have made concerning the male sample, this under-estimation of female mortality is highly plausible given the biases in late nineteenth-century Lisbon's statistical data (erroneous age declarations, under-recording of certain population categories in censuses, inaccuracies in 10-year age groups for counting annual deaths).

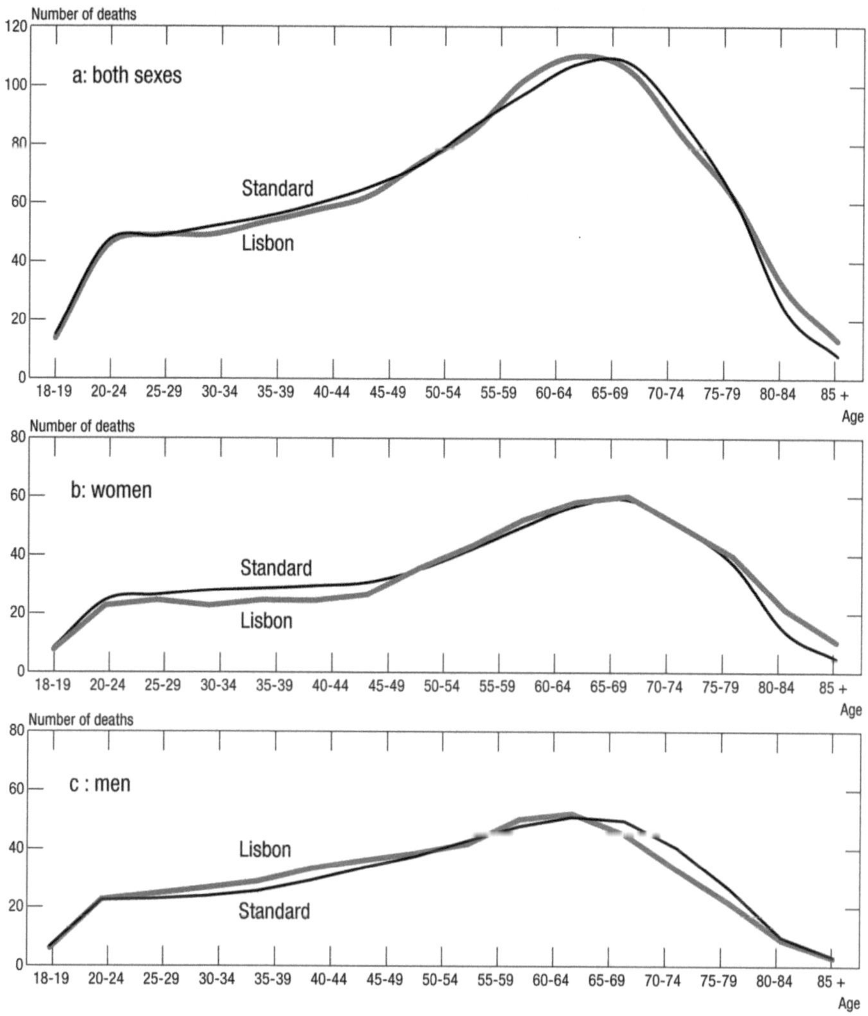

Fig. 4.7 Age distribution of deaths for Lisbon (1889) and our pre-industrial standard

suture closure coefficient of the reference population, male and female, after this modification. We shall henceforth denote it $P_{Lisbon1889}$.

4.1.2.3 Male–Female Weighting (for a Population of Both Sexes)

The notable difference between men and women in patterns of cranial suture closure is a good reason for preferring separate analyses by sex. However, so that

Table 4.2 Characteristics of the $P_{Lisbon1889}$ reference population

Suture closure coefficient (× 10)	18–19	20–24	25–29	30–34	35–39	40–44	45–49	50–54	55–59	60–64	65–69	70–74	75–79	80–84	85 +	Total
Contingency table, women																
0–2	6.2	15.7	16.6	7.0	8.4	8.7	9.8	8.2	4.7	5.1	7.6	3.8	6.2	0.9	1.1	110
3–9	0.9	4.0	3.7	10.2	8.4	7.1	7.4	8.9	10.9	8.4	4.6	9.4	6.2	2.8	9.2	102
10–18	0.4	1.6	2.3	0.5	3.0	4.9	4.9	6.2	7.0	11.8	13.7	20.7	10.8	5.7	7.9	101
19–27	0.4	0.8	1.4	2.7	1.8	2.2	2.9	7.5	7.8	16.9	12.2	15.0	13.9	9.4	5.6	101
28–40	0.0	0.8	0.0	2.7	2.4	1.1	1.0	4.1	7.8	11.8	19.8	13.2	13.9	15.1	13.3	107
Total	8	23	24	23	24	24	26	35	38	54	58	62	51	34	37	521
Contingency table, men																
0–8	4.9	17.9	12.2	11.7	11.6	12.8	7.4	3.5	12.4	0.0	3.9	0.0	1.1	0.0	0.0	99
9–15	0.4	3.6	6.4	6.5	6.5	7.3	13.8	12.7	7.4	9.4	2.6	8.1	6.8	1.9	2.9	96
16–21	0.4	0.5	2.3	4.1	5.8	4.6	5.3	8.1	9.9	9.4	11.6	8.1	7.9	3.3	4.4	86
22–28	0.4	1.0	2.9	2.9	2.2	4.6	6.4	9.2	7.4	14.1	18.0	12.1	12.4	5.2	5.8	105
29–40	0.0	0.0	1.2	1.8	2.9	3.7	3.2	4.6	4.9	17.2	18.0	14.8	6.8	5.6	2.9	88
Total	6	23	25	27	29	33	36	38	42	50	54	43	35	16	16	473
Contingency table, weighted men and women																
0–4	11.2	34.1	28.5	22.1	17.2	17.0	18.1	12.8	14.0	8.4	10.6	5.6	6.2	0.9	4.5	211
5–13	1.7	10.0	10.2	13.6	21.2	21.3	17.5	19.8	24.3	15.3	13.3	23.2	17.3	5.3	8.1	222
14–20	1.3	1.5	7.4	6.9	8.2	10.8	15.4	16.9	18.1	23.9	23.5	26.5	18.6	8.8	13.1	201
21–28	0.4	1.9	4.1	5.5	4.4	6.1	10.1	18.9	14.8	32.4	33.5	28.4	29.1	17.1	12.5	219
29–40	0.0	0.8	1.3	4.6	5.0	5.1	4.5	8.5	13.2	29.1	36.6	25.7	18.3	19.4	16.5	189
Total	15	48	52	53	56	60	66	77	84	109	117	109	90	52	55	1,042

Table 4.3 Conversion of mineralisation coefficients to stages by set of teeth

Stage	Teeth 11–18 or 41–48	Teeth 41–48	Teeth 14–17	Teeth 44–47	Teeth 14–17 and 44–47
I	4–5	4–5	1–2	1–2	2–4
II	6–13	6–12	3–6	3–6	5–12
III	14–16	13–17	7–9	7–8	13–17
IV	17–21	17–22	10–12	9–11	18–22
V	22–25	23–25	13	12–13	23–26
VI	26–29	26–29	14–15	14–15	27–30
VII	30–31	30–31	16	16	31–32
VIII	32	32	–	–	–

comparisons can be made with studies that generally use populations of both sexes combined, we propose a reference population to be used in these cases.

In our reference population, the number of males (473) is lower than the number of females (521). To prevent this imbalance – which reflects demographic reality at certain ages –[16] from applying "too many" female characteristics to the both-sexes sample, we have weighted the male sample to make it equal to the number of females. A multiplier of 1.101[17] is applied to each cell in the contingency table for the male sample (Table 4.2 at Springer Extra). Table 4.4 (at Springer Extra) shows the new distribution obtained for the "equally weighted" male sample, and Table 4.5 the consequent "both sexes" distribution.

4.1.2.4 Defining Stages of Suture Closure

Cranial suture closure is expressed by a coefficient which may have 41 values (0–40[18]). As in the case of annual age classes, it is awkward to work at this level of detail because of the random fluctuation observed in our sample. To achieve some degree of coherence in analysis, it is better to divide the suture closure coefficients into sub-sets, or "suture closure stages", to use Masset's term.

Various divisions have been proposed: seven stages (I–VII; Masset 1982, p. 171 and 178) or four (0–4; Bocquet-Appel and Masset 1995). Whereas the four stages (0–0.9; 1–1.9; 2–2.9; 3–4), regardless of sex, appear to be designed to achieve equal divisions, Masset's seven stages are governed by a more elaborate system, not spelt out by the author, and vary by sex to allow for the differences in suture closure between men and women (see above). In addition, they give disproportionate weight to the lower suture closure coefficients, since the first ten cover five of the seven stages proposed.

---

[16] At higher ages, women are more numerous than men.

[17] The multiplier $\alpha = 521/473$.

[18] Each suture segment is graded from 0 to 4. Divisions of 10th or even 50th may be used to calculate the coefficient. For the sake of convenience, we have multiplied these values by 10.

Table 4.4 Observed age distribution for 7 or 8 mineralisation stages

Teeth 11–18	Age in years																	
Coefficients combined	Stage	2	3	4	5	6	7	8	9	10	11	12	13	14	15	16	17	Total
4–5	I	9	7															16
6–13	II		35	39	34	4												112
14–16	III				16	18	4	1										39
17–21	IV					28	39	29	12	6								114
22–25	V							9	37	48	29	23	6					152
26–29	VI										14	27	28	22	14	4	3	112
30–31	VII												10	21	34	29	13	107
32	VIII															4	4	8
Total		9	42	39	50	50	43	39	49	54	43	50	44	43	48	37	20	660

Teeth 41–48	Age in years																	
Coefficients combined	Stage	2	3	4	5	6	7	8	9	10	11	12	13	14	15	16	17	Total
4–5	I	9	4															13
6–13	II		37	40	21													98
14–16	III				29	32	4											65
17–21	IV					17	40	27	15	6	2	2						109
22–25	V							12	36	44	27	17	4	4				144
26–29	VI									6	15	33	29	21	16	3		123
30–31	VII												9	20	31	28	16	104
32	VIII															6	4	10
Total		9	41	40	50	49	44	39	51	56	44	52	42	45	47	37	20	666

Teeth 14–17	Age in years																	
Coefficients combined	Stage	2	3	4	5	6	7	8	9	10	11	12	13	14	15	16	17	Total
1–2	I	9	7															16
3–6	II		34	39	31	10												114
7–9	III				19	32	31	8										90
10–12	IV					9	12	27	26	17	7	4	3	1				106
13	V							5	24	39	31	28	5	3	2			137
14–15	VI										5	20	32	21	6			84
16	VII												4	19	40	36	20	119
Total		9	41	39	50	51	43	40	50	56	43	52	44	44	48	36	20	666

Teeth 14–17	Age in years																	
Coefficients combined	Stage	2	3	4	5	6	7	8	9	10	11	12	13	14	15	16	17	Total
1–2	I	9	9															18
3–6	II		33	39	28	5												105
7–8	III				23	24	3											50
9–11	IV					21	36	22	8	3	1	3						94
12–13	V						4	17	43	50	34	26	11	5				190

(continued)

Table 4.4 (continued)

Teeth 14–17	Age in years																	
Coefficients combined	Stage	2	3	4	5	6	7	8	9	10	11	12	13	14	15	16	17	Total
14–15	VI									3	8	24	30	23	12	5		105
16	VII												1	17	34	32	20	104
Total		9	42	39	51	50	43	39	51	56	43	53	42	45	46	37	20	666

Teeth 14–17 and 44–47	Age in years																	
Coefficients combined	Stage	2	3	4	5	6	7	8	9	10	11	12	13	14	15	16	17	Total
2–4	I	9	7															16
5–12	II		35	38	27	4												104
13–17	III				23	28	6	1										58
18–22	IV					18	35	22	4									79
23–26	V						3	17	46	51	35	28	7	4				191
27–30	VI									3	7	24	31	19	8			92
31–32	VII												5	21	37	35	20	118
Total		9	42	38	50	50	44	40	50	54	42	52	43	44	45	35	20	658

Table 4.5 Distribution of deaths in Lisbon in 1889 according to official statistics

Age	% deaths	Applied to Lisbon population both sexes combined	% deaths at 2–17 years
0		1,902	–
1	0.4	737	–
2	0.5	212	24.6
3	0.3	144	16.7
4	0.2	104	12.0
5	0.4	58	6.7
6	0.3	42	4.9
7	0.3	34	3.9
8	0.3	28	3.2
9	0.3	26	3.0
10	0.2	20	2.3
11	0.2	19	2.2
12	0.1	23	2.7
13	0.1	26	3.0
14	0.1	28	3.3
15	0.1	28	3.3
16	0.1	35	4.1
17	0.1	35	4.1
Total 2–17 years		862	100

In an attempt to reduce these biases, we have opted for a division into equally weighted stages; each of our five stages corresponds to around 20 % of our sample subjects (see Table 4.2). This division is virtually identical to the results obtained by statistical distribution analysis of the observations in five groups (*k*-means clustering).

4.1.3 The $P_{Lisbon1889}$ Reference Population

This artificially reconstituted population $P_{\text{Lisbon1889}}$ will, therefore, be the basis for estimates of age at death for a set of buried adults. Table 4.2 shows its characteristics for men, women and both sexes combined, by stage and 5-year age group. (The distributions by suture closure coefficients and age groups are given in greater detail in Tables 4.2, 4.3, 4.4, and 4.5 at Springer Extra).

Compared with the Masset reference population (raw data, non-standardised population, corrected ages, five-stage division,), the $P_{\text{Reference}}$ population reveals the biological structure of the population. In particular, note the striking reduction in the weight of low-coefficient individuals (0–2 for men, 0–8 for women) in the composition of the youngest age groups. Another major difference between the two populations is perceptible in the distribution of coefficients of cranial suture closure for men over 60 and women over 70 (before any adjustment of the age structure). These differences reflect the addition of some 200 extra subjects aged over 60 (Fig. 4.8). However, the most significant modifications are the result of changes in the age structure of the $P_{\text{Reference}}$ population that were made to obtain $P_{\text{Lisbon1889}}$ (see Table 4.6 at Springer Extra and Fig. 4.9).

4.1.3.1 Advantages and Drawbacks of the New Reference Population

A reference population is never a neutral construct; great care must be exercised in constituting it, since its composition by sex, age and biological stage has a determining effect on subsequent anthropological findings. Our reference population is still imperfect due to possible errors of interpretation in the initial collection and the under-representation of certain biological features of suture closure. Our sample is still short of men (see Fig. 4.6) and under-represents young male subjects whose suture closure has not or hardly begun (coefficient 0–1). An age distribution observed from a larger number of individuals might turn out to be different from the one used in this study.

We believe that the care we have taken to ensure the representativeness of our reference population with respect to the “standards” of pre-industrial populations, in terms of biological development and health conditions, is necessary, although still open to improvement, if it is to be used with archaeological populations several centuries old.

The collections we propose as bases for anthropological analysis ($P_{\text{Reference}}$ and $P_{\text{Lisbon1889}}$) present a number of advantages over existing collections:

- The declared ages have been corrected by statistical methods (necessary and sufficient);

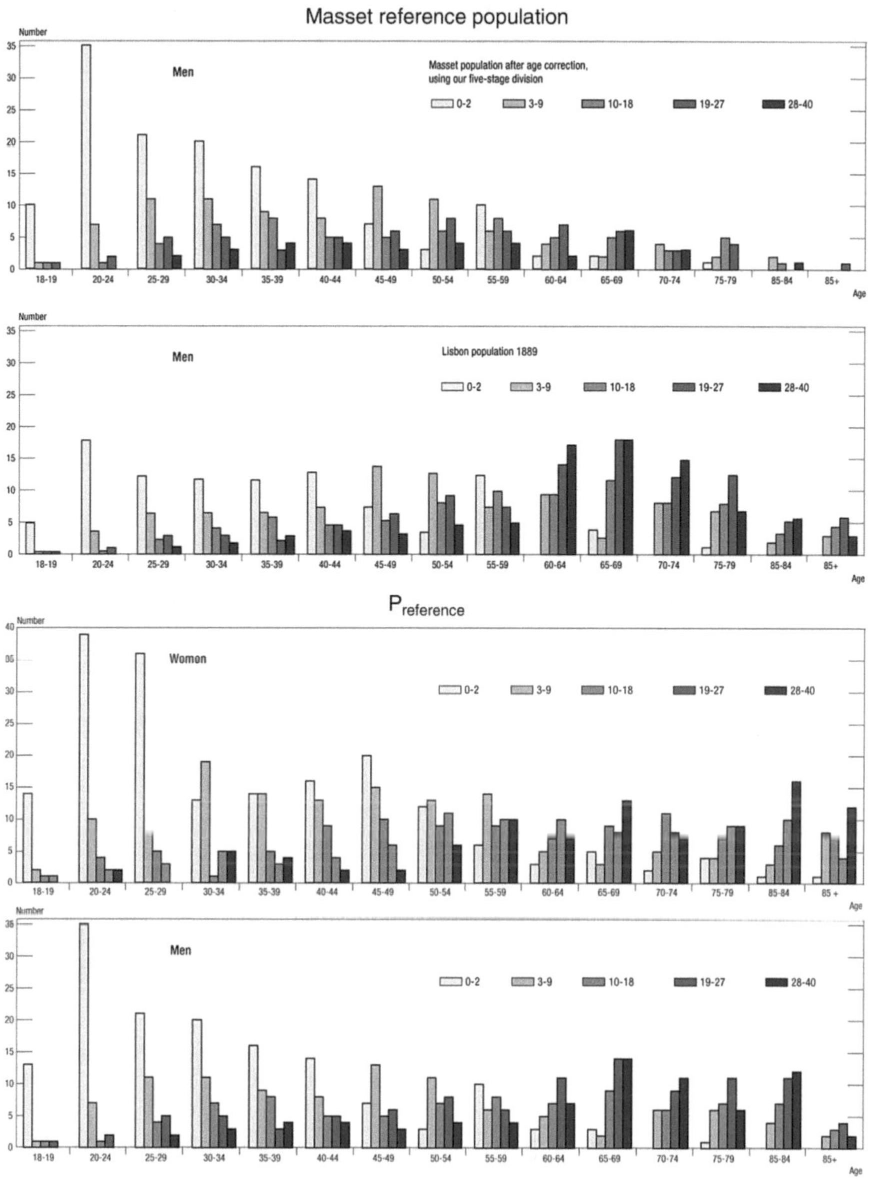

Fig. 4.8 Distribution by suture closure stage and 5-year age group of the individuals in the Masset reference population (*above*) and our ***raw*** reference population (*below*)

– The addition of some 200 individuals aged 60 and above gives greater consideration to the suture characteristics of the elderly and provides an observed sample that is slightly better balanced by age group.

Table 4.6 Reconstituted distribution of child deaths by age classes

Age class as published	Both sexes combined	% deaths (0–20 years)	% deaths (2–17 years)
00–01	1,902	52	
01–02	737	20	
02–05[a]	459	13	52
05–08[b]	133	4	15
08–12[c]	93	3	11
12–20[d]	304	8	22
Total 0–20 years	3,628	100	100
Estimate 2–17 years	875		

[a]Assuming 02–05 = actual ages 2–4
[b]Assuming 05–08 = actual ages 5–7
[c]Assuming 08–12 = actual ages 8–11
[d]Assuming 12–20 = actual ages 13–20

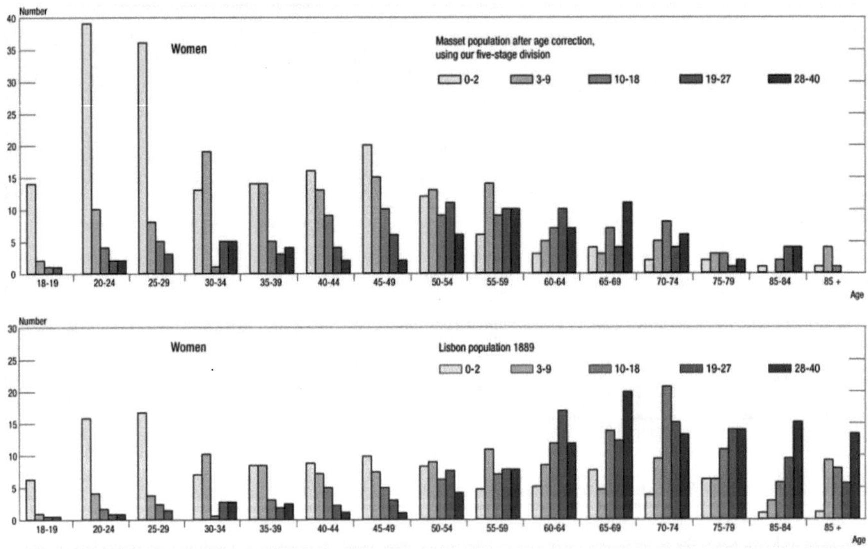

Fig. 4.9 Distribution by suture closure stage and 5-year age group of the individuals in Masset and our reference population ***adjusted*** to 1889 mortality in Lisbon ($P_{Lisbon1889}$)

- Visibly different biological processes between males and females (as far as our current evidence shows) mean that separate analysis by sex is to be preferred. However, for the study of both sexes combined, it is better to use a population in which males and females have been equally weighted. In our collection, where females are more numerous than males, not to correct would have given excessive weight to female biological characteristics.
- Using 5-year age groups gives greater precision to the estimates, while not ruling out a return to 10-year groups.

- The suture closure stages have been defined with care to ensure that an artificial division does not give certain age groups greater weight in the estimates than is actually observed.
- Rather than use an age and sex structure that does not correspond to a "traditional" demographic distribution ($P_{Reference}$), we have deliberately fitted our population to the mortality standards of pre-industrial populations so as to propose a reference matrix more appropriate for archaeological populations ($P_{Lisbon1889}$).[19] This was facilitated by the additional sample of men and women over 60.

4.2 Establishing a Reference Population for Juveniles

For children, as for adults, each method for estimating age uses a comparison collection specific to the age indicator observed. Within a given study it is best not to use age indicators that were not developed for the same reference population (or populations calibrated to be similar and consequently comparable), because the estimates obtained may diverge widely (see Chap. 3, Fig. 3.8).

On top of that there is a further problem: the most common methods using long bone measurements as age indicator propose indirect estimates in practice, without saying so explicitly (Sundick 1978; Fazekas and Kosa 1978; Alduc-Le Bagousse 1988). These authors construct correlation tables of age estimates from teeth (usually with Ubelaker's method (1989), with no allowance for margin of error) and the length of certain bones in a series of skeletons used as reference population. These tables are then used to estimate the age at death of juveniles for whom only post-cranial remains survive. This approach is subject to two biases: the variability of bone growth patterns and the poor quality of dental estimation methods.

More recently (Lalys 2002), tables have been more rigorously established from x-rays of the length of the ulna and the bones of the hand, but their use has been limited.

---

[19] Although it is essential to adapt a reference population to the pre-industrial model ($P_{Lisbon1889}$ and below, $P_{Maubuisson}$ and $P_{Antibes1890}$) when using the probability vector method, this is no longer the case when using estimation methods based on a constant distribution of stages by age group (see Chaps. 12 and 13). Nevertheless, careful thought must be given to the way reference collections are constituted before attempting to estimate the age at death of a buried population. Whatever methodology is used, the quality of the estimates depends on the representativeness of the biological characteristics observed in the reference collection (of known age and sex).

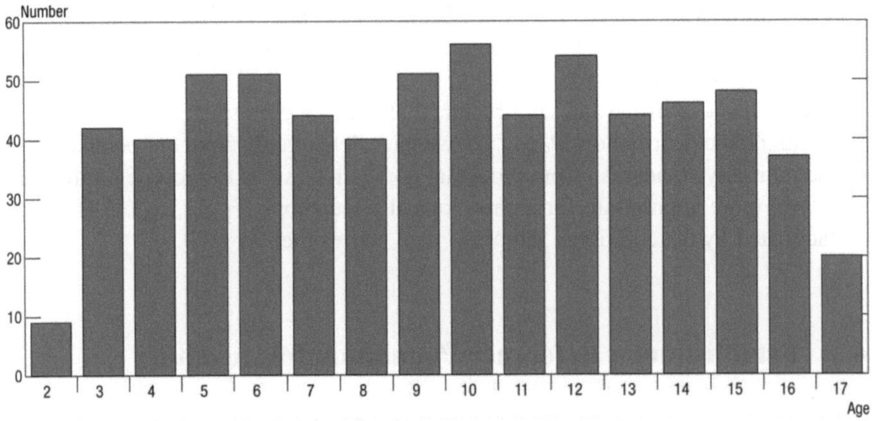

Fig. 4.10 Age distribution of the sample of orthopantomograms

4.2.1 Tooth Emergence and Mineralisation

We have expressed reservations about the use of tooth emergence to estimate the age of an archaeological population because the extent of tooth eruption is hard to estimate on archaeological remains, both for taphonomic reasons – the teeth need to be in the sockets – and the difficulty of discerning stages from dry bone.[20] We prefer to observe the process of dental mineralisation, which, in addition, is less subject to environmental constraints.

Unlike the situation with adults, there are no well-documented historical collections of children's bones. Juveniles are recorded in various collections such as those of the National Natural History Museum of the University of Lisbon, the Institute of Normal Anatomy of the Strasbourg Faculty of Medicine, the Museum of Anthropology of the University of Bologna, and many others (Ardagna 2004), but in numbers too disparate to be of use for statistical analysis. Earlier research is sometimes mentioned in the literature, but the results are hard to compare with current studies. The major series studied by Eduard Mühlreiter, for example, published by Theodore Emile de Jonge Cohen in 1920, gives the average age for the eruption of permanent teeth among children in Vienna, 1870–1890, but we have no information about how this was interpreted or even the nature of the sample. Similarly, children's declared ages are subject to the same errors as described for the adults in the Portuguese collections (although the range is smaller, there are preferred ages for children that cause the same distortions as for adults, Fig. 4.10). Finally, all the early studies are based on observing tooth eruption, which we have chosen to abandon in favour of dental mineralisation.

[20] Emergence is complete when the tooth breaks through the gum.

4.2.2 *Establishing a New Comparison Collection*

Over the past few years, with the help of stomatologists, dental surgeons and orthodontists practising mainly in the southeastern part of France,[21] we have collected a series of current orthopantomograms (dental panoramic radiographs) of children of known sex and age (in years and months). The reference sample chosen comprises 677 x-rays of children aged 2–17 years.[22] It comprises children born in southern France and North Africa.

This reference population is both homogeneous geographically – the Mediterranean basin – and heterogeneous socially and in terms of living conditions, so is well suited for representing a wide range of contexts.

Although the sample is not entirely age-weighted, it does comprise a statistically acceptable number of children in each age class, except for the two extremes. Increasing the number of 17-year-olds would have virtually no effect, given the extent of mineralisation observed at that age. On the other hand, an increase in the number of 2-year-olds might have repercussions for the contingency tables and the frequency matrices (see below), particularly if some children had a mineralisation coefficient higher than those in this sample (higher than 4).

4.2.3 *Number of Mineralisation Coefficients*

Examination of the tooth mineralisation process begins by noting for each item or x-ray, the mineralisation coefficient of each tooth: maxillary and mandibular, deciduous and permanent

In the method we propose, the number of coefficients is restricted to seven (Fig. 4.11) in order to minimise the inevitable problems of subjective interpretation that arise from more complex classifications (Moorrees et al. 1963a, b, proposed 16).

[21] Our warmest thanks go to all those who have forwarded to us the orthopantomograms necessary for establishing the reference population, friends, colleagues and practitioners, working in the Alpes-Maritimes *département*. In this last group, the following deserve special mention: Drs Kamilla-André and Terrasson (Cagnes-sur-Mer), Chaussy, Lachaud and Rücker (Cannes), Savoye (Le Cannet), Bougues (Marmande), Dossios, Favot, Jasmin, Mahler, Millet and Raybaud (Nice), and Alibert (Sophia Antipolis)

We are very grateful for the invaluable help provided by Eve-Line Boulle and Marième Bouali for data entry, Magali Belaigues-Rossard, Nicolas Lannoy and Magali Sucheki for statistical analysis. We also warmly acknowledge Arnaud Bringé, whose initial results were published in 2005 and 2006, for his close collaboration on this study, (Buchet et al. 2005, 2006c).

[22] In all, 715 dental x-rays of children aged 2–18 were analysed. We excluded from the sample the 17 individuals aged 18–20, in line with our biological distinction between juveniles and adults. A certain number of x-rays of children aged 2–17 were withdrawn (21 in all), usually where there had been one or more therapeutic extractions or where the extent of mineralisation diverged too far from the average distribution (recording errors or pathology).

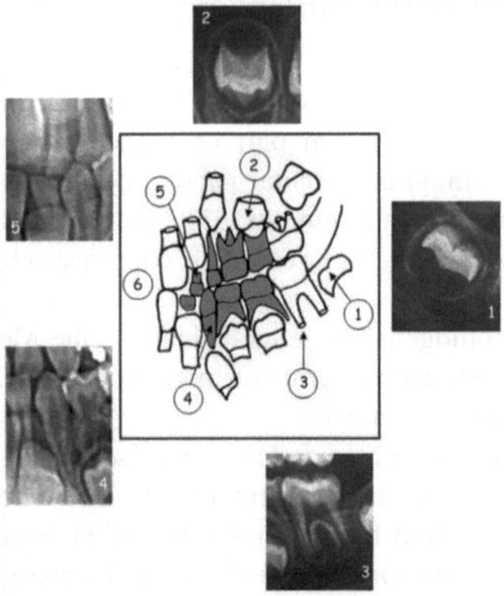

Mineralisation coefficient
0. Naturally absent (bud not yet formed)
1. Crown growing
2. Crown completed
3. Root growing
4. Root completed
5. Root resorption (completed)
6. Deciduous tooth fallen out naturally

Fig. 4.11 Tooth mineralisation coefficients: example of a child aged 6 years ± 4 months. *Source*: photographs, CEPAM; drawings and design, Luc Buchet

4.2.4 Possible Variability in Tooth Development

Using contemporary x-rays to produce a reference collection applicable to pre-industrial populations is not a straightforward task. However, on the basis of a comparative study of a Roman-period and a nineteenth-century population[23] which concluded that there had been no significant drift in the tooth mineralisation process (Saunders et al. 2000), we may assume that observations of contemporary teeth can be used to construct a reference population.[24] The authors of that study noted that, unlike mineralisation, the timing of tooth eruption appears to vary over time (because of environmental factors and improvements in health). In order to come as close as possible to the standard for pre-industrial populations and reduce the influence of the reference sample on the age estimates, its age distribution will be modified to fit that of a population subject to pre-industrial mortality rates (see below). Major dental observation series, such as that from nineteenth-century Vienna studied by Mühlreiter, and from twentieth-century Nice published by Muller-Bolla et al. in 2006, were not used because they focused on tooth eruption and not mineralisation.

[23] The Roman series (second-early third century AD) came from the Isola Sacra necropolis, 23 km west of Rome. It comprised some 2,000 individuals (of whom 800 children) of both sexes and all ages. The nineteenth-century series belongs to St Thomas's Anglican Church, Belleville, Ontario. It comprises 1,564 skeletons, of whom 282 children under 15. Of these, 229 were sufficiently well preserved for dental examination.

[24] A team of dentists and biostatisticians (Parner et al. 2001) investigated any possible trend over time in the eruption of permanent teeth by analysing two samples of Danish schoolchildren from 1969 to 1982. They observed a slight but statistically significant increase in mean age for both sexes and all teeth (with 95 % CI: 1.5 days per year for boys and 2.6 for girls). However, the interval between the two samples was short and the public health conditions in the years concerned cannot be compared to those of pre-industrial populations.

4.2.5 Selection of Teeth for Model Construction

To allow for archaeological conditions, where teeth are often lost after death, particularly deciduous teeth, a number of models are proposed. All are based on observing permanent teeth only.

Two of the proposed models are more appropriate for pathologists because they use all the teeth in a hemi-maxilla or hemi-mandible. For archaeological studies, where skeletons rarely have a full set of teeth, a selection has been made; three models are proposed using both the teeth that are most often present and those whose R^2 is higher than 0.8.

The first model involves the eight permanent maxillary teeth, the second the eight permanent mandibular teeth. The other three models use a smaller number of teeth: four permanent maxillary premolars or molars,[25] four permanent mandibular premolars or molars, four permanent maxillary or mandibular premolars or molars. The models using a limited number of teeth lose in statistical precision but come closer to actual archaeological conditions of differential conservation (see Tables 4.7 and 5.1).

4.2.6 Adapting the Reference Population

As in the case of adult age estimation, certain criteria had to be met, such as a sufficient number of observations to account for intra-population diversity, adequate representativeness of the sample with respect to living (and dying) conditions for pre-industrial populations, and division into mineralisation stages and age groups that are statistically and demographically useful.

We did not distinguish between boys and girls, because there is at present no satisfactory method for determining the sex of a child from its skeleton (see above).

The stage of tooth mineralisation is calculated for each individual by adding the mineralisation coefficients observed on the teeth of a hemi-maxilla or hemi-mandible, depending on the chosen model.[26] The stages of tooth mineralisation are the biological parameters used to estimate age at death for buried children.

As with the closure of cranial sutures for adults, various types of division may be envisaged for children's tooth mineralisation: a division into six, seven or eight stages depending on the case (Tables 4.3 and 4.4).

---

[25] Excluding the third molar, or wisdom tooth, because of the wide variation in its eruption and mineralisation.

[26] Since mineralisation occurs symmetrically ($0.92 > r > 1$) for permanent and deciduous teeth, either the left or right jaw can be used.

Table 4.7 Distribution by age and mineralisation stage according to the mortality by age observed in Lisbon in 1889

		Age in years																
Coefficient	Stage	2	3	4	5	6	7	8	9	10	11	12	13	14	15	16	17	Total
Teeth 11–18																		
4–5	I	162	18															180
6–13	II		91	79	30	3												203
14–16	III				14	12	2	1										29
17–21	IV					18	23	16	5	2								64
22–25	V							5	15	14	10	8	3					55
26–29	VI										5	10	13	11	6	3	4	52
30–31	VII												4	11	15	21	18	69
32	VIII															3	5	8
Total		162	109	79	44	33	25	22	20	16	15	18	20	22	21	27	27	660
Teeth 41–48																		
4–5	I	164	11															175
6–12	II		100	80	19													199
13–17	III				26	21	2											49
17–22	IV					11	24	15	6	1	1	1						59
23–25	V							7	14	12	9	6	2	2				52
26–29	VI									2	5	11	14	10	7	2		51
30–31	VII												4	10	14	21	22	71
32	VIII														1	4	5	10
Total		164	111	80	45	32	26	21	20	15	15	18	20	22	22	27	27	666
Teeth 14–17																		
1–2	I	164	19															183
3–6	II		92	80	28	6												206
7–9	III				17	20	19	4	1									60
10–12	IV					6	7	14	10	5	2	1	1	1				47
13	V							3	10	11	11	10	2	1	1			49
14–15	VI										2	7	15	10	3			37
16	VII												2	9	19	27	27	84
Total		164	111	80	45	32	26	21	20	16	15	18	20	21	23	27	27	666
Teeth 44–47																		
1–2	I	164	24															188
3–6	II		87	80	24	3												194
7–8	III				20	16	2											36
9–11	IV					14	22	12	3	1	0	1						53
12–13	V						2	9	17	14	12	9	5	3				71
14–15	VI									0	3	8	15	11	6	4		47
16	VII												0	8	16	24	27	76
Total		164	111	80	44	33	26	21	20	15	15	18	21	22	22	27	27	666
Teeth 14–17 and 44–47																		
2–4	I	162	18															180
5–12	II		91	79	24	3												197
13–17	III				20	18	4	1										43
18–22	IV					11	20	12	2									45

(continued)

Table 4.7 (continued)

		Age in years																
Coefficient	Stage	2	3	4	5	6	7	8	9	10	11	12	13	14	15	16	17	Total
23–26	V						2	9	18	14	12	10	3	2				70
27–30	VI									1	3	8	14	9	4			39
31–32	VII												2	10	18	27	27	84
Total		162	109	79	44	32	26	22	20	15	15	18	19	21	22	27	27	658

4.2.6.1 Adjusting the Sample to the Demographic Context

In order to give the reference population the demographic characteristics of pre-industrial populations, we

1. Used the Lisbon 1889 data, with the following assumption concerning the division into groups of published ages (Table 4.5):
2. Assumed that deaths were not distributed evenly within each age group but declined from year to year (Table 4.6).

The new reference population obtained is shown in Table 4.7.

4.2.6.2 Division into One-Year and Five-Year Age Classes

Various divisions may be proposed, depending on the required purpose:

- One-year age classes, to obtain a high degree of precision from the outset and to avoid "tweaking" the figures *ex-post* to find the likely distribution of deaths per year of age;
- Five-year age groups, more compatible with demographic models, enabling comparison with studies in historical demography or palaeodemography;
- "Atypical" age groups in demographic terms, relevant to the socioeconomic practices of the culture under study, particularly in Ancient and Mediaeval Europe (see Sect. 6.2.3)

The User may create more appropriate age groups if they so wish from the raw data supplied in Tables 4.7, 5.1, 5.2, 8.1, and 8.2 at Springer Extra and in the supplementary material available on the INED website.

To conclude, the establishment of a "reasoned" reference collection to estimate child age at death, using as sole age indicator the extent of tooth mineralisation, has made it possible to develop specific instruments. With the proposed models, the non-adult population can be distributed by a probabilistic method similar in its principle to that used for adults, thus providing statistical coherence for studies of archaeological populations.

4.3 Age Groups as Yet Inaccessible

4.3.1 Estimating Age at Death for Infants

For infants under the age of 2 or 3 years it is preferable to use other age indicators, because it would be very difficult to complete our reference collection from measures of tooth mineralisation. There is little chance of documenting the 0–2 age class because medical intervention on the teeth of infants is exceptional, apart from cases of major trauma for which access to x-rays is restricted.

However, for infants, the problem is less a matter of improving anthropological techniques for determining age at death than the small number of skeletons to which they can be applied, since they are systematically under-represented in cemeteries. The reasons for this are physiological (softer bones more prone to biochemical degradation in earth; Baud and Gossi 1980), taphonomic (Guy and Masset 1997) and religious/cultural (discriminatory funeral practices by child's age[27]; see Blaizot et al. 2003).

These biases seriously hamper a direct approach to this age group, which is of importance in detecting trends in demographic phenomena.

4.3.2 Estimating Age at Death for Adolescents and Young Adults

Another age group also raises identification problems that have not yet been fully resolved. Adolescents and young adults (approximate ages 15–25) belong by some biological criteria to the juvenile population although their way of life in pre-industrial societies placed them in the world of adults. Although in most contemporary Western countries the "age of physical maturity" corresponds to "legal coming of age", in ancient and medieval societies, the age of legal majority was much younger, at puberty. In anthropological terms this age is hard to identify because once a child's tooth mineralisation is virtually complete, at about 12, this biological indicator becomes less effective (higher margin of error).

For anthropologists, distinguishing between an adult population and a juvenile population (ages 18 or 25 depending on the criteria used[28]) is fairly easy. But more precise estimates of age at death in the 15–25 age group are harder to achieve.

[27] There are special places for infant graves, but archaeological digs do not always have the time, or luck, to reach them.

[28] Average age at skeletal maturity is indicated both by the fusion of the sphenoid and occipital and acceleration of the process of epiphyseal closure that marks the end of growth in a skeleton. Various authors place the age of maturity between 18 and 20 for the skull and 20 and 25 for the post-cranium (for example, the proximal humeral epiphysis is fused by age 25, like the distal epiphyses of the radius and ulna).

The methods developed for juveniles (tooth mineralisation stages) work up to about age 14–15 (appearance of the third molar); those proposed for adults (dental cementum rings or cranial suture closure) are appropriate for those aged 18–20 and over. If one day there is a real scientific consensus concerning the margin of error for age estimated from cementum rings, the same approach may perhaps be applied to both adolescents and young adults.

The techniques currently available for estimating age at death of adolescents and young adults do not provide a comprehensive vision of the processes that mark the end of growth. The currently available correlation tables based on the skeletal maturity of the post-cranium (fusion of proximal humeral epiphysis or distal epiphyses of the radius and ulna) do not provide the statistical precision we believe to be essential for a probabilistic approach to the age distribution at death.

This explains why anthropologists either ignore this age group (cf. a juvenility index that excludes individuals aged 15–19) or combine it with adults (cf. D_{20+} and above and D_{5+}, the denominators in the main age ratios used). But it is an important group both in social terms and for demographic trends: for the periods we study it was the age of "entry into adulthood", as the sociologists put it. We believe that it is crucial to understand their demographic behaviour more fully (migration, fertility, mortality); and at all events, it is a key challenge for palaeodemographic research in the years ahead.

Part II
Reconstructing Demographic Parameters

Chapter 5
Age at Death: Current Approaches and Methods

5.1 General Methodological Principles

After determining the age and sex of a set of exhumed individuals, the aim of the method is to obtain from among the various possible combinations the most likely distribution by groups of age at death. Whatever the archaeological population under study and whatever the age group, the following principles are recommended:

- **Use a reference population whose age structure is close to the standard for pre-industrial populations**

  (a) The population must be a reliable and well-documented instrument for comparison and should match, as far as possible, the biological standards of pre-industrial populations, both in their growth and ageing processes and in their exposure to the risk of dying at a given age.
  (b) However, it should be borne in mind that not all archaeological populations have necessarily the same age and sex structure as the reference population and that they do not all have the same mortality distribution. The frequent mortality crises that affected past populations radically modify period age-specific mortality (Séguy et al. 2006a; Séguy and Signoli 2007).

- **Maintain a probabilistic approach to age at death**

  (a) It is clearly established that determination of individual age at death, however useful it may be in pathology or an archaeological inventory, cannot be used for palaeodemographic study.
  (b) All that remains is a probabilistic approach to age. Many researchers have sought, and continue to seek, promising statistical methods (frequentist or Bayesian), all of which start from prior estimation of age at death of a set of buried individuals, immature or adult.

- **Seek compatibility with demographers' tools**

I. Séguy and L. Buchet, *Handbook of Palaeodemography*, INED Population Studies 2,
DOI 10.1007/978-3-319-01553-8_5, © Springer International Publishing Switzerland 2013

The objective is to propose palaeodemographic models likely to correspond to particular contexts, using criteria accessible from historical and archaeological data. In the current state of research, it is not possible to envisage a demographic study based directly upon archaeological data. The reason is that cemetery data have no connection with the tools traditionally used in demography.[1] It is therefore absolutely necessary to design new tools which take account of osteological constraints. It is by doing so that the French school diverged at an early date from the path taken by North American researchers and looked for innovative solutions to circumvent this obstacle.

5.2 Estimation by the "Probability Vector" Method

5.2.1 Outline of the "Probability Vector" Method

Claude Masset presented a statistical method known as the "probability method"[2] to distribute the entire buried adult population not by summing the ages of the individual subjects but based on the likely distribution by age group of all the buried individuals (Masset 1973b, 1982, 1989).

Using a reference collection in which the numbers in each age group had been artificially equalised, Masset established frequency matrices to connect a given biological stage (from I to VII) to all the age groups in which that stage is observed. These matrices include the probability for each skeleton of belonging to one of the seven age groups chosen, based on its degree of cranial suture closure. These are called "probability vectors" in French palaeodemographic studies.

The age indicator chosen by Masset was the degree of cranial suture closure for each individual, but the method can be applied to any other age indicator. His method was adapted by Catherine Bergot and Jean-Pierre Bocquet-Appel (1976), who used the head of the femur and humerus, Franziska Langenscheidt (1985) who used pubic symphysis, and Christian Theureau (1998), who applied it to four age indicators (endocranial suture closure, changes to the pubic symphysis, degree of mineralisation of the proximal ends of the femur and humerus), which he suggests should be used simultaneously with different weightings. The main results of Masset's thesis were presented in İşcan (1989a) and Bintliff and Sbonias (1999). His proposed method has mainly been used by French-speaking anthropologists in their site studies. The first application, proposed by Christian Simon in 1982 and again in 1987, concerned the Sézegnin cemetery (Switzerland). In 1989, Joël

[1] It is not possible to enter them directly into the life tables established by demographers.

[2] This method corresponds to what Daniel Courgeau in Chap. 12 calls the "Proportional Fitting Procedure (PFP)" or "ALK method" (Age Length Key), without the I for Iterative, because when these methods are applied in palaeodemography they are not really iterative and need to stop after the first loop.

Blondiaux studied five populations in Northern Gaul, and Véronique Gallien the Saint-Denis populations in 1992. In 1998, Luc Buchet proposed an overview of 14 populations mainly in the French regions of Nord-Pas-de-Calais, Basse-Normandie and Rhône-Alpes.

5.2.2 *A New Age Estimation for a Set of Buried Adults*

5.2.2.1 Probable Distribution by Age Group

Following the principle of Masset's "probability vector" method, we propose new frequency matrices (Tables 5.1 and 5.2) taken from the observed distribution of men and women by suture closure stage and selected age group (see Table 4.2). These matrices show for an individual at closure stage *X* the probability of belonging to each of the defined age groups from age 18 to "85+".

Several different matrices are proposed: male, female, both sexes (equal weight); 5-year and 10-year age groups (except the first).

The supplementary material available on the INED website contains the spreadsheets for automatically calculating the distribution of skeletons by 5-year or 10-year age groups.

5.2.2.2 Mean Age Associated with Probable Distribution by Age at Death

This probable distribution implies a probable mean age at death. In the present state of research we have no better way of calculating the mean age at death of a population of buried adults, although some advances are proposed in the prospective part of this book (Chap. 13).

It can, it however, be calculated from the distribution of deaths by 5-year rather than 10-year groups, which gives a slightly more precise estimate.

We suggest choosing the mean age at death of individuals over 20, which is one entry in our network of life tables. In this case, only 5-year groups can be used, setting apart the 18–19 age group.

5.2.3 *Estimating Age at Death of a Sample of Buried Immature Subjects*

5.2.3.1 Probable Distribution by Age Group

Compared with the problems encountered in determining adults' age at death, the estimation of children's age at death might appear more straightforward. In fact, the

Table 5.1 Frequency matrices with five stages and 15 5-year age groups (except first and last): male (equal weight), female and both sexes (equal weight)

Suture closure coefficient (external table) (× 10)	Closure stage	18–19 years	20–24 years	25–29 years	30–34 years	35–39 years	40–44 years	45–49 years	50–54 years	55–59 years	60–64 years	65–69 years	70–74 years	75–79 years	80–84 years	85+ years	Total
Female matrix, equal weight, five stages																	
0–2	A	5.66	14.30	15.10	6.32	7.64	7.93	8.92	7.49	4.23	4.60	6.94	3.42	5.62	0.86	1.0	100
3–9	B	0.87	3.96	3.62	9.97	8.24	6.95	7.22	8.75	10.65	8.27	4.49	9.21	6.06	2.78	9.0	100
10–18	C	0.44	1.59	2.27	0.53	2.96	4.84	4.83	6.09	6.88	11.64	13.54	20.37	10.66	5.58	7.8	100
19–27	D	0.44	0.80	1.38	2.66	1.79	2.17	2.93	7.51	7.71	16.78	12.14	14.94	13.83	9.39	5.5	100
28–40	E	0.00	0.75	0.00	2.50	2.24	1.02	0.92	3.85	7.25	11.04	18.55	12.30	13.00	14.13	12.4	100
Male matrix, equal weight, five stages																	
Suture closure coefficient (external table) (× 10)	Closure stage	18–19 years	20–24 years	25–29 years	30–34 years	35–39 years	40–44 years	45–49 years	50–54 years	55–59 years	60–64 years	65–69 years	70–74 years	75–79 years	80–84 years	85+ years	Total
0–8	A	4.91	18.01	12.29	11.82	11.68	12.92	7.46	3.48	12.43	0.00	3.88	0.00	1.14	0.00	0.00	100
9–15	B	0.39	3.72	6.66	6.72	6.79	7.63	14.33	13.18	7.71	9.76	2.68	8.39	7.05	1.96	3.03	100
16–21	C	0.44	0.60	2.72	4.80	6.78	5.36	6.19	9.43	11.56	10.96	13.53	9.43	9.24	3.85	5.10	100
22–28	D	0.36	0.98	2.78	2.81	2.08	4.38	6.08	8.81	7.09	13.45	17.22	11.57	11.88	4.95	5.57	100
29–40	E	0.00	0.00	1.33	2.01	3.31	4.19	3.63	5.26	5.65	19.64	20.57	16.89	7.74	6.45	3.32	100

Combined male/female matrix, equal weight, five stages																	
Suture closure coefficient (external table) (× 10)	Closure stage	18–19 years	20–24 years	25–29 years	30–34 years	35–39 years	40–44 years	45–49 years	50–54 years	55–59 years	60–64 years	65–69 years	70–74 years	75–79 years	80–84 years	85+ years	Total
0–4	A	5.32	16.13	13.50	10.47	8.14	8.03	8.56	6.07	6.61	3.99	5.00	2.67	2.93	0.45	2.13	100
5–13	B	0.74	4.50	4.58	6.14	9.54	9.60	7.90	8.90	10.93	6.88	5.99	10.46	7.81	2.40	3.63	100
14–20	C	0.65	0.76	3.70	3.43	4.08	5.38	7.67	8.41	8.99	11.91	11.72	13.18	9.27	4.37	6.50	100
21–28	D	0.19	0.88	1.88	2.50	2.00	2.80	4.62	8.61	6.74	14.77	15.26	12.94	13.29	7.81	5.70	100
29–40	E	0.00	0.43	0.68	2.45	2.65	2.72	2.38	4.51	7.00	15.41	19.42	13.61	9.69	10.31	8.75	100

Table 5.2 Frequency matrices with five stages (A, B, C, D, E) and seven 10-year age groups (except first and last): male (equal weight), female and both sexes (equal weight)

Female matrix, five stages

Suture closure coefficient (external table) (× 10)	Closure stage	18–29 years	30–39 years	40–49 years	50–59 years	60–69 years	70–79 years	80+ years	Total
0–2	A	35.06	13.96	16.85	11.72	11.54	9.03	1.84	100
3–9	B	8.45	18.20	14.17	19.39	12.76	15.27	11.75	100
10–18	C	4.30	3.48	9.67	12.97	25.18	31.03	13.37	100
19–27	D	2.62	4.45	5.10	15.22	28.92	28.77	14.93	100
28–40	E	0.75	4.74	1.94	11.10	29.60	25.30	26.56	100

Male matrix, equal weight, five stages

Suture closure coefficient (external table) (× 10)	Closure stage	18–29 years	30–39 years	40–49 years	50–59 years	60–69 years	70–79 years	80+ years	Total
0–8	A	35.2	23.5	20.4	15.9	3.9	1.1	0.0	100
9–15	B	10.8	13.5	22.0	20.9	12.4	15.4	5.0	100
16–21	C	3.8	11.6	11.6	21.0	24.5	18.7	9.0	100
22–28	D	4.1	4.9	10.5	15.9	30.7	23.4	10.5	100
29–40	E	1.3	5.3	7.8	10.9	40.2	24.6	9.8	100

Combined male/female matrix, equal weight, five stages

Suture closure coefficient (external table) (× 10)	Closure stage	18–29 years	30–39 years	40–49 years	50–59 years	60–69 years	70–79 years	80+ years	Total
0–4	A	35.0	18.6	16.6	12.7	9.0	5.6	2.6	100
5–13	B	9.8	15.7	17.5	19.8	12.9	18.3	6.0	100
14–20	C	5.1	7.5	13.0	17.4	23.6	22.4	10.9	100
21–28	D	3.0	4.5	7.4	15.3	30.0	26.2	13.5	100
29–40	E	1.1	5.1	5.1	11.5	34.8	23.3	19.1	100

margin of error makes it impossible to include in any predefined age class an individual child whose age spans more than one class. Some authors have proposed alternative solutions: as early as 1982 (pp. 89–90), Christian Simon suggested redistributing the children whose estimated age spans two age groups in line with the distributions observed in Sully Ledermann's life tables (1969, Network 100, e_0: 30.35 years). This idea was popularised by Pascal Sellier (1996) as the "principle of minimising anomalies in an archaic mortality pattern", whereby the 10–14 age group is considered to be the least exposed to the risk of death. The initial assumption of minimum risk at ages 10–14 is justified, but when some subjects are "reintroduced" into this age group, the equally strong probability that they belong to another group is overlooked. Bocquet-Appel (2005) appeals to the statistical properties of the normal distribution and, in doubtful cases, defines the subject's probability of belonging to one of the two groups. This more rigorous method is also more tedious to apply to large series.

We therefore considered it essential to approach age at death in probabilistic terms and to estimate the most probable distribution of deaths by age group in a population of buried children. The probability vector method, adapted to the child population, can also be used to estimate with greater statistical likelihood one of the estimators frequently used in palaeodemography – the juvenility index $\left(\frac{D_{5-14}}{D_{20+}}\right)$ – by giving the most probable value for the numerator.

With this improvement, the juvenility index remains a valuable indicator, albeit highly sensitive to the birth rate, which is unknown.

Using the tooth mineralisation stages observed from the 675 orthopantomograms whose distribution is corrected by the mortality distribution of pre-industrial populations (see Table 4.7), we established frequency matrices making it possible to distribute the deaths of immature subjects across 1-year age groups (ages 2–17) or 5-year groups (ages 2–4, 5–9, 10–14 and 15–17) (Table 8.4 at Springer Extra).

We also calculated the frequency matrices for distributing the child population around the central 5–14 age group (numerator of the juvenility index) (Table 8.5 at Springer Extra).

5.2.3.2 Other Possible Age Groups

In demography, age is divided in a regular, linear, but rather artificial fashion. Except for the first two age groups, 0–1 years and 1–4 years, generally studied separately, either 1-year or 5-year classes are used. By adopting the demographers' age divisions, palaeodemographers, as well as historical demographers and anthropologists, segment the lives of the populations they study in a way that is not necessarily the most appropriate. Although it is practical to use conventional age groups, and indeed necessary for comparative analysis, more specific behaviours may be masked. The apparent continuity of biological age suggested by demographers' divisions in fact conceals numerous discontinuities (see Sect. 2.2, Fig. 2.4).

Analyses based on a more socio-biological age division would certainly shed new light on the question (see Wrigley et al. 1997). Palaeodemographic observation could focus on age groups of special significance, both in terms of the sources and of the ancient and medieval societies of Western Europe. For example:

- A 2–4 age group: the period of dietary diversification during and after weaning[3] may cause gastro-intestinal disorders;
- A 5–8 age group: end of early childhood, the child leaves the world of its mother or nurse; first contacts with the outside world and its risks (poor awareness of danger) and first tasks (helping the family). In biological terms, the immune system is established and the child is less susceptible to childhood diseases;

[3] On the basis of isotope analyses, Estelle Herrscher (2003) estimates that the mediaeval children buried at Saint-Laurent, Grenoble, were weaned between ages 2.6 and 3.3 years.

- A 9–13 age group: end of childhood (according to the ideas of the period) and entry into the world of adults, both in religious terms – confirmation[4] – and social terms (legal majority). It is also a period when the child may need to leave the home or family to study or work (risk of work accidents);
- A 14–15 to 17–18 age group: young adults by life style, but still adolescent in bone growth (teeth mature around age 18 and bone growth not complete until age 25 or so).

However, to take account of biological observations based on x-rays of subjects in the reference population, the proposed division needs to be slightly adjusted. Of course, this new division will doubtless be modified in response to future study results.

Possible age groups

- 2 years and below
- 3-4 years
- 5-7 years
- 8-12 years
- 13-15 years
- 16-17 years[5]

Using frequency matrices to estimate the distribution of deaths by age group makes it easier to define both their amplitude and their segmentation.

We therefore established the various matrices corresponding to the above age groups (Table 8.6 at Springer Extra), and they were first used in an archaeological context for the Lisieux and Yvoire sites (Buchet and Séguy 2008).

The use of new age groups that better match children's physiological development and their place in ancient and mediaeval societies opens the way to a better understanding of the key stages of children's lives – and their survival – in historical periods, and allows finer interpretation of sets of archaeological skeletons. However, one unknown will always remain: how reliable is the study sample? Because the distribution by age at death observed in cemeteries also, and perhaps mainly, reflects the natural conditions of conservation of the bones and the care taken when the remains are removed.

[4] Since the Middle Ages, confirmation has occurred at various ages, usually between 12 and 14 years.

[5] Using the age of 17 as the upper limit for the immature population makes it possible, in a site study, to link the results obtained with those calculated for the adult population (ages 18 and above) with the probability vector method.

5.2.3.3 Mean Age at Death for These Distributions

As for the adult population, it is possible to calculate the mean age at death for the various age distributions at death cited above. However, note that unlike adults' mean age at death, which represents the average years of life of those who survived the rigours of childhood, the mean age at death calculated for the child population corresponds to the average life of children who fell victim to those same rigours. Demographically speaking, the two indicators are not equivalent.

The mean age at death of children dying between ages 2 and 17 is calculated from the death distribution by 1-year rather than 5-year age classes (as is the mean age of children dying between ages 5 and 14, which we intended to include in the mortality models. See below.).

5.2.3.4 Conclusion

A way had to be found for palaeodemographers to avoid the pitfalls of either overlooking the margin of error by wrongly supposing that the positive and negative errors cancel out, or of proposing an arbitrary redistribution of individuals, particularly the immature, whose estimated age overlaps two classes. The probabilistic approach we have adopted avoids the biases linked to the determination of individual ages.

The methods proposed offer enhanced precision while evading a statistical difficulty; they also have the major advantage of harmonising the palaeodemographic techniques used for adults and the immature. Even if the biological indicators are different, the methods of analysis now use the same concepts: probable distribution of deaths by age group and calculation of the mean age at death for that distribution.

By focusing on observation by age class – 1-year for children and 5-years for adults[6] – and subdividing the degrees of growth and biological maturation, our method further improves precision. This gives researchers the opportunity to propose other, larger, divisions of biological stages or age groups.

However, the process of creating an "ideal" reference population is far from over. The variety among its individuals (growth and ageing processes, resistance to stress and infection) requires a more complex approach than the deductive one adopted here (if the hypothesis is true, the result deduced from it is necessarily also true). The nature of the data we possess should guide us towards a Bayesian approach in which the initial assumptions, expressed as prior probability

[6] Using 1-year age classes for adults whose ages cover 77 years (from ages 18 to 94) would have given only a dozen individuals on average per age class. Together with the fine detail of the biological observations, segmented into 41 suture closure coefficients, this would have caused an unacceptable fragmentation of the data.

distributions, are compared with the observed data to provide posterior probability distributions, which represent our knowledge once the observations have been made.

5.2.4 Adaptations to Particular Contexts

Our earlier work[7] showed that when the age structure of the population using the cemetery (burying population) was known from records, the frequency matrix could be adjusted so that it kept the biological (suture) characteristics of the initial reference population and its age distribution complied with the mortality distribution relevant to the site under study, applying the principle of matrix redistribution.

Using, where possible, the age structure of the mother population makes it possible to avoid the random variations of the palaeodemographic sample, since these are all contained in the initial structure.

The pre-industrial mortality model may also be inappropriate in certain contexts, such as the study of a socially advantaged population (like the nuns of the Royal Abbey of Maubuisson, see Sect. 11.2) or an epidemic of plague that decimates a population with no distinction of age. In such cases, it is better to adopt the model closest to the actual experience of these populations, on condition that the lifestyle of the archaeological population is itself close to that of the historical population taken as a reference.[8]

At this stage in research, three mortality models applicable to palaeodemographic contexts[9] are available: for a particularly advantaged female population (like the Maubuisson nuns), for a population exposed to an epidemic or other disaster that strikes without distinction of age or sex (like that observed in Martigues during the 1720–1721 plague), and for a port city with a large number of young adult immigrants, as in Antibes (see corresponding age at death distributions in Table 8.7 at Springer Extra).

One of the aims of current research is to pursue the development of these models and extend the scope of their application. Whatever method of estimation is used (vector or frequentist), the more demographic information is available about the population under study, the more accurate and reliable will be the estimated age distribution of deaths.

---

[7] The first applications concerned archaeological sites recent enough for us to have accurate demographic information on the mother population (Buchet and Séguy 1999, 2003; Buchet et al. 2003; Signoli et al. 2005).

[8] Chapter 13 also shows how specific situations can be simply and effectively taken into account by the estimation method it proposes.

[9] In the sense of proven demographic situations that can be re-used in similar contexts.

5.3 "Estimator" Method

5.3.1 *Outline of the "Estimator" Method*

This was the first response provided by French palaeodemographers to the problems of individual ages, via a new index providing demographic information on the buried population (Bocquet-Appel and Masset 1977, 1996).

Bocquet-Appel and Masset noticed that the ***juvenility index*** could be obtained from cemetery data, with a slight adaptation to allow for the under-recording of children under 5 in most cemeteries. They proposed, therefore, to define the palaeodemographic juvenility index as the ratio of children aged 5–14 to "adults" in the sense of all individuals aged over 20 years $\left(\frac{D_{5-14}}{D_{20+}}\right)$, often abbreviated as JI.

To avoid imposing on archaeological populations a mortality model marked by contemporary factors (see Chap. 6), the authors took care to select a set of life tables corresponding to specific characteristics observed in populations with traditional demographic patterns: a juvenility index above or equal to 0.100 and a ratio of more than 2 between deceased individuals aged 5–9 and those aged 10–14. After removing the tables containing blatant anomalies, the authors had a sample of 40 tables,[10] from Geneva in 1625 to Nicaragua in 1940, corresponding to populations with mainly agricultural lifestyles and little access to modern medicine.

On the basis of these tables they established statistical correlations between the juvenility index and certain demographic parameters: life expectancy at birth, infant probability of dying and 0–5 mortality rate. These regressions, known as palaeodemographic estimators, were published in 1977, and later corrected to allow for the regression margin of error (Masset and Parzysz 1985)

5.3.1.1 Advantages and Limitations of the "Estimators"

These "estimators" take into account specific features of the mortality of pre-industrial populations and are based on indicators directly related to osteological data. The juvenility index can be calculated quite simply and used to fill the gaps in the archaeological record and provide an estimate of the number of children under 5, rarely found in excavated cemeteries. In this sense, the estimators are a valuable tool. French-speaking palaeodemographers saw this and were quick to adopt it, but always with a critical eye.

A number of difficulties in using the estimators were soon reported (Sauter and Simon 1980; Bocquet-Appel and Masset 1982, 1985, 1995, 1996; Masset and Parzysz 1985; Murail and Sellier 1995), notably regarding the number of deceased individuals. First, we can never be sure that no adult or child skeleton is missing

[10] Half the tables before the twentieth century and some more contemporary ones concern countries that had not yet begun their demographic and public health transition. The sample was extended to 45 tables in 2002 (Bocquet-Appel 2002).

from the excavation or the anthropological examination. Second, the definition of the 5–14 age class raises two problems: not only are 5–14-year-olds not always well represented in cemeteries, but their attribution to a particular class is difficult in the case of those near to the age limits because of the margin of error (e.g. 5 years +/– 9 months).[11] It is also hard to detect[12] and interpret growth rates; the number of children dying at ages 5–14 is highly sensitive to variations in population growth and a high juvenility index may be due either to a stationary population with high mortality (and low life expectancy at birth) or to an expanding population.

5.3.1.2 Other Proposed Indicators

The estimator method was well received by North American palaeodemographers, who looked for other ratios of deceased individuals, either to find fertility parameters (Buikstra et al. 1986) or to obtain a comparative indicator. Mary Jackes (1986, 1988, cited by Konigsberg et al. 1989) proposed using the mean of three mortality rates[13] from ages 5 to 20 (${}_5q_5$; ${}_5q_{10}$; ${}_5q_{15}$; denoted MCM: Mean Childhood Mortality) to compare results from one site to another. This index correlates closely with the juvenility index and provides further information, but still exhibits the same shortcomings: it is highly sensitive to the population growth rate and the random fluctuations of small samples.

5.3.2 "Revised Estimators"

Conscious of the methodological limitations of estimators, the method's authors suggested adding another indicator to the juvenility index: *adult mean age at death*, while continuing to take account of the population growth rate (Bocquet-Appel and Masset 1996; Bocquet-Appel 2002). They also looked for a new way to estimate the age structure of the buried population independently of its initial distribution, generally unknown, and thus to obtain an unbiased measure of the mean age at death of a set of adults.

[11] The problem can be solved by applying the properties of a normal distribution: starting from estimated age and its standard deviation, many software packages can calculate the likelihood of belonging to one age class or the other.

[12] Bocquet-Appel and Masset propose a relationship based on the juvenility index $\left(\frac{D_{5-14}}{D_{20+}}\right)$ and the senility index $\left(\frac{D_{60-\omega}}{D_{20+}}\right)$. Unfortunately, the latter is hard to define by anthropological analysis. Since there are currently no criteria for assessing the growth rate of buried populations or for distinguishing its sign, only assumptions can be made, which is why the authors proposed regressions for various annual growth rates.

[13] Calculated like Halley's table by considering all individuals deceased after age $x + a$ to be survivors at age x. Thus ${}_5q_5$ represents the ratio $\left(\frac{D_{5-9}}{D_{5+}}\right)$.

5.3.2.1 An Attempt to Adapt the "IPFP" Method

They settled on a mathematical method (Iterative Proportional Fitting Procedure, IPFP[14]) which requires a frequency matrix correlating biological stages of ageing and age groups. Their idea was to estimate the most probable distribution of age at death after successive iterations. However, because of the numerous empty age classes produced by this method (see Chap. 12), they preferred to stick with the calculation of mean age at death even though its manner of estimation was methodologically false.

In 1996, Bocquet-Appel and Masset published the regression equations linking their anthropological parameters (adult mean age at death and juvenility index[15]) to such demographic indicators as life expectancy at birth, life expectancy at age 20, infant probability of dying, juvenile probability of dying, population growth rate, taken from a sample of 45 life tables. These regressions can only provide coherent palaeodemographic information if the estimator a_{20} (mean adult age at death) is correctly calculated. This point is made in Lyle W. Konigsberg and Susan R. Frankenberg's criticism (2002) of this method.

For our part, we have adapted the calculation program by restricting the number of iterations to one, as imposed by the limits of the method. These more reliable results have been tested for various archaeological sites (Buchet et al. 2004, 2006b).

5.3.2.2 A New Indicator: P_{5-19}

More recently, Bocquet-Appel (Bocquet-Appel 2002; Bocquet-Appel and Bacro 2008) has proposed a new indicator, which he denotes ${}_{15}P_5$, namely, deaths at ages 5–19[16] divided by deaths at age 5: $\frac{D(5-19)}{D5+}$.

Compared with the juvenility index, which does not include deaths between 15 and 19, this new indicator has the advantage of being a continuous variable. It avoids the difficulty of the cut-off point at age 14, which is hard to identify biologically (whereas the mature population, around ages 18–19, is easier for anthropologists to identify). Like the juvenility index or Jackes's Mean Childhood Mortality, it leaves out the youngest subjects, for whom archaeological observation is often incomplete.

However, as an input to mortality models (see Sect. 8.2), P_{5-19} does not appear to provide better estimates than the juvenility index.

---

[14] Although this name does not correspond to the mathematical method used (see Chap. 12).

[15] Adult mean age at death correlates weakly with the earliest probabilities of death, unlike the juvenility index. Combining two age indicators that contain different information provides a better estimate of demographic parameters.

[16] We prefer to denote it P_{5-19} to avoid any confusion with the probability of survival at ages 5–19.

5.3.3 *Brainstorming*

Palaeodemographers were long at odds over the matter of individual ages, but the debate has now shifted to the question of how to interpret the matrix allocating the reference population to biological stages and age groups, a matrix used to estimate the age at death of an archaeological population as a whole. Either we consider the distribution of age groups by suture closure stages (probability vector method) or we consider the distribution of stages by age group, following the principle first stated by Masset in 1982 (with his *Approx* program, written with Bocquet-Appel), and by Konigsberg and Frankenberg in 1992. In Chap. 12 of the prospective section of this book, Daniel Courgeau provides a critical overview of these two approaches, and then in Chap. 13, with Henri Caussinus, explores the possibilities of the second proposal.

The methodological advances suggested by Courgeau and Caussinus do not challenge either the principle or the choice of palaeodemographic indicators. To obtain a more accurate picture of the distributions by age at death of pre-industrial populations, we carried out a specific study of this question. In this way, palaeodemographers will have the mortality models best suited to the demographic contexts of their historical populations.

Chapter 6
Current Demographic Models

The purpose of palaeodemographic analysis is to understand the population as it was, within a given socio-environmental context, where the individuals it comprised formed a dynamic group marked by births, deaths and sometimes migrations, and when all we have, at best, is the bones of those who died.

This is no easy task, because most of these demographic measurements are inaccessible to palaeodemographers, who, with few exceptions, have no way of knowing the sex and age distribution of the population studied. Nor can they usually calculate the probabilities of dying or life expectancy at birth, which are the standard input parameters for life tables, without making assumptions about the completeness of the sample. This is because, for the biological reasons we discuss above, the numerator (how many people died between ages x and $x + a$) always has a margin of error that affects the calculations, especially for adults; and, for archaeological reasons, we can never be sure that we have all the components of the denominator (those present at age x). Palaeodemographers are thus forced to make assumptions to underpin their analyses.

6.1 Review of Basic Hypotheses in Palaeodemography

Whatever the conditions of the site, whether a dig is partial or exhaustive, any attempt to go beyond determining the sex and age at death of a set of exhumed skeletons requires accepting the following assumptions:

- The relationship observed in our reference populations between a biological indicator of age and membership of a given age group is constant over time and space (biological uniformity hypothesis) and independent of the size of the observed sample;
- The palaeodemographic sample, i.e. the set of individuals whose sex has been determined and whose biological indicator of age has been measured, is

I. Séguy and L. Buchet, *Handbook of Palaeodemography*, INED Population Studies 2, DOI 10.1007/978-3-319-01553-8_6,

representative of the entire population buried in this place (in other words, all age groups are properly represented);
- There are close biological, socioeconomic, health and demographic links between the exhumed population and the buried population.

These assumptions are necessary if we are to use equations (population models) that relate a population structure by sex and age to demographic behaviours that depend on age (fertility, migration, mortality). The deaths by age, or age group, observed by palaeodemographers directly result from interactions between the structure of a population and its vital parameters.

6.2 Population Models

The demographer Alfred J. Lotka (1934, 1939) proposed a theoretical model to connect various components of demographic analysis. He showed that, whatever its initial age structure, a closed population – with no migration – exposed indefinitely to constant fertility and mortality, tends towards a constant age structure and constant growth rate. This is the concept of the stable population, also called Malthusian population, referring to the theory of the Reverend Thomas Malthus, who had foreshadowed this idea one and half centuries earlier.

Using these established mathematical relationships, it is sufficient to know one or two parameters to estimate any missing, or biased, demographic variables. These models led to the emergence of a new discipline, demographic projection, which is making steady progress thanks to new developments in mathematics, statistics and information technology.

These population models are also of great interest to palaeodemographers, who can address the dynamics of archaeological populations via the observed distribution of ages at death. But the conclusions they reach will vary widely, depending on their initial hypotheses and the degree of complexity attributed to the demographic phenomena.

6.2.1 The Stable Population Concept in Palaeodemography

Since Gyula Acsádi and János Nemeskéri in 1970, the hypothesis of a stable population with zero growth, known as a stationary population, has been accepted by most palaeodemographers, who consider that over long periods of use of a burial ground (generations or even centuries), the traditional mechanisms regulating population will have maintained a growth rate close to zero, reflecting a balance between births and deaths (Dupâquier 1972; Bideau 1983). In such a situation, not only is the age structure constant, but so are population size and numbers of births and deaths. A population of this sort is governed entirely by the law of mortality.

For palaeodemographers, this means that the age-sex distribution of the buried population is identical to that of the structure of the living population. It is then a simple matter to calculate the various demographic parameters that correspond to the theoretical population found in the life table.

In reality, a hypothesis of this sort is hard to defend, because it bears little relation to the living conditions of pre-industrial populations, which were regularly subject to mortality crises that strained their ability to recover. The dynamic model of moderate growth, occasionally disturbed by mortality crises, as proposed by Jean-Noël Biraben (1969, 1979) and modelled by John Komlos and Sergey Nefedov (2002), most certainly comes closer to the realities of pre-industrial demographics. Results are seriously biased if a population is considered as stationary when in fact it is not.

Demographers were quick to (re)introduce the notion of population growth into their models. Ansley J. Coale and Paul Demeny in 1966 (followed by Coale et al., in 1983) proposed an extensive set of stable populations corresponding to various levels and structures of mortality and various fertility behaviours (see below). palaeodemographers (Bennett 1973; Weiss 1973; Bocquet-Appel and Masset 1977; Valkovics 1982; Sattenfield and Harpending 1983; Johansson and Horowitz 1986; Henneberg and Steyn 1994; Bocquet-Appel and Masset 1996) have also turned to stable populations, even if the estimation of a growth rate, whether positive or negative, from reliable osteological indicators[1] is as yet an unresolved problem.

It must not be forgotten, however, that stable populations are also virtual, since growth rates are seldom constant for long enough to allow a population to achieve a "stable" form. The theory of stable populations (Bourgeois-Pichat 1966) reaches its limits in the case of small human groups (several hundred individuals), where random fluctuations distort the conclusions. The same is true when the chronological sequence observed is too short (a few decades): the population dynamics are subject to too many short-term variations and cannot settle to a "stable" pattern.

6.2.2 Semi-stable and Quasi-stable Populations

As a result, the concepts of semi-stable and quasi-stable population proposed by Jean Bourgeois-Pichat (1990, 1994) may well be of use in palaeodemography.

---

[1] Some authors have suggested tracking changes in the age distribution of deaths in order to measure the growth rate (Valkovics 1982); others (Longacre 1976; Cohen 1977) have based their calculations on population estimates from various periods, but their results are not robust enough to serve as a grounding for palaeodemographic hypotheses.

At present, most studies compare the distribution of ages at death obtained for a stationary population ($r = 0$) with that resulting from a non-zero growth rate, either chosen arbitrarily or estimated (see examples in Chap. 11).

A semi-stable population is one which, at any given time, coincides with the stable population that corresponds to the indicated fertility and mortality parameters. Once its mortality and fertility have been determined, a semi-stable population immediately achieves a stable state, whereas, by the same hypothesis, a stable population takes some time to do so. The necessary condition here is that the population maintain a constant age structure over time. Bourgeois-Pichat (1990, p. 823) observed that the age structure of populations in developing countries varied little, although their growth parameters (fertility, mortality) were not constant, and proposed applying the formulae established for stable populations to these real populations.

The same concept can be applied to theoretical populations corresponding to mortality models: a quasi-stable population is defined by a constant fertility function and mortality that varies within a set of model tables. According to Bourgeois-Pichat (1994, p. 170), the set of stable populations calculated from model life tables corresponds very closely to the various stages of the quasi-stable population associated with this set.

Furthermore, Bourgeois-Pichat (1990, p. 831) adapted the system of semi-stable populations to the imperfect data from developing countries (frequent biases are inaccurate age declaration and underestimated number of children). Many observed populations do develop in a manner close to the model of semi-stable populations, especially when fertility varies little.

If duration is replaced by the point in time t, these population models make it possible to use the properties of stable populations in palaeodemography and are able to accept short-term variations.

6.2.3 *Migration*

The previous models assume that the populations are closed, with no inward or outward migration. Some demographers (Preston and Coale 1982) have also developed models that take account of the age-dependent migration function.

The age structure of an open population is determined by the numerical growth rate of each of its age groups at a given point in time (natural increase and net migration). Samuel H. Preston and Ansley J. Coale (1982, p. 253) provide mathematical equations that relate the age structure of a population to its various vital parameters under the various population models proposed (stationary, stable, open).

Although these models are more complex to handle, they are not without interest in palaeodemography. Given the historical facts, we cannot ignore migration.

6.3 Various Mortality Regimes

To monitor developments in population health, it was the custom for many years, and remains so in some cases, to focus primarily on mortality measures (mainly using general indicators such as mean age at death and life expectancy at birth). However, the age distribution of the deceased is the result of both the age structure of the living population (shaped by fertility, mortality and migration, if any) and the mortality law.

This focus on mortality is partly due to the study material itself, i.e. the remains of the deceased, and also to the appeal of model life tables, useful tools for "getting the dead to talk", although it is sometimes forgotten that they also include the "fertility" dimension (see Coale and Demeny's models). Model life tables reveal the link between an observed distribution by age at death and the vital parameters of the associated theoretical population, if we assume an underlying population model.

Life tables have been part of the demographer's toolkit since the eighteenth century, when the first "political arithmeticians" (Graunt 1662; Halley 1693) established the relationship between age and the probability of dying. At first, these calculations of mortality were mainly used by actuaries, but they gradually drew the attention of demographers, who established the figures and calculation methods, and laid down rules of interpretation.[2]

6.3.1 Model Life Tables

A model table is a mathematical model for estimating a level of mortality at age x, the unknown, from one or two known parameters in the observed table, by using observations taken from other populations at a given place and date.

Most model tables available are constructed from data observed in developed or developing countries, and on regression methods correlating one or more mortality indicators with the age-specific probability of dying. For a given value of the input parameter (for example, q_x, or e_0), the model table can be used to obtain the probable timing of mortality as it emerges from observations taken from a set of existing life tables. All model life tables assume a close link between mortality and age.

Model tables were developed for two purposes: to fill gaps in the demographic record of mortality by sex and age, and to calculate demographic forecasts. Various estimation formulae have been proposed, using variable inputs, sample reference tables generally based on contemporary observations, and mathematical formulae both simple and complex.

[2] For example, the role of causes of death in variations in mortality.

6.3.2 Main Life Tables Used by Demographers

6.3.2.1 United Nations Life Tables

The first life tables constructed for the purpose of demographic observation were the UN tables (1955), which set out to estimate mortality from the infant probability of dying ($_1q_0$), on the assumption that there was a relation between two successive probabilities.[3] These tables were based on data from developed countries and were criticised (Gabriel and Ronan 1958) for their chosen input (one of the least reliable probabilities) and the calculation method used (biased by estimating a linked succession of probabilities). Their principle was nonetheless applied in work by Coale and Demeny (1966), Coale et al. (1983), Ledermann (1969), Brass (1971), the OECD (1980) and the United Nations (1982).

6.3.2.2 Coale and Demeny's Models

Coale and Demeny's tables (1966, extended and improved in 1983 with the help of Barbara Vaughan) differed from the earlier ones by recording regional differences in mortality (four regional models[4] corresponding roughly to geographical groupings, covering only the developed countries, i.e. European and of European settlement). They also used a more complex form of calculation and input into the models was by life expectancy at birth.[5]

These models were a response to the two main objections made to the UN tables (a single mortality regime and regressions calculated from the infant probability of dying) and also had the advantage of being a two-parameter system: one parameter is the regional model chosen (empirical choice of mortality structure) and the other the input into the model (choice of a level of mortality). However, in practice, users generally take the West model (Coale and Demeny's general pattern model) in the absence of any closer knowledge of the mortality to be estimated.

---

[3] For a detailed presentation of the various mortality models, see Bourgeois-Pichat 1994, pp. 41–86.

[4] The East family (Austria, Czechoslovakia, Germany, Hungary, Northern and Central Italy, Poland: 31 tables) is characterised by high infant mortality and high mortality at ages 50+. The North family (Iceland, Norway, Sweden: 9 tables) has low infant mortality and low mortality at ages 50+. The South family (Southern Italy and Sicily, Spain, Portugal: 22 tables) has high under-5 mortality, low mortality at ages 40–60 and high mortality at ages 65+. The West family (130 tables from the other 22 countries, mainly in Europe, plus Australia and New Zealand, Canada and the United States, Israel, Japan, Taiwan and the white population of South Africa) comes close to the general mortality model observed in the preliminary phase.

[5] Life expectancy at birth: a general mortality index measuring the mean number of years a newborn would live if exposed throughout its life to the mortality conditions observed in its year of birth.

These models require no calculation: all the table parameters are given, including the associated stable populations (which may explain their success). They have, however, been criticised on two points: first, the preponderant weight of "European" tables, which makes them ill-suited to certain regions in the world,[6] and second, the input into the tables (e_0) is not the index used to construct the tables (e_{10}), causing a bias in the calculation of the probability of dying.

6.3.2.3 Ledermann's Life Tables

Using principal component analysis to determine the most useful indicators for establishing model life tables (Ledermann and Bréas 1956), Sully Ledermann proposed a series of sets of model tables in 1969 (seven single entry and three dual entry), based on the observations from 157 tables which were largely those used to construct the UN tables.

Ledermann's life tables make it possible to approach mortality without the biases caused by the method used to calculate the UN life tables, and avoid the need for a specific input, contrary to the UN and Coale and Demeny tables. For each of the proposed inputs, they also provide estimates of those inputs (so that the user can calculate the life table associated with any value of the input index) and 95 % variance limits for age specific probabilities of dying.

A few quibbles have been raised: the fact that calculation by successive regressions causes a wider margin of error at the end of the chain (as errors are accumulated at each probability) is partly outweighed by the possibility of using the set closest to the initial data. Ledermann's tables do not allow for the possibility of differing mortality structures, as do Coale and Demeny's proposed regional models.

6.3.2.4 Brass' Logit Model

The logit model used by William Brass (Brass and Coale 1968) is not based on a set of observed tables but on a relationship that can be used to construct a life table for a given population from a known table taken as a reference table (also known as a standard). His method is based on the idea that the ratio of probabilities of dying between two tables will continually rise or fall with age (Pressat 1995). This means that a table defined by its survivors at age x can be related to the survivors at the same age in a reference table by a linear regression in which the constant determines the level of mortality and the intercept the relationship between child mortality and adult mortality (see Chap. 9).

---

[6] In the 1983 revision, Coale, Demeny and Vaughan took account of this objection and included some tables for Africa. In 1976, Samuel H. Preston had already proposed introducing a fifth "non-Western" pattern based on life tables from Latin America.

The importance of the choice of reference table was pointed out by Brass, among others, in 1971, and he proposed a "European" standard, close to the UN and Ledermann model tables and Coale and Demeny's West model; and in 1975, an African standard to adjust the data from countries where child mortality is still high.

The main weakness of this method – inaccurate estimation of the proportion of deceased at high ages – was corrected by Douglas C. Ewbank et al. (1983), who proposed introducing two further parameters, κ and λ, for the impact on the youngest and oldest age groups.

6.3.2.5 Life Tables for Developing Countries

Returning to the tradition of life tables, the OECD (1980)[7] and UN (1982)[8] set out to compile a set of reliable mortality tables for developing countries, so that models could be developed for places where mortality was still high.

The methods they used were inspired by the work of Coale, Demeny and Ledermann: identification of sub-groups (in the UN tables), principal component analysis, regressions of logarithms or logits of probabilities of dying, publication of tables ranked by life expectancy at birth, publication of associated stable populations.

6.3.3 Main Mortality Models Used by Palaeodemographers

6.3.3.1 Traditional Mortality Models

In the 1970s, the models established by Acsádi and Nemeskéri (1970) and Weiss (1973), among others, were popular among researchers. But because these tables were directly based on archaeo-anthropological data (determination of individual sex and age of exhumed skeletons) and ethnological data where age is not a reliable variable either, they contained intrinsic errors. Some researchers still continue to construct them or use those of Weiss (Lovejoy et al. 1985b; Corruccini et al. 1989; Alekseeva and Buzhilova 1997; Drusini et al. 2001; Budnik et al. 2004).

Others have thought it more prudent to use contemporary life tables, such as those of Coale and Demeny, Ledermann and Brass' logits (1971, European

[7] The OECD used a corpus of 104 life tables from developing countries, divided into four groups by mortality by age (the fifth region was an aggregate of the four). For each group the OECD also published a standard table. However, the poor quality of the input data limits the validity of these model tables.

[8] The UN only used 36 tables from 22 developing countries in Latin America, Asia and Africa, also divided into four "geographical" regimes and a general mortality pattern. But what they gained in terms of quality they perhaps lost in terms of representativeness.

standard)[9] to address the fertility of buried populations or correct the biases in the anthropological sample, although these tables have long been criticised for their inability to reflect the demographic features of pre-industrial populations. Renée Pennington (1996) more wisely preferred to use the African standard (Brass 1975) to model the demographic consequences of even a slight reduction in early childhood mortality in the Neolithic period.

While Masset and Bocquet-Appel's palaeodemographic estimators include the mortality features of pre-industrial populations and use indicators directly related to osteological data, they do not provide all the life table parameters. Some authors (Sellier 1996; Guillon 1998; Castex 2005; Castex et al. 2005; Bizot et al. 2005; Crubézy et al. 2006) have consequently combined the palaeodemographic estimators with Ledermann's model tables in order to reconstruct the entire mortality curve, in total contradiction with their initial hypothesis of a mortality by age specific to pre-industrial populations.

With the improvement in survey data quality, interest in these "traditional" mortality models, constructed from a large number of observations for predictive purposes, has waned in recent years. The quality of current survey data has not only rendered indirect estimation methods less useful but has also made it possible to develop analysis techniques based on individuals. This shift is also perceptible in palaeodemography.

6.3.3.2 Parametric Models

William Siler's mathematic model uses the "hazard function", based on the idea of exponential increase (or decrease) in mortality. He used it first for animal populations (Siler 1979) and then successfully applied it to human populations in 1983.[10] By adding a fourth parameter to the Gompertz-Makeham model, Siler achieved a much better fit for the mortality of pre-industrial populations, where the probability of dying was extremely high in infancy and rapidly declined thereafter.

This model was applied by Timothy B. Gage to non-human primates (Gage and Dyke 1986; Gage 1998) and in anthropological demography (Gage 1988, 1989, 1990; Gage and Mode 1993). Kathleen A. O'Connor (1995), O'Connor et al. (1997) was the first to then apply Gage's work to palaeodemography.

Box 6.1 gives a more detailed presentation of the various parametric models. The interested reader may also consult the general description by James W. Wood et al. (1992a).

[9] To our knowledge, the life models for developing countries (UN 1982; OECD 1980) have not been used in palaeodemography, either due to unawareness of their potential utility, or because researchers consider that the causes of death among developing country populations are too remote from the health conditions of pre-industrial populations.

[10] Mortality by age displays very similar patterns for all mammals. This observation has made it possible to apply to human populations a mathematical model originally designed for primates (Siler 1983).

Box 6.1 The Main Probability Distribution Specifications of Mortality Models in Palaeodemography

Daniel Courgeau

In demographic event history analysis, a certain number of time functions are used to specify the mortality of a population, or indeed the occurrence of any other phenomenon, often called failure. For more details see *Event history analysis in demography* (Courgeau and Lelièvre 1992).

Main Functions Used

An individual's failure time is taken to be a random variable T, greater than or equal to zero, whose distribution is to be examined. This may be specified in various ways, three of which are presented here.

The *survivor function,* for mortality, is the probability that failure time T will be later than or equal to a given date t:

$$S(t) = P(T \geq t)$$

The *probability density function* is then the limit when $dt \rightarrow 0$ of the probability that time T falls within $[t, t + dt]$ divided by dt:

$$f(t) = \lim_{dt \to 0} \frac{P(t \leq T \leq t + dt)}{dt} = -\frac{dS(t)}{dt}$$

The *hazard function* specifies the instantaneous rate of failure conditional upon survival to time t:

$$\lim_{dt \to 0} \frac{P(T < t + dt | T \geq t)}{dt} = \frac{f(t)}{S(t)} = -\frac{d\ln S(t)}{dt}$$

For palaeodemographic estimates, the probability density function $f(t)$ is used mainly for continuous time models, and differences between survival functions for discrete time models.

Main Models

Only a few models are used here for mortality. They depend on the type of mortality under study.

(continued)

Box 6.1 (continued)

(a) *Gompertz model*

This model was published in 1825 to characterise the mortality of adults, which increases exponentially with age. Its hazard function is expressed by:

$$h(t) = \lambda\rho\exp(\rho t).$$

The corresponding probability density function is

$$f(t) = \lambda\rho\exp[\rho t + \lambda(1 - e^{\rho t})],$$

and the survivor function

$$S(t) = \exp[\lambda(1 - e^{\rho t})].$$

The probability of dying between ages t_1 and t_2 is measured by the difference

$$S(t_1) - S(t_2).$$

(b) *Gompertz-Makeham model*

This was proposed in 1860 to take account of accidental deaths, which are independent of age. It also applies to adults. Its hazard function is expressed

$$h(t) = \lambda_1 + \lambda\rho\exp(\rho t).$$

Its probability density function is

$$f(t) = (\lambda_1 + \lambda\rho\exp(\rho t))\exp[-\lambda_1 t + \lambda(1 - e^{\rho t})]$$

and survivor function

$$S(t) = \exp[-\lambda_1 t + \lambda(1 - e^{\rho t})].$$

(c) *Siler model*

This was proposed in 1979 for mortality at all ages. Its hazard function is expressed

$$h(t) = \lambda\rho_1 e^{-\rho_1 t} + \lambda_1 + \lambda_2\rho_2 e^{\rho_2 t}.$$

(continued)

Box 6.1 (continued)

Its probability density function is

$$f(t) = (\lambda\rho_1 e^{-\rho_1 t} + \lambda_1 + \lambda_2\rho_2 e^{\rho_2 t})\exp[-\lambda(1 - e^{-\rho_1 t}) - \lambda_1 t + \lambda_2(1 - e^{\rho_2 t})]$$

and survivor function

$$S(t) = \exp[-\lambda(1 - e^{-\rho_1 t}) - \lambda_1 t + \lambda_2(1 - e^{\rho_2 t})].$$

(d) *Weibull model*

This was proposed in 1951 and has been used by various authors in palaeodemography to model adult mortality with a parameter $\lambda > 1$. Its hazard function is expressed by:

$$h(t) = \lambda\rho(\rho t)^{\lambda-1}.$$

Its probability density function is

$$f(t) = \lambda\rho(\rho t)^{\lambda-1}\exp\left[-(\rho t)^{\lambda}\right]$$

and survivor function

$$S(t) = \exp\left[-(\rho t)^{\lambda}\right].$$

However, although the hazard function does increase with age, it increases more slowly than with the Gompertz model.

6.3.3.3 Maximum Likelihood Estimation

Maximum Likelihood Estimation (MLE, see Box 12.1, Chap. 12) has a long history but has only been used in palaeodemography in recent years, in association with the use of event-history models. This began with Richard R. Paine (1989) who attempted to find the best fit between the distribution of ages at death established from skeletons[11] and the theoretical distribution associated with life tables (in particular, Coale and Demeny's). Other applications followed in order to determine the demographic parameters resulting from this corrected distribution of ages at death (Konigsberg and Frankenberg 1992; Siven 1991a, b; Skytthe and Boldsen 1993).

[11] I.e. the individual ages estimated from osteological remains.

But on top of the problems already mentioned in determining individual age at death, others arise from the use of a possibly imperfect mortality model.

Work by James W. Wood et al. (1992a) to measure populations' health conditions drew attention to the problem of non-stationary populations and to two less familiar biases, those of selective mortality and population heterogeneity. Population heterogeneity is a well-known phenomenon in demography (Vaupel et al. 1979) and mortality analysis now usually takes account of the resistance or susceptibility to death and disease of each individual in the population under study. These considerations enabled Wood et al. (2001, 2002) to propose a "latent trait method" for estimating the distribution of ages at death, taking account of the non-stationary and heterogeneous nature of the population, but at the cost of quite complex equations.

6.3.3.4 Other Models for Estimating Distribution of Ages at Death

In recent years, other mathematical models have been proposed for reconstructing the age-sex structure of a buried population, taking account of the probabilistic constraints of age determination (Müller et al. 2001; Love and Müller 2002), and for testing the validity of their palaeodemographic results by comparing the distribution by age of buried populations not with stable populations but rather with populations characterised by variable fertility and mortality rates (Bonneuil 2005).

More recently still, some researchers have examined the specific features of the mortality of pre-industrial populations (Forfar 2006; Luy and Wittwer-Backofen 2005; 2008; Séguy et al. 2006b, 2008), distinguishing mortality during "normal" periods with that of demographic crises (Bocquet-Appel and Bacro 2008).

6.4 Should Population Dynamics Be Modelled by Fertility or by Mortality?

To monitor human health over time, it was long the practice, and sometimes still is, to focus on mortality, mainly via the average indicators of mean age at death and life expectancy at birth. In the 1970s, as the global population surged, researchers began to consider that fertility had a much stronger impact than mortality on the population pyramid, and hence on the number of people liable to die between two ages.[12]

[12] The idea is that fertility affects the number of births in a year and contributes to widening or narrowing the base of the population pyramid, gradually altering its profile over time. Conversely, mortality, which affects all ages in roughly the same way from 1 year to the next (except for demographic crises), has a more moderate effect on population structure.

These two ways of addressing population dynamics can be identified in the work of palaeodemographers. For example, Robert McCaa, Mary Jackes and Jean-Pierre Bocquet-Appel start from a certain level of fertility, defined by the growth rate of their population, to determine the distribution of ages at death that results from a mortality regime fixed in advance.

For our part, we prefer to consider that the old demographic regime did not enable populations to achieve very high growth rates over the long term. This was because their high fertility was quickly outweighed not only by high "normal" infant and child mortality rates but also by recurrent and sometimes severe mortality crises. Although age-specific fertility rates lie within fairly stable limits from 1 year to the next, age-specific mortality rates can vary considerably in terms of both intensity and of the most severely affected age groups.[13]

Under these conditions, pre-industrial populations could only respond through weaker social control of marriage (and associated births) and the survival reflex whereby couples replaced children who had died young. Equally, short-term episodes of rapid demographic increase or decline should not be overlooked.

[13] Only non-selective epidemics, such as plague, have little or no effect on the age-sex structure of the surviving population (Séguy et al. 2006b).

Chapter 7
Finding the Right Models for Pre-industrial Populations

Contemporary model tables, or mathematical models based on current conditions, are not, in our view, the best tools for describing pre-transitional human behaviour. The theory of demographic transition (Landry 1934; Davis 1945; Notestein 1945; Chesnais 1986) described this shift from a traditional regime, in which high fertility and mortality were roughly balanced, to one of low birth and death rates. Depending on time and place, this shift has taken from a few years to a couple of centuries. It is a consequence of social and economic development, and progress in public health and medicine that improve children's chances of survival. Lower infant and child mortality leads to a reduction in the birth rate.[1] The time lag between the two events (lower mortality generally precedes lower fertility) produces a population boom throughout the transition period, until equilibrium is restored between births and deaths. Lower fertility and mortality cause both bottom-up population ageing (as the base of the age pyramid narrows) and top-down ageing (as people live longer).

This theory invalidates the hypothesis of a continuity of demographic behaviour from pre-historic times to the present which underpins the use of contemporary models to study populations of the past.

Specific models must therefore be developed for pre-industrial populations, along the lines of Jean-Pierre Bocquet-Appel and Claude Masset's "estimators". The aims are threefold:

- To use a sufficiently large corpus of life tables statistically representative of mortality among pre-industrial populations;
- To propose inputs easily accessible from historical or osteological sources;
- To take account of the growth rate, either positive or negative, of the population.

[1] Whether mortality reduction precedes or follows fertility decline is still subject to discussion. In some cases, it appears that they have been simultaneous or even reversed (Coale and Watkins 1986).

I. Séguy and L. Buchet, *Handbook of Palaeodemography*, INED Population Studies 2,
DOI 10.1007/978-3-319-01553-8_7,

7.1 Constraints To Be Included in the Models

7.1.1 A Mortality Pattern Specific to Pre-industrial Populations

Whatever their lifestyles or the latitudes where they live, most populations that have not started their demographic transition present very high rates of fertility and child mortality. Their pattern of mortality by age differs noticeably from what is now observed in developed countries (Figs. 7.1 and 7.2): the risk of dying is very high during the first years of life, declines after age 5 and gradually rises again after age 20. This means a low life expectancy at birth, very similar to life expectancy at age 20.

The specific features and variability of the demographic behaviour of European populations from Antiquity to the Middle Ages are largely unknown. Consequently, by using as wide a basis as possible for our models we can claim a certain universality, even though we must accept a wider margin of error.

7.1.2 Models Constructed from Inputs Accessible to Palaeodemographers

We have already pointed out that, unlike historical demography based on pre-statistical written evidence, osteological data, because of the margin of

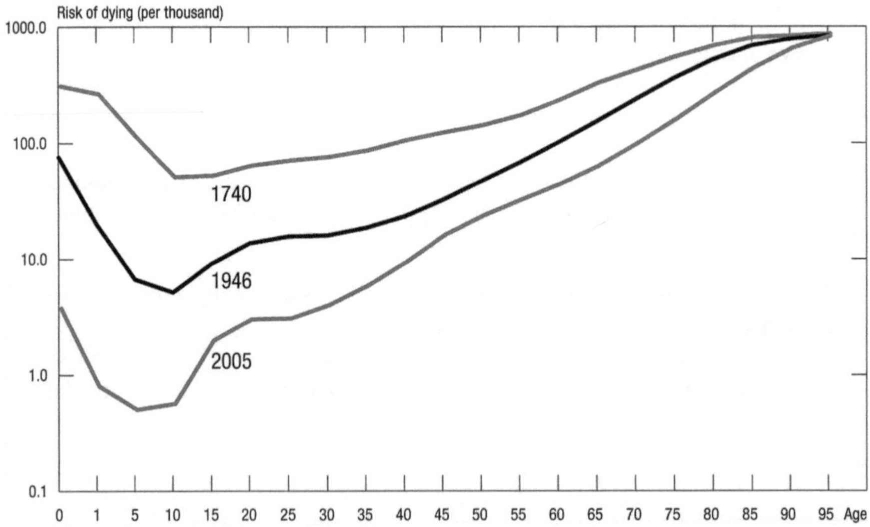

Fig. 7.1 Risk of dying at a given age in France, 1740–2005 (both sexes). *Sources*: Yves Blayo 1975; France Meslé and Jacques Vallin 2001

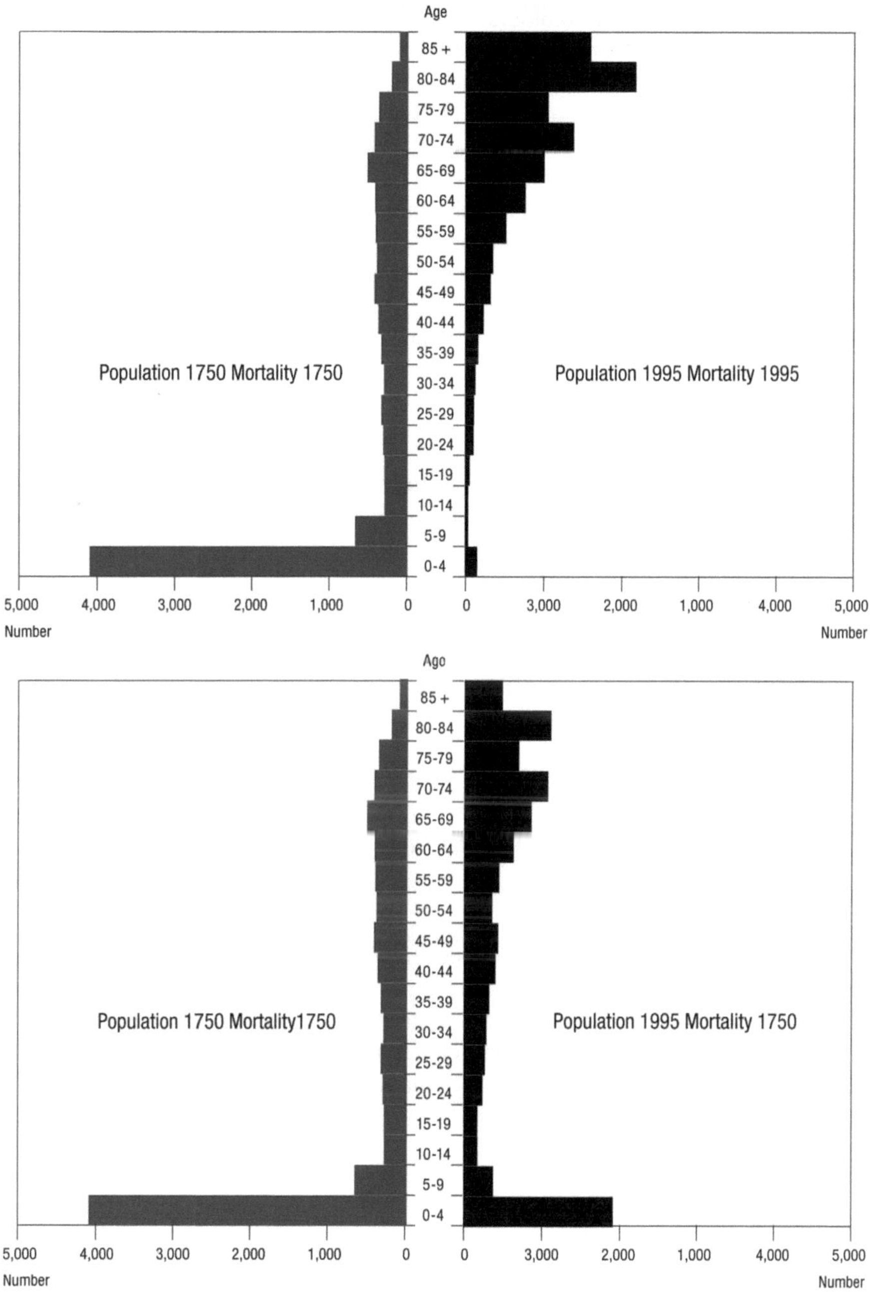

Fig. 7.2 Distribution of age at death in France, 1740 and 1995, according to population structure and mortality distribution. *Key*: *Top*, observed distribution (i.e. corresponding to the 1750 and 1995 population pyramids); *bottom*, distribution of deaths if the mortality by age of 1740–1749 had applied to France in 1995 (cf. Fig. 7.1)

uncertainty associated with biological evidence, cannot easily be used to calculate probabilities of dying. It is crucial, therefore, to provide specific inputs for mortality models intended for palaeodemographers. For the models to work properly, these inputs must be the variables on which the linear regressions are established.

Certain archaeological constants (such as the virtually systematic under-representation of children under 5 in burial grounds), the various ways of determining age (different for children and adults) and the clear segmentation of the mortality curve between children and adults,[2] all make it necessary to use variables representative of these two age groups. Variables must be chosen that provide information separately about the under-20s[3] (excluding ages ages 0–4 because of their under-representation) and adults (over-20s or over-18s depending on the age indicators used).

7.1.3 Margin of Error and Palaeodemographic Sample Size: Two Variables to Consider

The results proposed must take account of both the quality of the estimates, as measured by the residual standard deviation, and the size of the palaeodemographic sample. The margin of error of the models is generally minor, but the small size of some palaeodemographic samples, which must be considered, may affect the expected results.

7.1.4 Growth Rate Is Difficult to Measure, but Cannot Be Ignored

As we saw in the previous chapter, the hypothesis of a zero-growth (so-called "stationary") population was long accepted by palaeodemographers for the reason that over the long periods of burial ground use (several generations, even centuries), the regulation mechanisms of traditional populations must have maintained a growth rate close to zero, reflecting the balance between births and deaths.

---

[2] Research has shown that child and adult mortality in the past could vary independently and that the higher the general rate of mortality the weaker the correlation between the adult and child rates (Woods 1993; Wrigley et al. 1997).

[3] Anthropologically speaking, maturity begins at around age 18 (see above). Demographically, we could have taken that threshold if we had been working with 1-year age groups. Given the unreliability of pre-statistical data, we have preferred to work with 5-year age groups, and set the transition point between immature and mature according to the age groups adopted (15–19 and 20–24 years).

However, recent findings in historical demography have undermined this hypothesis: the image of a population where moderate growth is occasionally upset by crises of excess mortality surely comes closer to the realities of the traditional demographic regime (high mortality and high fertility).

The stable population hypothesis is based on the idea of natural growth due solely to the balance between births and deaths; it assumes a total lack of migration (closed population hypothesis). However, in the past as in the present, population dynamics are partly due to migration (inward and outward), which reinforces or counteracts the effects of natural increase. These movements must therefore be taken into account, even if measuring them from reliable osteological or archaeological evidence will raise some problems of its own.

7.2 Modelling

7.2.1 Collecting the Sample of Observed Tables

Starting from an initial corpus of about a 1,000 tables taken from statistical yearbooks and demographic publications up to 1997, we compiled a purposive sample of 292 life tables[4, 5] corresponding to populations presenting the features defined for pre-industrial populations in terms of lifestyles (farming populations, little urbanisation, no modern medical care) and mortality (high infant and child mortality).

Data presentation was harmonised, some probabilities (generally for the oldest age groups) had to be estimated, and the tables were tested to detect and eliminate those containing inconsistencies. The sample of selected tables then underwent statistical analysis[6] in order to group tables with similar characteristics in terms of both level and structure of mortality, and for differential male/female mortality. The results of these analyses were used to individualise the corpus of tables to be

[4] This work was aided by Stéphane Renard, for data entry, and, at various stages of the statistical study, by Magali Belaigues-Rossard, Luc di Benedetto, Paul Bermier, Bertrand Buffière, Annie Carré, Nadège Couvert, Benoît Haudidier and Carole Perraut, on short-term contracts at INED. Our thanks for all their help. We also wish to warmly acknowledge the work of Arnaud Bringé who oversaw and guided all the stages in this lengthy research project.

[5] The "both sexes" sample comprises 292 life tables, the "male" sample 290 and the "female" sample 286.

[6] Two main methods were used: random k-means clustering and automatic table classification. These discriminant analyses were based on a series of indicators specially defined to characterise the "oldest" life tables: discriminating variables chosen to measure changes in mortality structure: $\frac{{}_1q_0}{{}_4q_1}, \frac{{}_5q_5}{{}_5q_{10}}, \frac{{}_{20}q_{20}}{{}_{20}q_{40}}, \frac{{}_1q_0}{e_{20}}$ and $\frac{e_{5-15}}{e_{20}}$; discriminating variables chosen to distinguish changes in level: e_{5-15} and e_{20-40}; juvenility index $\left(\frac{D_{5-14}}{D_{20-\omega}}\right)$ and the ratio $\frac{D_{5-9}}{D_{10-14}}$ (where $D =$ number of deceased in the table); variables chosen to distinguish differential mortality by sex: ${}_1q_{10}$, e_{20}, e_{20-40} and e_{40-60}.

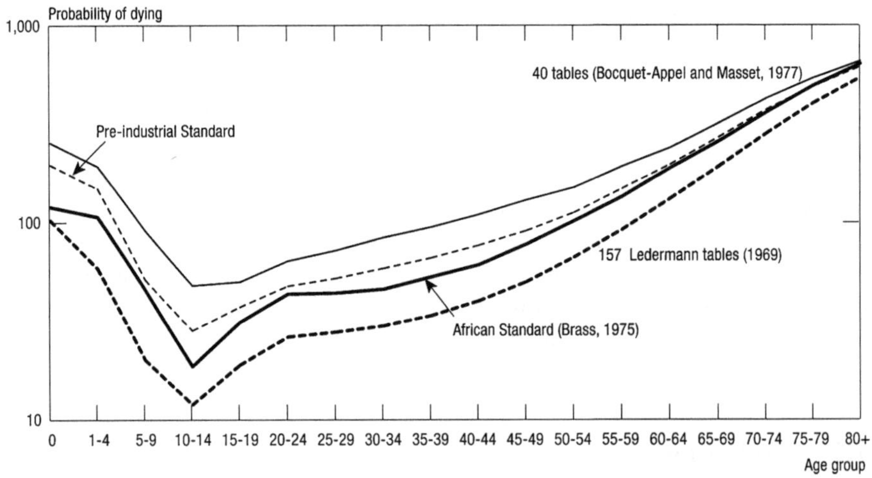

Fig. 7.3 Mortality regimes underlying various models

chosen for the modelling: 167 "both sexes" tables,[7] 139 "female" tables and 147 "male" (the selected tables are listed in Table 8.8 at Springer Extra). The mean pattern in this sample of tables (Fig. 7.3) is noticeably different from those of existing models.[8] It comes closer to the values of the African standard (Brass 1975) than those of Bocquet-Appel and Masset (1977) or Ledermann (1969).

7.2.2 *Choice of Mathematical Model*

With this sample, a large number of linear regressions were run, relating a given parameter to all the estimated probabilities of dying. Two mathematical models were used: the logarithmic model (described in Chap. 8) and the logistical model (Chap. 9).

The two models were used to estimate levels of mortality at various ages x, by reference to data observed at other times and places (i.e. the 167 reference tables). They involve accepting the hypothesis that all the populations to which these models are applied pertain to pre-industrial mortality patterns.

[7] The sample comprises historical tables from industrialised countries (seventeenth to nineteenth centuries) and contemporary tables from developing countries (end-nineteenth to twentieth centuries), all selected according to the level of the demographic indicators cited above. No sub-group stands out: the average level of mortality in the developing countries is significantly very close to that observed in the historical tables of the industrialised countries (Student's T-test: $T = 0.023$, significance level: 0.05) except for three probability values (${}_4q_1$, ${}_5q_5$, ${}_5q_{65}$). The same holds for mortality structure: no significant difference emerges between the two samples, except for three probability values (${}_4q_1$, ${}_5q_5$, ${}_5q_{65}$).

[8] Not all Coale and Demeny's source tables have been published, so they could not be included in this comparison.

7.2.3 Choice of Entry Parameters

Since the preceding value and following value methods cannot be used with palaeodemographic data, only entry parameters compatible with reliable osteological indicators were chosen, such as:

- Juvenility index $JI = \frac{D_{5-9}}{D_{10-14}}$ denoted “JI”
- Bocquet-Appel (2002)’s proposed ratio $P = \frac{D_{5-19}}{D_{20-\omega}}$, “P”
- Mean adult age at death[9] “a_{20}”
- Mean age of children deceased between ages 5 and 14[10] “a_{5-14}”

Care must be taken to ensure that the palaeodemographic variable estimated from osteological indicators and used to enter the model is not considered as strictly identical to the demographic variable calculated from life table data and used for various regressions, although we cannot measure the discrepancy (except for special cases: Signoli et al. 2005).

By convention, we shall call the palaeodemographic variable “estimated *JI*”, “estimated *P*” or “estimated a_{20}”.

7.2.4 Including the Population Growth Rate

In the absence of sufficient data to provide sub-samples for various growth rates, we artificially created the stable populations associated with each of the tables in our samples.[11] For that purpose, we calculated the distribution of death by age (and survivors by age) from the mortality distribution of the observed population and a growth rate lying within an interval of −3 % and +3 %, by 0.25 % increments,[12] making 25 values including zero.

[9] Method for calculating this variable in the tables: $[(22.5 \times d_{20-24}) + (27.5 \times d_{25-29}) + (32.5 \times d_{30-34}) + (37.5 \times d_{35-39}) + (42.5 \times d_{40-44}) + (47.5 \times d_{45-49}) + (52.5 \times d_{50-54}) + (57.5 \times d_{60-64}) + (62.5 \times d_{65-69}) + (67.5 \times d_{70-74}) + (72.5 \times d_{75-79}) + (77.5 \times d_{80-84}) + (88 \times d_{80-\omega})]/d_{20-\omega}$

Method for calculating this variable for the osteological series: see Chap. 6.

[10] Method for calculating this variable in the tables: $[(7.5 \times d_{5-9}) + (12.5 \times d_{10-14})]/d_{5-14}$

Method for calculating this variable for the osteological series: see Chap. 6.

[11] However, the growth rates of the populations for which we have life tables were not always specified in the publications. By convention, we have considered that their growth rates were very low, and have constructed our stable populations on that assumption.

[12] In this way we obtain a network of 24 stable populations associated with each table in our sample (namely 167 × 24 = 4,008 tables for the “both sexes” sample; 147 × 24 = 3,528 tables for our “male” sample and 139 × 4 = 3,336 tables for the “female” sample).

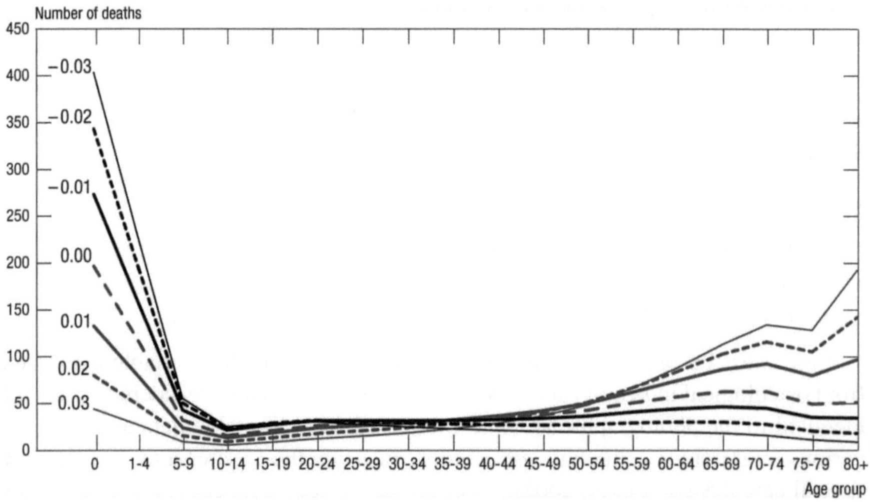

Fig. 7.4 Distribution of age at death for various values of *r*. N.B. The example is taken from the mean table of the "both sexes combined" pre-industrial standard

In practice, the range of demographic situations affecting pre-industrial populations was much narrower,[13] with the exception of mortality crises and the subsequent periods of recovery – or continued depression. In these special but recurring cases, the population losses might also exceed, even substantially, the limits set here.

We thus have a wide range of stable populations that enable us to measure the relationship between the various demographic parameters, mortality distribution and age structure of the population. Figure 7.4 illustrates the variations in distribution of age at death that result from a fixed mortality distribution and variable growth rates, and Fig. 7.5 the ensuing modifications in population structure.

7.2.5 Quality and Presentation of the Models

To obtain the best model, we analysed the studentised residuals and the adjusted R^2 value of each regression before excluding outlying tables. The exclusion procedure

[13] Under the old demographic regime, long-term population growth was hindered not only by high mortality (particularly among infants and children) but also by frequent mortality crises (see above). In such situations, growth rates of 2.5 % or 3 %, as seen in contemporary observations (Pison 2007) are hardly plausible in the long term. For example, Bocquet-Appel and Masset (1977) proposed ten growth rates (plus and minus 2 %; 1 %; 0.5 %; 0.2 %; 0.1 %); Henneberg and Steyn (1994) proposed growth rates in increments of 0.5 from −1 % to +5 %; and Bocquet-Appel and Masset (1996) used growth rates from −2.5 % to +2.5 % by increments of 0.25.

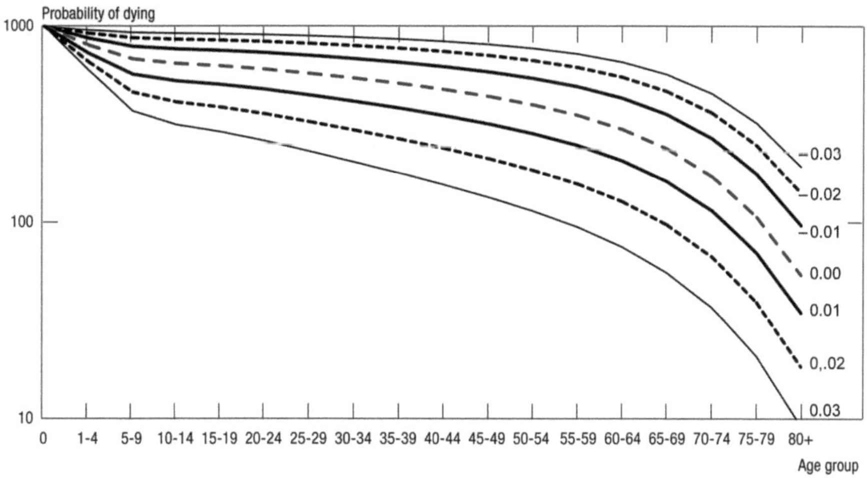

Fig. 7.5 Distribution of survivors by age for various values of r. N.B. The example is taken from the mean table of the "both sexes combined" pre-industrial standard

complies with fixed rules that limit operator subjectivity and enable the program to be re-run (see Chap. 8, note 2). The significance level of the model's parameters was also tested automatically in order to retain only those whose p-value was lower than 0.005.

Each estimate comes with the value of the residual standard deviation and adjusted R^2; both of these can be use to assess the quality of the proposed model.

There are models for each sex and both sexes combined to allow for difficulties in determining the sex of children. Due to the number of models proposed (75 per parameter), we can give only an overview here for various values of r, but the reader may consult the source documents (in the Tables available at Springer Extra).

Chapter 8
Model Tables for Pre-industrial Populations

8.1 Logarithmic Model

This model relates the logarithm of probabilities of dying to a given variable expressed either as a logarithm or a pure number. Estimation by the successive probability method uses the model proposed by Sully Ledermann (1969). The method consists of constructing a series of regression equations relating each probability to the preceding probability (or the following one, for inverse regressions). In this way it avoids the bias due to chain estimation, because the entry value can only be used for reliable estimation of the immediately following value, and the correlation rapidly weakens for later values.

This principle has been adapted to palaeodemographic data, with equations relating the log probabilities of dying to a given palaeodemographic indicator (X) expressed as a pure number or its logarithm. The mathematical model is of the following form:

$$\log_a q_{(x+a)} = a_0 + a_1 X + \varepsilon\,(\pm 2\sigma)$$

or

$$\log_a q_{(x+a)} = a_0 + a_1 \log X + \varepsilon(\pm 2\sigma)$$

where ε is a white Gaussian noise (or stochastic error).

The first results of this work were published in 2006 and 2008 with help from Magali Belaigues-Rossard, Paul Bernier, Nadège Couvert, Carole Perraut and Arnaud Bringé (Séguy et al. 2006b, 2008). Here we present the final version of the mortality models for palaeodemographers, and a presentation currently in preparation will include variables of use to historical demographers.

I. Séguy and L. Buchet, *Handbook of Palaeodemography*, INED Population Studies 2,
DOI 10.1007/978-3-319-01553-8_8, © Springer International Publishing Switzerland 2013

8.1.1 Preliminary Hypotheses for the Multiple Linear Regression

To model the demographic parameters associated with a fixed mortality distribution and chosen growth rate, the simplest solution would have been a multiple linear regression of the following type:

$$\log_a q_{(x)} = a_0 + a_1 X + a_2 r + \varepsilon$$

where X is the palaeodemographic variable, r the growth rate and ε white Gaussian noise.

However, the collinearity that affects the equation variables (Table 8.11), since X is calculated as a function of r, would make such a model unstable. It was preferable, therefore, to propose models for each chosen value of r, operating independently from each other.

This network of model tables makes it possible to start from one or more palaeodemographic variables and estimate part or all of the life table for each chosen value of r. This amounts to having a panel of stable populations – or quasi-stable if we assume that the stable state is achieved instantaneously – associated with each of the observed mortality distributions.

8.1.2 Preliminary Hypotheses for the Simple Linear Regression

For the model to operate properly, certain initial hypotheses need to be validated (Poulain 1981, pp. 69–71):

In the regression

$$\log_a q_{(x+a)} = a_0 + a_1 X + \varepsilon$$

where $\log_a q_{(x+a)}$ is the dependant variable, X the explanatory variable, a_0 and a_1 the parameters of the model, and ε the stochastic error,

- The variables X and $\log_a q_{(x+a)}$ must be observed without measurement error;
- The relations between X and $\log_a q_{(x+a)}$ must be linear (linearity hypothesis, cf. Fig. 8.1);
- The distribution of the error term ε is normal (residual normality hypothesis);
- The error term ε must have a zero mean and constant variance with respect to each of the variables $\log_a q_{(x+a)}$ and X (homoscedasticity hypothesis);
- These same values of ε must be independent of each other (zero covariance);
- No spatial or temporal auto-correlation.

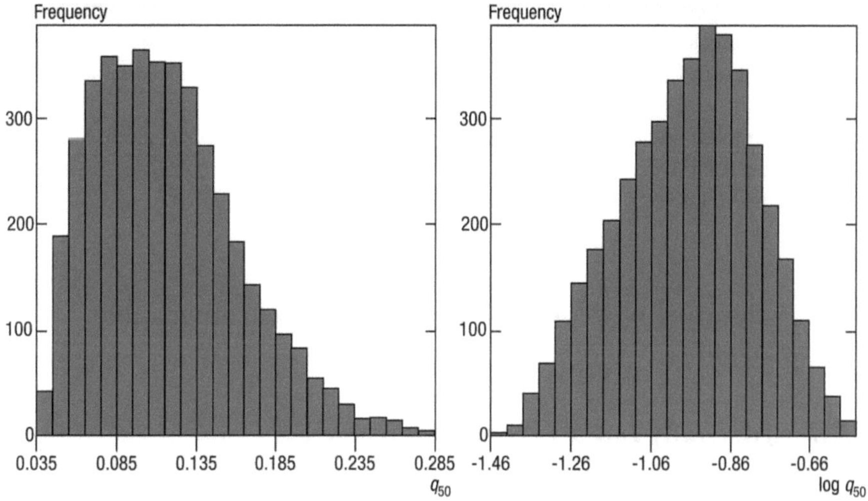

Fig. 8.1 Observed distributions of the value and logarithm of the probability of dying q_{50}. *Key*. *Left*, the distribution of the probability of dying at ages 50–54 ($_q q_{50}$) in the chosen sample of tables is approximately log-normal; *right*, the distribution of the logarithm of this probability ($\log_5 q_{50}$) is approximately normal

All of these conditions have been verified for the model presented below, using a software procedure of preliminary variable analysis, automatic exclusion of outlying tables and quality controls of the proposed models.[1]

8.1.3 Quality of Proposed Regressions

For all the proposed regressions

- The p-values associated with the parameter estimates are all very low (approximately 10^{-17}), showing that each of the estimated parameters is significantly non-zero;
- The adjusted correlation coefficients (R^2) are of good quality. We chose the 0.8 threshold for our models. Where the regressions have an R^2 below that threshold,

[1] The exclusion of outlying tables is not based on a simple visual examination of the values of the studentised residuals, an inevitable source of error, but on a program that guarantees rigorous selection. Outlying observations are analysed via Residual Student with constraints fixed as a function of the value of R^2, a measure of model quality. The greater R^2, the weaker the constraints on Residual Student, so that the model quality takes account of data uncertainty, which becomes an increasingly important factor as the model is refined. The significance tests of the model's parameters are also based on an automatic program that tests the p-value associated with each of these coefficients so that only models with a p-value below 0.05 are retained. A further constraint has been added to the model: a non-zero constant to ensure comparability between models.

they are greyed out to signal lower reliability for the estimate (e.g. in Table 8.13). All the regressions for each indicator, including those we did not choose because of their poor R^2, are available in the supplementary material on the INED website;
- The estimated standard deviation of each model is acceptable. It can be used to calculate relatively precise confidence intervals for the probabilities estimated by the models.

8.2 Estimating of Probabilities from Palaeodemographic Variables

8.2.1 *Juvenility Index* (JI) *and Indicator* P

The linear regression uses the logarithm of these variables, whose distribution is approximately log-normal. The models are

$$\log_a q_{(x+a)} = a_0 + a_1 X + \varepsilon\ (\pm 2\sigma)$$

and

$$\log_a q_{(x+a)} = a_0 + a_1 \log P(\pm 2\sigma)$$

with

$JI = \frac{D_{5-14}}{D_{20-\omega}}$ and $P = \frac{D_{5-19}}{D_{5-\omega}}$. These indicators vary noticeably as a function of the growth rate (Fig. 8.2).

For a set growth rate, the two indicators correlate well, particularly with the logarithms[2] of the early probabilities ${}_1q_0$ to ${}_5q_{10}$ (see Table 8.1).[3]

However, whatever the value of r, the juvenility index provides a better quality of estimate for the first three probabilities than indicator P; conversely, indicator P markedly improves the estimate for ${}_5q_{10}$, as long as the population is growing ($r \geq 0$). The estimate for ${}_4q_1$ is still very poor with indicator P ($R^2 \leq 0.8$, whatever the growth rate considered). It is better to use the juvenility index, where R^2 is always greater than 0.8 (Tables 8.2 and 8.3).

Although the two indicators correlate less well with adult mortality, they can be used to estimate some age-specific probabilities of dying, as long as a lower regression quality is accepted ($R^2 \geq 0.78$) and preference is given to a situation of

[2] Use of logarithms is justified by the log-normal distribution of values for the juvenility index.

[3] Since anthropological determination of sex is still a difficult task, we present here only a sample of the modelling done on the "both sexes" sample. All the regressions are available in the additional material on the INED website.

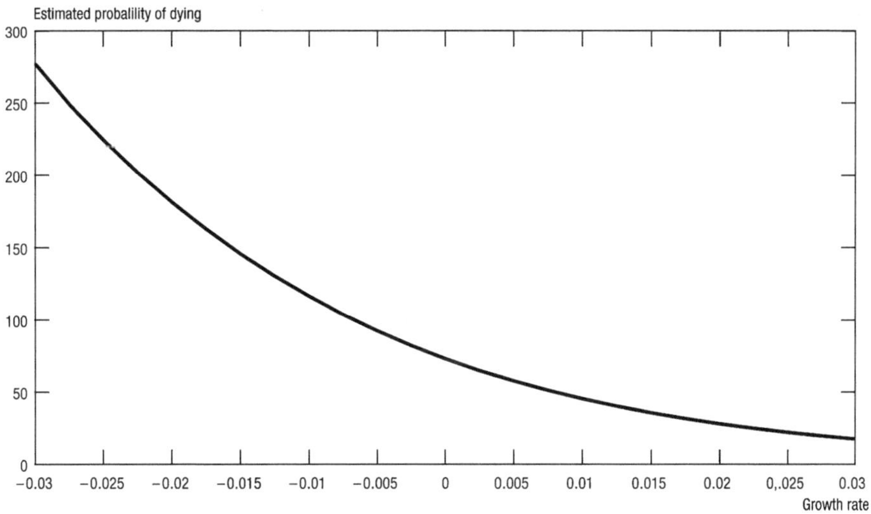

Fig. 8.2 Estimation of $_5q_5$ as a function of the growth rate for a juvenility index of 0.127

Table 8.1 Correlation between inputs and growth rate

Variable	a_{5-14}	a_{20+}	*JI*	log *JI*	*P*	log *P*
Growth rate	−0.322	−0.852	0.762	0.881	0.86	0.898

population decline ($r \leq -0.015$). Under these conditions, indicator *P* will estimate $_5q_{40}$, $_5q_{45}$, $_5q_{50}$ and even $5q30$ in extreme cases. The juvenility index, on the other hand, can be used to estimate $5q30$ when $r \leq -0.0025$ (Table 8.9 at Springer Extra).

However, along with the quality of the estimator as a demographic variable (i.e., calculated from the parameters of the observed life tables), consideration must also be given to the palaeodemographic indicator taken from anthropological observation.

The indicator "estimated *P*" has two advantages over the juvenility index, which has been variously criticised (see Chap. 5): it is easier to observe (the maturity threshold is more accurately determined anthropologically) and it covers all the ages in the exhumed population above 5 years. It remains to be verified from historically documented series whether these advantages justify opting for "estimated *P*" instead of "estimated *JI*".

8.2.2 Mean age of Deceased Children Aged 5 to 14: a_{5-14} (Both Sexes)

We looked for a palaeodemographic indicator that would not reproduce the failings of the juvenility index (particularly the archaeological inventory of 5–14-year-olds). The mean age of children dying at ages 5–14 is a data point accessible to anthropologists (Buchet et al. 2006b, and this book).

Table 8.2 Estimation model for parameter *JI*. Both sexes combined for various values of *r*

$\text{Log}_a q_x$	Explanatory variable	a_0	a_1	Adjusted R^2	$\sigma\text{Log}_a q_x$	Final number of tables	Value of *r*
$\text{Log}_1 q_0$	Log *IJ*	−0.335	0.441	0.823	0.042	129	+0.010
$\text{Log}_1 q_4$	Log *IJ*	−0.365	0.573	0.808	0.054	128	
$\text{Log}_5 q_5$	Log *IJ*	−0.387	1.065	0.979	0.030	163	
$\text{Log}_5 q_{10}$	Log *IJ*	−0.879	0.787	0.886	0.054	163	
$\text{Log}_1 q_0$	Log *IJ*	−0.318	0.437	0.822	0.042	129	+0.0075
$\text{Log}_1 q_4$	Log *IJ*	−0.327	0.580	0.803	0.056	131	
$\text{Log}_5 q_5$	Log *IJ*	−0.346	1.052	0.977	0.032	164	
$\text{Log}_5 q_{10}$	Log *IJ*	−0.846	0.780	0.895	0.052	162	
$\text{Log}_1 q_0$	Log *IJ*	−0.310	0.423	0.808	0.042	129	+0.0050
$\text{Log}_1 q_4$	Log *IJ*	−0.316	0.565	0.804	0.055	129	
$\text{Log}_5 q_5$	Log *IJ*	−0.304	1.043	0.974	0.034	165	
$\text{Log}_5 q_{10}$	Log *IJ*	−0.811	0.777	0.904	0.050	161	
$\text{Log}_1 q_0$	Log *IJ*	−0.285	0.427	0.810	0.043	131	+0.0025
$\text{Log}_1 q_4$	Log *IJ*	−0.294	0.557	0.801	0.055	129	
$\text{Log}_5 q_5$	Log *IJ*	−0.264	1.029	0.970	0.036	166	
$\text{Log}_5 q_{10}$	Log *IJ*	−0.779	0.770	0.901	0.051	163	
$\text{Log}_1 q_0$	Log *IJ*	−0.269	0.422	0.809	0.043	131	0.0000
$\text{Log}_1 q_4$	Log *IJ*	−0.300	0.530	0.807	0.052	124	
$\text{Log}_5 q_5$	Log *IJ*	−0.224	1.017	0.967	0.037	166	
$\text{Log}_5 q_{10}$	Log *IJ*	−0.747	0.761	0.910	0.048	162	
$\text{Log}_1 q_0$	Log *IJ*	−0.252	0.416	0.807	0.043	131	−0.0025
$\text{Log}_1 q_4$	Log *IJ*	−0.272	0.529	0.804	0.053	125	
$\text{Log}_5 q_5$	Log *IJ*	−0.185	1.004	0.964	0.039	166	
$\text{Log}_5 q_{10}$	Log *IJ*	−0.714	0.754	0.914	0.047	162	
$\text{Log}_1 q_0$	Log *IJ*	−0.243	0.403	0.818	0.040	126	−0.0050
$\text{Log}_1 q_4$	Log *IJ*	−0.263	0.513	0.805	0.052	123	
$\text{Log}_5 q_5$	Log *IJ*	−0.146	0.991	0.961	0.041	166	
$\text{Log}_5 q_{10}$	Log *IJ*	−0.679	0.750	0.922	0.045	161	
$\text{Log}_1 q_0$	Log *IJ*	−0.229	0.398	0.810	0.041	127	−0.0075
$\text{Log}_1 q_4$	Log *IJ*	−0.245	0.505	0.801	0.052	123	
$\text{Log}_5 q_5$	Log *IJ*	−0.107	0.978	0.957	0.043	166	
$\text{Log}_5 q_{10}$	Log *IJ*	−0.646	0.744	0.929	0.043	160	
$\text{Log}_1 q_0$	Log *IJ*	−0.211	0.394	0.825	0.040	124	−0.0100
$\text{Log}_1 q_4$	Log *IJ*	−0.227	0.497	0.799	0.053	123	
$\text{Log}_5 q_5$	Log *IJ*	−0.070	0.965	0.953	0.045	166	
$\text{Log}_5 q_{10}$	Log *IJ*	−0.614	0.736	0.931	0.042	160	

Since the distribution of a_{5-14} values is approximately log-normal, the linear regression introduces the logarithm of the variable as follows

$$\log_a q_{(x+a)} = a_0 + a_1 \log a_{5-14} (\pm 2\sigma).$$

However, this indicator does not display a close linear correlation with the population growth rate, because dispersion is too high. It produces low correlations

Table 8.3 Estimate model for parameter *P*. Both sexes combined for various values of *r*

Log $_aq_x$	Explanatory variable	a_0	a_1	Adjusted R^2	σ Log $_aq_x$	Final number of tables	Value of r
Log $_1q_0$	Log *P*	−0.2106	0.6369	0.808	0.044	127	+0.0100
Log $_1q_4$	*Log P*	*−0.2686*	*0.7478*	*0.763*	*0.057*	*120*	
Log $_5q_5$	Log *P*	−0.1391	1.4785	0.908	0.063	162	
Log $_5q_{10}$	Log *P*	−0.6406	1.1641	0.909	0.048	162	
Log $_5q_{30}$	Log *P*	−0.9081	0.4676	0.806	0.030	93	
Log $_1q_0$	Log *P*	−0.1926	0.6269	0.817	0.043	125	+0.0075
Log $_1q_4$	*Log P*	*−0.2598*	*0.7236*	*0.759*	*0.057*	*120*	
Log $_5q_5$	Log *P*	−0.1193	1.4319	0.906	0.064	161	
Log $_5q_{10}$	Log *P*	−0.6221	1.1318	0.912	0.047	162	
Log $_5q_{30}$	Log *P*	−0.8942	0.4621	0.808	0.030	94	
Log $_1q_0$	*Log P*	*−0.0155*	*0.5727*	*0.762*	*0.046*	*130*	+0.0050
Log $_1q_4$	Log *P*	−0.1846	0.6080	0.808	0.044	126	
Log $_5q_5$	*Log P*	*−0.2511*	*0.7003*	*0.754*	*0.058*	*120*	
Log $_5q_{10}$	Log *P*	−0.0997	1.3869	0.903	0.065	160	
Log $_1q_0$	Log *P*	−0.1863	0.5760	0.805	0.042	123	+0.0025
Log $_1q_4$	*Log P*	*−0.2712*	*0.6506*	*0.708*	*0.062*	*125*	
Log $_5q_5$	Log *P*	−0.0801	1.3438	0.900	0.066	159	
Log $_5q_{10}$	Log *P*	−0.5994	1.0541	0.914	0.047	163	
Log $_1q_0$	Log *P*	−0.1799	0.5559	0.805	0.042	122	0.0000
Log $_1q_4$	*Log P*	*−0.2132*	*0.6739*	*0.641*	*0.076*	*141*	
Log $_5q_5$	Log *P*	−0.0607	1.3033	0.893	0.068	159	
Log $_5q_{10}$	Log *P*	−0.5801	1.0266	0.915	0.047	163	
Log $_1q_0$	*Log P*	*−0.1818*	*0.5314*	*0.776*	*0.045*	*127*	−0.0025
Log $_1q_4$	*Log P*	*−0.2044*	*0.6527*	*0.635*	*0.077*	*141*	
Log $_5q_5$	Log *P*	−0.0411	1.2647	0.886	0.070	159	
Log $_5q_{10}$	Log *P*	−0.5606	1.0001	0.915	0.047	163	
Log $_1q_0$	Log *P*	−0.1661	0.5195	0.800	0.043	122	−0.0050
Log $_1q_4$	*Log P*	*−0.2589*	*0.5839*	*0.649*	*0.069*	*131*	
Log $_5q_5$	Log *P*	−0.0214	1.2279	0.878	0.072	159	
Log $_5q_{10}$	Log *P*	−0.5411	0.9748	0.914	0.047	163	
Log $_1q_0$	Log *P*	−0.1648	0.4969	0.800	0.042	120	−0.0075
Log $_1q_4$	*Log P*	*−0.2507*	*0.5663*	*0.643*	*0.069*	*131*	
Log $_5q_5$	Log *P*	−0.0015	1.1928	0.870	0.075	159	
Log $_5q_9$	Log *P*	−0.5214	0.9505	0.913	0.047	163	
Log $_5q_{50}$	Log *P*	−0.2891	0.6054	0.812	0.045	121	
Log $_1q_0$	*Log P*	*−0.1811*	*0.4699*	*0.720*	*0.050*	*133*	0.0100
Log $_1q_4$	*Log P*	*−0.2424*	*0.5496*	*0.636*	*0.070*	*131*	
Log $_5q_5$	Log *P*	0.0209	1.1625	0.859	0.078	160	
Log $_5q_{10}$	Log *P*	−0.5015	0.9273	0.911	0.048	163	
Log $_5q_{50}$	Log *P*	−0.3045	0.5681	0.802	0.046	124	

Note: The threshold chosen for the models is 0.8. The regressions (in italics and smaller characters) have an R^2 below this threshold, so the estimates are less reliable (this convention holds for the following tables)

Table 8.4 Palaeodemographic estimation model – parameter $a_{5–14}$, – both sexes combined, for various values of r

Log $_aq_x$	Explanatory variable	a_0	a_1	Adjusted R^2	σ Log $_aq_x$	Final number of tables	Value of r
Log $_1q_4$	$a_{5–14}$	2.1425	−0.3277	0.68	0.0611	117	−0.0075
Log $_5q_5$	$a_{5–14}$	3.4949	−0.5225	0.75	0.0863	130	−0.0075
Log $_1q_4$	$a_{5–14}$	2.1475	−0.3287	0.68	0.0611	117	−0.0050
Log $_5q_5$	$a_{5–14}$	3.5037	−0.5242	0.75	0.0863	130	−0.0050
Log $_1q_4$	$a_{5–14}$	2.1527	−0.3298	0.68	0.0611	117	−0.0025
Log $_5q_5$	$a_{5–14}$	3.5128	−0.5260	0.75	0.0863	130	−0.0025
Log $_1q_4$	$a_{5–14}$	2.1581	−0.3309	0.68	0.0611	117	0.0000
Log $_5q_5$	$a_{5–14}$	3.5223	−0.5278	0.75	0.0863	130	0.0000
Log $_1q_4$	$a_{5–14}$	2.1637	−0.3320	0.68	0.0611	117	0.0025
Log $_5q_5$	$a_{5–14}$	3.5321	−0.5297	0.75	0.0862	130	0.0025
Log $_1q_4$	$a_{5–14}$	2.1695	−0.3331	0.68	0.0611	117	0.0050
Log $_5q_5$	$a_{5–14}$	3.5423	−0.5316	0.75	0.0862	130	0.0050
Log $_1q_4$	$a_{5–14}$	2.1755	−0.3343	0.68	0.0611	117	0.0075
Log $_5q_5$	$a_{5–14}$	3.5528	−0.5335	0.75	0.0862	130	0.0075
Log $_1q_4$	$a_{5–14}$	2.1817	−0.3355	0.68	0.0611	117	0.0100
Log $_5q_5$	$a_{5–14}$	3.5636	−0.5355	0.75	0.0862	130	0.0100

with infant and child probabilities of dying (0.41 for $_1q_0$; 0.68 for $_4q_1$; 0.75 for $_5q_5$, both sexes[4]) and there is no correlation with the other probabilities of dying.

The variable a_{5-14} is consequently not a very useful parameter for modelling the mortality of buried populations (Table 8.4): composite indicators like *JI* and *P* are greatly to be preferred.

8.2.3 *Mean Adult Age at Death* (a_{20})

This variable is a composite indicator of adult mortality that is ill-suited for describing the variety of situations throughout this major part of life. However, in the current state of research it is the only relatively reliable palaeodemographic indicator at our disposal (although its method of calculation from a probable distribution by age group involves some variability).

Given the dispersion of a_{20} values (as calculated for each of our reference tables), the linear regression can be based directly on the value of the variable. Its form is

$$\log_a q_{(x+a)} = a_0 + a_1 + a_{20}(\pm 2\sigma).$$

[4] For women, the only possible estimate is for $_5q_5$ (with $R^2 = 0.71$); for men, no satisfactory correlation can be found between this indicator and any of the 18 probabilities of dying (with $R^2 \geq 0.68$).

Mean adult age at death (a_{20}) correlates well with adult and childhood probabilities of dying for the various chosen values of r. It has reasonable explanatory power for the probabilities of dying from ${}_5q_{10}$ to ${}_5q_{65}$, although regression quality noticeably declines for the last two probabilities (${}_5q_{70}$ and ${}_5q_{75}$), which should be estimated in some other way (see below). The infant and childhood probabilities of dying (${}_1q_0$, ${}_4q_1$, ${}_5q_5$) generally correlate poorly with mean adult age at death and are therefore not given in Tables 8.10, 8.11, and 8.12 (at Springer Extra). The ${}_5q_5$ estimate has a satisfactory R^2, if r is positive, but the quality of fit is countered by a very high standard deviation. It remains difficult to estimate ${}_5q_{15}$ for any value of r: to optimise the model a significant number of outlying tables need to be excluded. At present, we have no other osteological indicator to replace one of the three variables presented here for estimating ${}_5q_{15}$ (research is underway).

8.3 Regressions from Probabilities of Dying

These models with palaeodemographic input parameters do not always provide access to all the 18 probabilities. To avoid this bias, while remaining within the same mortality regime and mathematical model, we suggest filling in the missing estimates by using our regressions based on the relationship between probabilities of dying. These models are of more particular interest to historical demographers, whose sources can be used, subject to certain assumptions, to calculate probabilities of dying by sex and age. The models can usefully operate as supplements to the "palaeodemographic" models where regression quality declines (as in the examples given above). Since the probabilities have a roughly log-normal distribution, the regressions are run on the logarithms of the probability values.

For estimation from the preceding probability, the model is as follows:

$$\log_a q_{(x+a)} = a_0 + a_1 \log_a q_x (\pm 2\sigma).$$

And from the following probability:

$$\log_a q_{(x)} = a'_0 + a'_1 \log_a q_{(x+a)} \ (\pm 2\sigma)$$

with ${}_aq_x$ being the probability of an individual aged x dying in the age interval $(x, x + a)$, and σ the residual standard deviation.

8.3.1 *Regressions from the Preceding Probability*

The method consists of constructing a series of regression equations linking each probability to the preceding one. This avoids the problem of bias in chain

Table 8.5 Model for estimation by preceding probability, both sexes combined

Log ${}_{a}q_{x}$	Explanatory variable	a_0	a_1	Adjusted R^2	σ Log ${}_{a}q_{x}$	Final number of tables	Value of r
Log ${}_{1}q_{4}$	*Log ${}_{1}q_{0}$*	*−0.116*	*1.039*	*0.74*	*0.0733*	*148*	0
Log ${}_{5}q_{5}$	*Log ${}_{1}q_{4}$*	*−0.397*	*1.092*	*0.75*	*0.0976*	*147*	
Log ${}_{5}q_{10}$	Log ${}_{5}q_{5}$	−0.685	0.673	0.81	0.0663	158	
Log ${}_{5}q_{15}$	*Log ${}_{5}q_{10}$*	*−0.536*	*0.595*	*0.69*	*0.0686*	*137*	
Log ${}_{5}q_{20}$	Log ${}_{5}q_{15}$	−0.136	0.829	0.86	0.0482	154	
Log ${}_{5}q_{25}$	Log ${}_{5}q_{20}$	0.007	0.971	0.90	0.0422	157	
Log ${}_{5}q_{30}$	Log ${}_{5}q_{25}$	−0.087	0.894	0.90	0.0396	157	
Log ${}_{5}q_{35}$	Log ${}_{5}q_{30}$	−0.055	0.913	0.91	0.0381	155	
Log ${}_{5}q_{40}$	Log ${}_{5}q_{35}$	−0.035	0.911	0.89	0.0390	154	
Log ${}_{5}q_{45}$	Log ${}_{5}q_{40}$	−0.004	0.931	0.91	0.0358	154	
Log ${}_{5}q_{50}$	Log ${}_{5}q_{45}$	−0.008	0.898	0.92	0.0317	155	
Log ${}_{5}q_{55}$	Log ${}_{5}q_{50}$	0.017	0.893	0.92	0.0316	153	
Log ${}_{5}q_{60}$	Log ${}_{5}q_{55}$	−0.084	0.741	0.91	0.0243	152	
Log ${}_{5}q_{65}$	Log ${}_{5}q_{60}$	0.000	0.810	0.90	0.0241	144	
Log ${}_{5}q_{70}$	Log ${}_{5}q_{65}$	−0.014	0.729	0.81	0.0268	156	
Log ${}_{5}q_{75}$	Log ${}_{5}q_{70}$	0.029	0.790	0.83	0.0209	143	

estimation: the input probability can only be used to provide an acceptable estimate of the immediately following one; the correlation rapidly declines for later probabilities. This model has the advantage of providing access to any probability in the life table. It is sufficient to know the probability of dying between two ages. It cannot be directly used with palaeodemographic indicators, but serves to fill in the missing estimates, especially for higher ages. This estimate is independent of the population growth rate because it is based on the relationship between two successive probabilities ("$r = 0$" in the "value of r" column in the following tables).

The preceding probability provides very acceptable adult probabilities of dying (after age 20, whether for both sexes or each sex separately). The estimation of childhood probabilities of dying (${}_{4}q_{1}$ and ${}_{5}q_{5}$; and also ${}_{5}q_{15}$ for both sexes, and ${}_{5}q_{10}$ for the female model) remains a difficult task. The quality of the model ($R^2 \leq 0.8$, see lines in italics and smaller characters in Tables 8.5, 8.6, and 8.7) is affected by the extreme diversity of observed situations. For these ages, it is better to estimate probabilities of dying either from more specific parameters (see above) or from the following probability.

8.3.2 Regressions from the Following Probability

The preceding probability method cannot, by definition, provide a fit for the first one, ${}_{1}q_{0}$. The infant probability of dying needs to be estimated from the following probability or probabilities (${}_{4}q_{1}$ or ${}_{5}q_{5}$), bearing in mind that the best estimates

Table 8.6 Model for estimation by preceding probability, female

Log ${}_aq_x$	Explanatory variable	a_0	a_1	Adjusted R^2	σ Log ${}_aq_x$	Final number of tables	Value of r
Log ${}_1q_4$	*Log* ${}_1q_0$	*−0.289*	*0.777*	*0.57*	*0.0735*	*121*	0
Log ${}_5q_5$	*Log* ${}_1q_4$	*−0.329*	*1.136*	*0.66*	*0.1169*	*127*	
Log ${}_5q_{10}$	*Log* ${}_5q_5$	*−0.676*	*0.663*	*0.77*	*0.0742*	*120*	
Log ${}_5q_{15}$	Log ${}_5q_{10}$	−0.319	0.741	0.80	0.0615	118	
Log ${}_5q_{20}$	Log ${}_5q_{15}$	−0.010	0.922	0.91	0.0392	127	
Log ${}_5q_{25}$	Log ${}_5q_{20}$	−0.017	0.948	0.92	0.0387	132	
Log ${}_5q_{30}$	Log ${}_5q_{25}$	−0.055	0.916	0.92	0.0362	130	
Log ${}_5q_{35}$	Log ${}_5q_{30}$	−0.030	0.940	0.90	0.0405	135	
Log ${}_5q_{40}$	Log ${}_5q_{35}$	−0.056	0.913	0.92	0.0352	131	
Log ${}_5q_{45}$	Log ${}_5q_{40}$	0.026	0.980	0.93	0.0334	130	
Log ${}_5q_{50}$	Log ${}_5q_{45}$	−0.046	0.871	0.94	0.0283	127	
Log ${}_5q_{55}$	Log ${}_5q_{50}$	−0.028	0.852	0.93	0.0293	128	
Log ${}_5q_{60}$	Log ${}_5q_{55}$	−0.024	0.808	0.93	0.0240	127	
Log ${}_5q_{65}$	Log ${}_5q_{60}$	0.025	0.842	0.93	0.0222	124	
Log ${}_5q_{70}$	Log ${}_5q_{65}$	0.004	0.754	0.84	0.0266	133	
Log ${}_5q_{75}$	Log ${}_5q_{70}$	0.045	0.814	0.91	0.0162	118	

come from ${}_5q_5$ rather than ${}_4q_1$ because ${}_4q_1$ displays considerable variability and its values may be higher than those of ${}_1q_0$.[5]

The model for estimation of ${}_1q_0$ is thus expressed

$$\log_1 q_0 = a_0 + a_1 \log_5 q_5 + \varepsilon$$

and for ${}_4q_1$

$$\log_4 q_1 = a'_0 + a'_1 \log_5 q_5 + \varepsilon$$

The modelling has been done for both sexes combined and each sex separately, but only the "both sexes" estimate is relevant in palaeodemography (save in exceptional cases). For the female model, note that the ${}_4q_1$ estimate is of poor quality. This may be due to lower data quality or to the extreme variety of situations. The following tables (Tables 8.8, 8.9, and 8.10) also provide the ${}_5q_5$ estimate, based on the following probability ${}_5q_{10}$, using the same estimation model as before:

$$\log_5 q_5 = a_0 + a_1 \log_5 q_{10} + \varepsilon$$

[5] If ${}_1q_0 \prec {}_4q_1$: the results obtained for the model parameters come very close to those found in the general model. They are only slightly better. If ${}_1q_0 \prec {}_4q$: the model is more precise (according to the data at our disposal), but must be treated with precaution because it is based on a smaller number of tables (N = 11).

Table 8.7 Model for estimation by preceding probability, male

Log ${}_aq_x$	Explanatory variable	a_0	a_1	Adjusted R^2	σ Log ${}_aq_x$	Final number of tables	Value of r
Log ${}_1q_4$	*Log* ${}_1q_0$	*−0.218*	*0.935*	*0.74*	*0.0685*	*121*	0
Log ${}_5q_5$	*Log* ${}_1q_4$	*−0.338*	*1.149*	*0.67*	*0.1340*	*141*	
Log ${}_5q_{10}$	Log ${}_5q_5$	−0.575	0.763	0.90	0.0593	132	
Log ${}_5q_{15}$	Log ${}_5q_{10}$	−0.562	0.579	0.81	0.0545	116	
Log ${}_5q_{20}$	Log ${}_5q_{15}$	−0.024	0.889	0.90	0.0423	138	
Log ${}_5q_{25}$	Log ${}_5q_{20}$	0.050	1.010	0.88	0.0483	139	
Log ${}_5q_{30}$	Log ${}_5q_{25}$	−0.033	0.942	0.91	0.0406	143	
Log ${}_5q_{35}$	Log ${}_5q_{30}$	0.015	0.965	0.92	0.0383	144	
Log ${}_5q_{40}$	Log ${}_5q_{35}$	0.011	0.933	0.93	0.0335	136	
Log ${}_5q_{45}$	Log ${}_5q_{40}$	0.016	0.928	0.94	0.0290	140	
Log ${}_5q_{50}$	Log ${}_5q_{45}$	−0.030	0.870	0.93	0.0296	143	
Log ${}_5q_{55}$	Log ${}_5q_{50}$	−0.002	0.874	0.93	0.0265	138	
Log ${}_5q_{60}$	Log ${}_5q_{55}$	−0.038	0.792	0.94	0.0218	134	
Log ${}_5q_{65}$	Log ${}_5q_{60}$	0.002	0.810	0.92	0.0201	135	
Log ${}_5q_{70}$	Log ${}_5q_{65}$	−0.020	0.710	0.91	0.0162	130	
Log ${}_5q_{75}$	Log ${}_5q_{70}$	0.031	0.785	0.84	0.0187	127	

Table 8.8 Model for estimation by following probability, both sexes combined

Log ${}_aq_x$	Explanatory variable	a_0	a_1	Adjusted R^2	σ Log ${}_aq_x$	Final number of tables	Value of r
Log ${}_1q_0$	Log ${}_5q_5$	−0.183	0.407	0.808	0.043	128	0
Log ${}_1q_4$	Log ${}_5q_5$	−0.142	0.547	0.802	0.057	134	
Log ${}_5q_5$	Log ${}_5q_{10}$	0.485	1.140	0.850	0.099	125	

Table 8.9 Model for estimation by following probability, males

Log ${}_aq_x$	Explanatory variable	a_0	a_1	Adjusted R^2	σ Log ${}_aq_x$	Final number of tables	Value of r
Log ${}_1q_0$	Log ${}_5q_5$	−0.158	0.411	0.812	0.048	117	0
Log ${}_1q_4$	Log ${}_5q_5$	−0.292	0.447	0.805	0.051	105	
Log ${}_5q_5$	Log ${}_5q_{10}$	0.540	1.170	0.900	0.074	131	

Table 8.10 Model for estimation by following probability, females

Log ${}_aq_x$	Explanatory variable	a_0	a_1	Adjusted R^2	σ Log ${}_aq_x$	Final number of tables	Value of r
Log ${}_1q_0$	Log ${}_5q_5$	−0.158	0.411	0.812	0.048	117	0
Log ${}_1q_4$	Log ${}_5q_5$	−0.292	0.447	0.805	0.051	105	
Log ${}_5q_5$	Log ${}_5q_{10}$	0.540	1.170	0.900	0.074	131	

Compared with the preceding probability method, this model improves the estimates for the first two probabilities, but does not achieve the expected level of quality ($R^2 > 0.9$), so we turned to specific inputs, particularly the juvenility index (see below) to fit mortality at young ages.

8.4 Multi-input Models

As we have seen, single inputs only provide information about certain segments of the curve, certain age groups. A dual-input model would cover a wider field, provided that the explanatory variables were not collinear. However, the variables chosen for palaeodemographic models cross-correlate strongly (correlation coefficient greater than 0.70), which would make any model unstable (Table 8.11). The only variable that could be associated with the others is a_{5-14}, which, as we have seen, is ineffective for estimating mortality. Dual-input models cannot therefore be used with the inputs we have specifically defined for the palaeodemographic approach.

We can avoid the problem by proposing the series of regressions that "best" estimates the mortality of buried populations from ages 0 to 80 plus. If for each probability of dying we take the input with the best adjusted R^2 and smallest standard deviation, it is possible to define the "best" palaeodemographic model for a given value of population growth rate (Tables 8.13, 8.14, and 9.1 at Springer Extra and supplementary material on INED website).

For example, the first probability ($_1q_0$) can be estimated from the various palaeodemographic indicators proposed (see above), but since *JI* fits best – countered by a slightly greater standard deviation – it will be preferred for the both sexes models, whatever the population growth rate. In the separate sex models we propose, although determining the sex of buried infants remains a problem, indicator *P* sometimes provides a better estimate than the juvenility index (Tables 8.14 and 9.1 at Springer Extra). However, the differences are minor: slight improvements in the second decimal place for R^2 and standard deviation, and a larger number of tables contributing to the estimation, may swing the

Table 8.11 Cross-correlation of the various inputs (both sexes combined and all values of *r*)

Variables	a_{5-14}	a_{5-14}	*JI*	Log *JI*	*P*	Log *P*
a_{5-14}	1.000	0.305	−0.342	−0.441	−0.259	−0.323
a_{5-14}		1.000	−0.793	−0.755	−0.889	−0.808
IJ			1.000	0.915	0.968	0.975
Log *IJ*				1.000	0.926	0.887
P					1.000	0.530
Log *P*						1.000

Table 8.12 Model for young adult migration, female

Log $_aq_x$	Explanatory variable	a_0	a_1	Adjusted R^2	σ Log $_aq_x$	Final number of tables	Remarks	Value of r
Log $_1q_0$	Log IJ	−0.2673	0.4676	0.846	0.039	109	or Log P	+0.0025
Log $_1q_4$	*Log IJ*	*−0.5384*	*0.3325*	*0.757*	*0.037*	*87*	*or Log P*	
Log $_5q_5$	Log IJ	−0.2488	1.0559	0.948	0.048	138		
Log $_5q_{10}$	Log P	−0.5948	1.0364	0.905	0.049	127	or Log IJ	
Log $_5q_{15}$	a_{20}	0.3483	−0.0290	0.816	0.055	114		−0.0050
Log $_5q_{20}$	Log $_5q_{15}$	−0.0103	0.9219	0.914	0.039	127		
Log $_5q_{25}$	a_{20}	0.5626	−0.0298	0.825	0.057	125		
Log $_5q_{30}$	a_{20}	0.6295	−0.0301	0.875	0.046	130		
Log $_5q_{35}$	a_{20}	0.5606	−0.0304	0.933	0.032	136		+0.0025
Log $_5q_{40}$	a_{20}	0.5394	−0.0292	0.909	0.037	132		
Log $_5q_{45}$	a_{20}	0.6800	−0.0308	0.921	0.035	131		
Log $_5q_{50}$	a_{20}	0.5708	−0.0272	0.892	0.038	129		
Log $_5q_{55}$	a_{20}	0.4988	−0.0239	0.871	0.037	124		
Log $_5q_{60}$	a_{20}	0.3785	−0.0193	0.816	0.038	121		
Log $_5q_{65}$	a_{20}	0.3192	−0.0159	0.815	0.031	113		
Log $_5q_{70}$	Log $_5q_{65}$	0.0037	0.7536	0.844	0.027	133		
Log $_5q_{75}$	a_{20}	0.2382	−0.0097	0.834	0.018	84		

Table 8.13 Model for young adult migration, male

Log $_aq_x$	Explanatory variable	a_0	a_1	Adjusted R^2	σ Log $_aq_x$	Final number of tables[a]	Remarks	Value of r
Log $_1q_0$	Log IJ	−0.2645	0.4247	0.813	0.048	118	or Log P	+0.0025
Log $_1q_4$	Log IJ	−0.3893	0.4709	0.808	0.052	110		
Log $_5q_5$	Log IJ	−0.2554	1.0380	0.973	0.039	146		
Log $_5q_{10}$	Log IJ	−0.7601	0.7868	0.905	0.058	146	or Log P	
Log $_5q_{15}$	a_{20}	0.1920	−0.0275	0.834	0.048	114		−0.0100
Log $_5q_{20}$	a_{20}	0.0424	−0.0227	0.812	0.043	113		
Log $_5q_{25}$	a_{20}	0.5106	−0.0299	0.824	0.054	134		
Log $_5q_{30}$	a_{20}	0.6175	−0.0310	0.896	0.042	134		
Log $_5q_{35}$	a_{20}	0.6742	−0.0331	0.931	0.035	141		+0.0025
Log $_5q_{40}$	a_{20}	0.7996	−0.0337	0.931	0.034	142		
Log $_5q_{45}$	a_{20}	0.7442	−0.0310	0.903	0.039	144		
Log $_5q_{50}$	a_{20}	0.6620	−0.0278	0.907	0.034	136		
Log $_5q_{55}$	a_{20}	0.6055	−0.0248	0.869	0.037	131		
Log $_5q_{60}$	a_{20}	0.4102	−0.0192	0.811	0.034	135		
Log $_5q_{65}$	a_{20}	0.3089	−0.0151	0.820	0.027	108		
Log $_5q_{70}$	Log $_5q_{65}$	−0.0199	0.7100	0.911	0.016	130		
Log $_5q_{75}$	Log $_5q_{70}$	0.0308	0.7853	0.840	0.019	127		

Note: The value of "estimated a_{20}" reflects the mortality of both native-borns and migrants. Even with an identical risk of dying in the age group affected by the migration – here ages 15–35 – the number of deaths observed in that age group will be modified. It will be higher than for a closed population with no immigration (or lower with no emigration). The resulting mean adult age at death is also affected by the number of migrants

Table 8.14 Model for mature adult migration, male

Log $_aq_x$	Explanatory variable	a_0	a_1	Adjusted R^2	σ Log $_aq_x$	Final number of tables	Remarks	Value of r
Log $_1q_0$	Log IJ	−0.2645	0.4247	0.813	0.048	118		+0.0025
Log $_1q_4$	Log IJ	−0.3893	0.4709	0.808	0.052	110		
Log $_5q_5$	Log IJ	−0.2554	1.0380	0.973	0.039	146		
Log $_5q_{10}$	Log IJ	−0.7601	0.7868	0.905	0.058	146		
Log $_5q_{15}$	a_{20}	0.0757	−0.0274	0.832	0.048	119		
Log $_5q_{20}$	a_{20}	0.0952	−0.0253	0.815	0.046	121		
Log $_5q_{25}$	a_{20}	0.4856	−0.0316	0.871	0.046	134		
Log $_5q_{30}$	a_{20}	0.6350	−0.0335	0.902	0.042	143		
Log $_5q_{35}$	a_{20}	0.6742	−0.0331	0.931	0.035	141		
Log $_5q_{40}$	a_{20}	0.7833	−0.0344	0.930	0.034	142		+0.0075
Log $_5q_{45}$	a_{20}	0.6943	−0.0310	0.909	0.037	139		
Log $_5q_{50}$	a_{20}	0.6620	−0.0278	0.907	0.034	136		+0.0025
Log $_5q_{55}$	a_{20}	0.6055	−0.0248	0.869	0.037	131		
Log $_5q_{60}$	a_{20}	0.4102	−0.0192	0.811	0.034	135		
Log $_5q_{65}$	a_{20}	0.3089	−0.0151	0.820	0.027	108		
Log $_5q_{70}$	Log $_5q_{65}$	−0.0199	0.7100	0.911	0.016	130		
Log $_5q_{75}$	Log $_5q_{70}$	0.0308	0.7853	0.840	0.019	127		

balance towards indicator P. The difference between the two estimates remains slight and one may decide, for convenience' sake, to estimate the four probabilities of dying at ages 0–14 from the juvenility index.[6]

A further probability – $_5q_{70}$ – is difficult to determine from palaeodemographic variables. This is due partly to the quality of the observed life tables (the final probabilities may have been extrapolated when the source tables stopped at age 70), and partly to the capability of the composite index used (mean age at death for the over-20s) to account for mortality at advanced ages. The best estimate is never that provided by models with specific inputs but rather one based on the relationship between two successive probabilities. To estimate mortality after age 70, it is preferable to use the model based on the preceding probability and one or both of the variables $_5q_{65}$ and $_5q_{70}$,[7] independently of the population growth rate chosen (the influence of the growth rate on the relationship between two successive probabilities is negligible: the value of the preceding probability is the determining parameter). In some cases the variable a_{20} can be used to provide a direct estimate of the final probability of dying.

Some persistent difficulties in the estimation of certain probabilities must be pointed out, particularly for the female probability of dying at ages 1–4 ($_1q_4$) where

---

[6] This remark is based solely on demographically observed data, without consideration for biases that may affect the method of calculating these indicators from osteological data.

[7] Using an estimated probability introduces a slight bias. However, this chain estimation is restricted to the immediately preceding probability and is unlikely to distort the construction of a life table up to age 80 and over.

$r \leq 0$, and the "both sexes" probability of dying ages 15–19 (${}_5q_{15}$) where $r \leq 0.05$ (lines in italics and smaller characters in Tables 8.13 and 8.14 at Springer Extra). Although the first difficulty has few consequences for palaeodemography, where children are generally not differentiated by sex, the second is more problematic. In both cases these difficulties are probably due to breaks in the mortality curve at key ages, which are partly concealed under the age-group approach. In the current state of source evidence, this bias cannot easily be corrected.

8.5 Models with Migration

It is no easy task to identify migrants within a group (see Sect. 2.4), although it can be done. This is why we set out to include in our models the possibility of inward or outward migration by sex and age.

As in the case of population growth, it is not (as yet) possible to determine a precise migration rate by age from osteological data. We propose various models, which can be modulated according to the assumptions suggested by historical and archaeological context. Constructing palaeodemographic models that include migration in the same way as population growth or varying health conditions is made easier by the fact that our regression equations operate independently of each other. It is easy to combine them to obtain the entire mortality curve and the demographic components of the theoretical population associated with it.

These calculations are based on using an instantaneous growth or migration rate ρ, i.e. a growth rate established for the entire period considered, by the formula

$$\rho = \frac{\log(Pt_n) - \log(Pt_0)}{n}$$

where Pt_0 is the burial ground population at the start of the period; Pt_n the population at the end of the period; and n the length of the period.

We give two examples here: one concerning the emigration of young men and women (aged 15–35) and the other the immigration of men only at a later age (aged 40–50). The first relates to temporary or permanent population movements during the great clearances in France in the eleventh and twelfth centuries; and the other to the policy of granting land to Roman army veterans sent out to newly conquered provinces.

To use these models we must assume that the migration occurs over relatively short periods of time,[8] otherwise the population pyramid will gradually evolve, and at some point, if the growth or decline affects all age classes, the population will tend towards a stable state. However, Thomas J. Espenshade et al. (1982) do show

[8] It is always possible to attribute a growth rate to the youngest age classes, if it is considered that adult migration had a short-term effect on births.

that if the annual number of migrants and their age distribution remain constant over time and the fertility of immigrant women remains below replacement level, then in the long term the population tends towards a stationary form.

8.5.1 Models for Emigration of Young Adults of Both Sexes

In the examples given below (Tables 8.12 and 8.13), we assume that the initial population had a very moderate growth rate ($r = 0.0025$) and that twice as many young men as young women emigrated. Any other parameters can naturally be proposed.

8.5.2 Models for Immigration of Mature Men

In this example, we use the same assumption of $r = 0.0025$ in the host population and assume a male immigration rate of 0.0075 in the 40–49 age class. The equations for reconstructing the life table associated with these parameters are given I n Table 8.14.

8.6 Conclusions and Recommendations

These models, like existing life tables, can be used to reliably reconstruct the mortality of populations for which we have only fragmentary information. The best observations can be used to deduce the missing data and reconstruct the full mortality curve.

Using the properties of stable populations, of which stationary populations are a special case, it is easy to find the main demographic characteristics of the theoretical population associated with the calculated table.

Compared with the mortality models generally used by palaeodemographers (see Chap. 6), our work has a number of advantages:

- The fact that all the observed tables belong to the set of populations that had not yet begun their demographic transition brings our models closer to the characteristic mortality pattern of pre-industrial populations. There is no need to speculate about a "mean" fertility rate or a general mortality regime (e.g. Coale and Demeny's West model) to approximate the mortality of archaeological populations;
- Compared with contemporary mortality models, our regressions are established on variables directly accessible from osteological data. It is unnecessary to

calculate unlikely probabilities of dying in order to use our models in palaeodemographic study;

- Compared with the life tables proposed by Bocquet-Appel and Masset, our regressions are based on an extensive corpus of life tables statistically representative of pre-industrial populations. A large number of tables are used in each regression, considerably reducing the margin of error associated with each estimate.

Like other models, ours include growth assumptions to capture the dynamics of archaeological populations. Given the difficulty of collating a large enough number of tables observed under various growth regimes, we had to work with the stable populations associated with each of the 167 observed tables. This method of construction makes it impossible to enter the growth rate directly into the regression linking the palaeodemographic variable to the logarithm of the probability of dying. For that reason, the end user will have to choose for each regression a given growth rate of between +3 % and −3 %. In addition to the rigorous nature of the calculation, this system has the advantage of flexibility.

Not least, our models have the advantage of specifying the quality of each regression, thereby enabling users to choose the best indicator or mathematical model (see Chap. 9).

The pre-conditions for using our models are those that also apply to contemporary model tables: the mortality of the region under study must correspond to that of the chosen model. In other words, use of our models involves accepting the assumption that the mortality of archaeological populations was very similar to that observed for pre-industrial populations. In addition, the mortality under study must not concern a period that includes accidental events with severe demographic consequences (war, famine, epidemic) (Box 8.1).

Box 8.1 Annual and Instantaneous Growth Rate

Daniel Courgeau

First we assume that the growth rate remains constant throughout the period under consideration.

Let $P(t_0)$ be the initial population at time t_0 and $P(t_n)$ that population at time t_n.

Let us suppose we are working on an annual basis. The annual growth rate r is defined by the formula

$$P(t_1) = (1 + r)P(t_0).$$

so that after n years we have

$$P(t_n) = (1 + r)^{\mathrm{n}}P(t_0).$$

This means that a population of 1,000 individuals with an annual growth rate of 0.016 will grow in 100 years to 4,891.

(continued)

Box 8.1 (continued)

If we work with continuous time, the instantaneous growth rate ρ is defined by the formula

$$dP(t) = \rho P(t)dt$$

or more simply

$$\frac{dP(t)}{P(t)dt} = \rho$$

showing that the derivative of the natural logarithm of $P(t)$ is constant and equal to ρ, and that therefore this logarithm is proportional to the time, or more simply that

$$P(t_n) = P(t_0)e^{\rho n}.$$

This means that the same population as above with an instantaneous growth rate of 0.016 will grow in 100 years to 4,953 individuals. The difference is not great but with higher rates it can become quite large.

The relationship between the two rates for the same initial and final observed populations is

$$1 + r = e^{\rho}.$$

In palaeodemography, we work not on populations but on deaths, and further assumptions are needed. If these populations have an invariable mortality and age structure throughout the period under consideration, then the deaths observed in year t, $D(t)$ can be expressed as a function of the annual growth rate r, of the death rate m, which remains constant, and of the population $P(t)$:

$$D(t_n) = d \times P(t_n) = d \times P(t_0) \times (1 + r)^{\text{n}}$$

and

$$D(t_0) = d \times P(t_0).$$

It follows from the preceding relationship between r and ρ that

$$\rho = \frac{1}{n}\ln\left(\frac{D(t_n)}{D(t_0)}\right).$$

(The reader will find applications to archaeological examples in chapter 11.)

Chapter 9
Definition and Exploration of a Pre-industrial Standard

The idea that the probability of dying at a certain age could be estimated from an observed probability at another age in the same population gave rise to a long tradition of model life tables[1] (see Chap. 8) which have gradually been refined to take better account of the most relevant parts of the curve and the best-fitting variables.

The idea that two life tables could be linked by a linear regression, taking account of the logits of the cumulative probabilities, opened the way to the use of relational models. These are more flexible and provide a means to circumvent some of the difficulties encountered in estimating certain probabilities, particularly at key points on the mortality curve. We therefore decided to use the properties of the logit model proposed by William Brass (Brass and Coale 1968) for application to pre-industrial populations.

By adapting to demography a method developed for medical experiments on animals, Brass (1971) demonstrated a linear relation between the logit of survivors of age x in a life table and the logit of survivors of age x in any other life table. This relation makes it possible to construct any life table from two parameters in any known table, and also, by taking a table as reference and varying the parameters α and β of the linear regression, to produce a large number of theoretical tables that can be used to explore potential situations.

Brass's model, initially designed to track changes in the mortality of Western populations, was rapidly adapted to other standards (André Lambert 1973, for Latin

This work owes much to the initial approaches of Magali Belaigues-Rossard in 1999–2000. We are also grateful to Brahim Ahmedou for writing a macro for automatically generating new tables and to Arnaud Bringé for the statistical validation.

[1] After Frank W. Notestein's pioneering work in 1944 (cited by Josianne Duchêne 1999, p. 155), the first set of model tables was that proposed by the UN (1955, 1956), which started from the statistical link between two successive probabilities and estimated the entire mortality curve from one indicator (in this case, the infant probability of dying).

I. Séguy and L. Buchet, *Handbook of Palaeodemography*, INED Population Studies 2,
DOI 10.1007/978-3-319-01553-8_9,

America; Brass 1975, for African populations) as a means to adjust the data from other countries where statistics might be deficient or prone to error.

Like the model tables based on the observation of a large number of data, the prime purpose of relational models was to study populations for whom the demographic data were too biased. Although over the last fifteen or so years, improvements in civil registration and census data in developing countries have reduced the need for these tools in contemporary demography, the same is not true for palaeodemography. The data from cemeteries and burial grounds are irremediably incomplete and biased, albeit in highly variable proportions. The development of relational models that can be used for pre-industrial populations is still an ongoing task and the search for the most "meaningful" demographic indicators just as essential.

9.1 The Brass Method

This is based on the observation of the risk of dying between birth and a given age. It is used to deduce the parameters of one life table from any other by simple linear regression of the logits of cumulative probabilities for a given age in both tables.

The equation is

$$\text{logit}\, {}_xq_0 = \alpha + \beta\, \text{logit}\, {}_xq_0{}'$$

${}_xq_0$: the probability of dying before age x in the observed table[2]
${}_xq_0{}'$: the probability of dying before age x in the reference table

$$\text{with logit}\, {}_xq_0 = 0.5 * \ln\left(\frac{{}_xq_0}{1 - {}_xq_0}\right)$$

following the relationship defined by Brass (1971, pp. 73–74)[3]

[2] The various ${}_xq_0$ are calculated as follows:
${}_5q_0$ is obtained from ${}_1q_0$ and ${}_4q_1$, such that ${}_5q_0 = {}_1q_0 + {}_4q_1 - ({}_1q_0 \times {}_4q_1)$
${}_{10}q_0$ is obtained from ${}_5q_0$ and ${}_5q_5$ such that ${}_{10}q_0 = {}_5q_0 + {}_5q_5 - ({}_5q_0 \times {}_5q_5)$
${}_{15}q_0$ is obtained from ${}_{10}q_0$ and ${}_5q_{10}$ such that ${}_{15}q_0 = {}_{10}q_0 + {}_5q_{10} - ({}_{10}q_0 \times {}_5q_{10})$ etc.

[3] Brass introduces a fixed constant into the usual transformation:

$$\text{logit}\, {}_xq_0 = \ln\left(\frac{{}_xq_0}{1 - {}_xq_0}\right) = \ln\left(\frac{1 - l_x}{l_x}\right)$$

This fixed constant is intended to counterbalance the weight of the logarithmic transformation of the l_x (probabilities of surviving from birth to age x). The two terms in the equation are both very small when $x = 0$ and $l_x = 1$. On the logit scale this amounts to modifying parameter α (which must be multiplied by 0.5) and leaving parameter β unchanged.

This model can be used to measure the difference in level (α) and gradient (β) between two tables:

- Parameter α measures the distance between the survival curves, i.e., the differences in the levels of mortality. If $\alpha > 0$, the level of mortality in the observed table is higher than that in the reference table (the greater α, the lower the expectancy of life at birth); if $\alpha < 0$, the level of mortality in the observed table is lower than that in the reference table;
- Parameter β represents the age structure of mortality, the relationship between the mortality of the young and that of adults. If $\beta > 1$, mortality increases faster with age in the observed table than in the reference table; conversely, if $0 < \beta < 1$, youth mortality is higher in the observed table than in the reference table. The greater β, the higher the proportion of survivors before median age at death in the reference table, and conversely after median age.

The model proposed by Brass (Brass and Coale 1968; Brass 1971, 1975) is not therefore based on a set of model tables but on the linear relationships between a life table taken as reference (also called standard) and an observed life table. The linear regressions obtained between the logits of the ${}_xq_0$ probabilities in the reference table and those in the observed table are used to fit the latter's data to the standard. However, a model does not operate solely as a way of comparing two life tables, but can also be used to reconstruct missing values in the observed population, starting from the ${}_xq_0'$ logits in the reference life table. The linear regression establishes the distance of each ${}_xq_0$ from the probability of dying at the same age in the reference table.

Unlike the model tables constructed on the previous model

$$\log_a q_{x+a} = a_0 + a_1 \log_a q_x$$

(see Chap. 8), which could only be constructed from the complete series of probabilities of dying in the observed tables, The Brass method merely requires knowledge of at least two probabilities between ages 0 and x. All the others are deduced by assuming logit linearity. It is also more flexible to use under "normal" conditions, because the model can be entered from any probability of dying, unlike the model tables proposed in the previous chapter where the user is constrained by entry parameters fixed in advance. However, the constraints of osteo-archaeological data considerably restrict the possibilities of using it in this way.

9.2 Necessary Adaptations for Palaeodemographic Data

As outlined above, the Brass method is not directly applicable to our data because first, estimation of probabilities of dying is still problematic, and second, the mortality regime adopted by Brass does not correspond to what we have observed for average pre-industrial populations.

We therefore needed to adapt the model to our data, and, first of all, to define the table to serve as reference table, because the Brass method requires that we remain close to the defined standard, so that the proposed variations for parameters α and β do not fall outside certain intervals (generally set between -1.5 and 1 for α and between 0.5 and 1.5 for β, see below).

To stay as close as possible to the mortality regime of pre-industrial populations, we chose the mean of all the life tables selected in each of our three samples (male, female and both sexes combined; Table 9.1). By convention, this mean mortality regime is called the "pre-industrial standard". Taking the mean table seemed to us a better compromise than selecting from among the hundred or so tables in the sample the one that came closest to the observed mean (as was done by Brass 1971, 1975).

Compared with the African standard, the pre-industrial standard presents significant differences, both in terms of mortality levels (mainly that of under-5s and, to a lesser extent because of the scale of values, the 15–60 age group; after age 60, the two curves converge) and of the increase in risk with age (note that the excess accidental mortality among young adults visible on the African standard does not occur in the pre-industrial standard) (Table 9.2). The African standard comes closer to the characteristics of contemporary mortality despite the still extremely high levels of infant and child mortality. A further difference is the much higher median age at death (50.96 years versus 38.43 years for our reference table for the two sexes combined).

9.3 Pre-industrial Standard

9.3.1 Characteristics

The pre-industrial standard is defined by the mean value (with standard deviation) of probabilities of dying by age, observed in each of the life table samples selected previously. For example, the mean of the 167 tables for both sexes combined is the pre-industrial standard, both sexes combined (Fig. 9.1); the mean of the 147 male tables is the pre-industrial standard for males (Fig. 9.2); and the mean of the 139 female tables the pre-industrial standard for females (Fig. 9.3). Next, we calculated the parameters of the life tables associated with each of these three pre-industrial standards (Table 9.3).

9.3.2 Differential Mortality

For all the observed tables, the gender differential in mortality is not greatly marked (less than 5 % in the risk of dying in each age group), except for a slight male excess mortality at ages 45–65, where it may be as high as 10 % (Fig. 9.4).

Table 9.1 Parameters of the "pre-industrial standard" life table and values of the associated entries: both sexes combined, male, female

Pre-industrial standard	Both sexes combined					Male					Female				
Age groups	Q_x	Standard deviation of Q_x or of variable	D_x	S_x	E_x	Q_x	Standard deviation of Q_x or of variable	D_x	S_x	E_x	Q_x	Standard deviation of Q_x or of variable	D_x	S_x	E_x
0–1	0.200	0.057	200	1,000	34.34	0.203	0.056	203	1,000	34.24	0.189	0.047	189	1,000	34.98
1–4	0.150	0.058	119	800	41.77	0.142	0.051	114	797	41.86	0.149	0.049	121	811	42.02
5–9	0.052	0.025	35	681	46.93	0.050	0.025	34	683	46.63	0.053	0.025	37	690	47.20
10–14	0.029	0.013	19	646	44.35	0.028	0.012	18	649	43.97	0.031	0.013	20	653	44.72
15–19	0.038	0.014	24	627	40.58	0.035	0.013	23	631	40.17	0.039	0.016	24	633	41.05
20–24	0.049	0.016	29	603	37.09	0.049	0.016	29	608	36.56	0.048	0.017	30	609	37.62
25–29	0.054	0.018	31	574	33.85	0.053	0.019	31	579	33.31	0.055	0.019	32	579	34.40
30–34	0.060	0.020	32	543	30.64	0.058	0.020	32	548	30.03	0.062	0.020	33	547	31.26
35–39	0.068	0.022	35	511	27.43	0.066	0.023	34	516	26.72	0.067	0.021	35	514	28.14
40–44	0.079	0.024	37	476	24.24	0.082	0.027	39	482	23.45	0.075	0.023	36	479	25.00
45–49	0.093	0.027	41	439	21.09	0.102	0.032	46	443	20.31	0.084	0.027	37	443	21.81
50–54	0.115	0.032	46	398	18.00	0.127	0.036	50	397	17.33	0.105	0.033	43	406	18.58
55–59	0.152	0.042	53	352	15.02	0.165	0.043	57	347	14.48	0.139	0.039	50	363	15.47
60–64	0.202	0.042	61	299	12.26	0.217	0.046	64	290	11.84	0.191	0.045	60	313	12.57
65–69	0.275	0.049	66	238	9.74	0.291	0.052	66	227	9.44	0.266	0.053	67	253	9.94
70–74	0.381	0.057	66	172	7.49	0.394	0.054	63	161	7.29	0.369	0.057	69	186	7.63
75–79	0.492	0.063	53	107	5.57	0.516	0.063	50	97	5.41	0.484	0.068	57	117	5.64
80–84	0.657	0.138	36	54	3.55	0.664	0.109	31	47	3.51	0.655	0.131	40	60	3.58
85+			18	0				16	0				20	0	
JI	0.089	0.0466				0.087	0.0438				0.093	0.0444			
a_{20+}	57.49	3.7340				56.90	3.8232				58.06	3.9395			
P	0.114	0.0413				0.110	0.0411				0.118	0.0421			

Table 9.2 Parameters of the reference table, both sexes combined, and probabilities of dying of the African standard

Pre-industrial standard (mean of the 167 tables of the sample both sexes combined)						African standard
s_0	1,000	$d_{(0,1)}$	200	$_1q_0$	0.200	0.150
s_1	800	$d_{(1,5)}$	120	$_4q_1$	0.150	0.095
s_5	681	$d_{(5,10)}$	35	$_5q_5$	0.052	0.025
s_{10}	645	$d_{(10,15)}$	18	$_5q_{10}$	0.029	0.019
s_{15}	627	$d_{(15,20)}$	24	$_5q_{15}$	0.038	0.031
s_{20}	603	$d_{(20,25)}$	29	$_5q_{20}$	0.049	0.043
s_{25}	574	$d_{(25,30)}$	31	$_5q_{25}$	0.054	0.044
s_{30}	543	$d_{(30,35)}$	32	$_5q_{30}$	0.060	0.046
s_{35}	510	$d_{(35,40)}$	35	$_5q_{35}$	0.068	0.052
s_{40}	476	$d_{(40,45)}$	37	$_5q_{40}$	0.079	0.062
s_{45}	438	$d_{(45,50)}$	41	$_5q_{45}$	0.093	0.078
s_{50}	397	$d_{(50,55)}$	46	$_5q_{50}$	0.115	0.100
s_{55}	352	$d_{(55,60)}$	53	$_5q_{55}$	0.152	0.137
s_{60}	298	$d_{(60,65)}$	60	$_5q_{60}$	0.202	0.187
s_{65}	238	$d_{(65,70)}$	66	$_5q_{65}$	0.275	0.261
s_{70}	173	$d_{(70,75)}$	66	$_5q_{70}$	0.381	0.361
s_{75}	107	$d_{(75,80)}$	53	$_5q_{75}$	0.492	0.493
s_{80}	54	$d_{(80,85)}$	36	$_5q_{80}$	0.657	0.636
s_{85}	19	$d_{(85+)}$	19	$_5q_{85}$		0.786

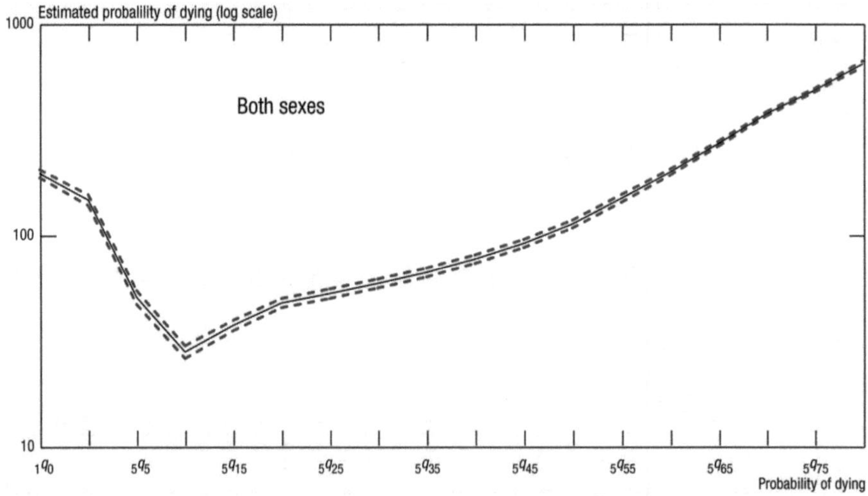

Fig. 9.1 Pre-industrial standard: probability of dying by age *Key*: Mean value and minimum and maximum values with a 95 % confidence interval

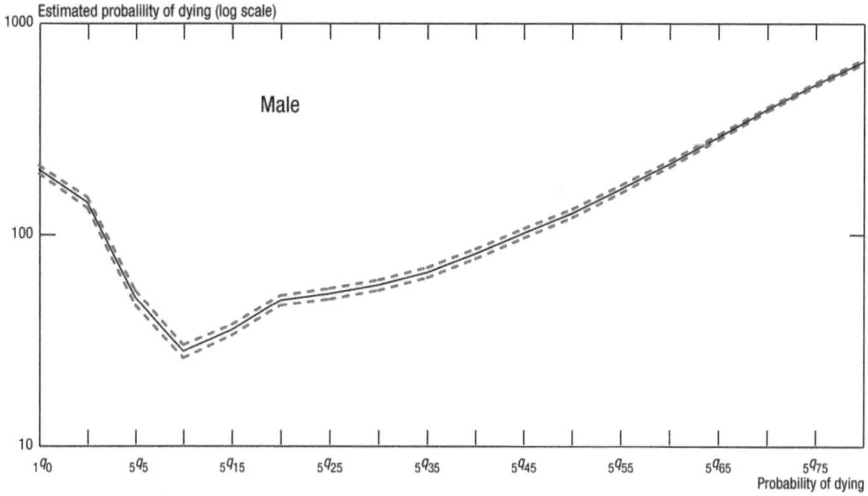

Fig. 9.2 Pre-industrial standard: probability of dying by age *Key*: Mean value and minimum and maximum values with a 95 % confidence interval

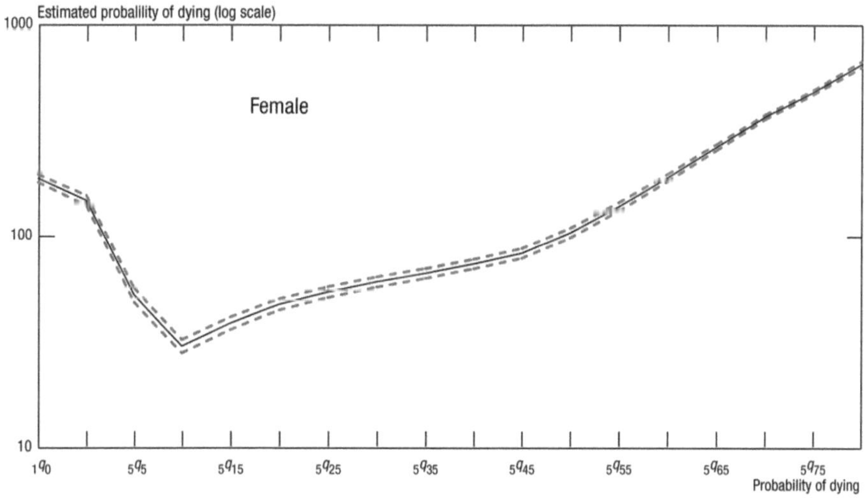

Fig. 9.3 Pre-industrial standard: probability of dying by age *Key*: Mean value and minimum and maximum values with a 95 % confidence interval

After specifying the characteristics of the mean table for each sample (both sexes, male and female), details must now be given for the variables necessary for the model, i.e. the series of values for $_xq_0$ and associated logits (Table 9.3).

A further point to be borne in mind is the number of parameters deemed necessary for our relational model. Demographers soon realised that the complexity

Table 9.3 Parameters of the "male", "female" and "both sexes" reference tables (pre-industrial standards)

$_5q_x$	F	M	Both sexes	$_5q_0$	F	M	Both sexes	Logit $_xq_0$	F	M	Both sexes
$_5q_0$	0.310	0.317	0.319	$_5q_0$	0.310	0.317	0.319	$_5q_0$	−0.400	−0.385	−0.378
$_5q_5$	0.053	0.050	0.052	$_{10}q_0$	0.347	0.351	0.355	$_{10}q_0$	−0.317	−0.307	−0.299
$_5q_{10}$	0.031	0.028	0.029	$_{15}q_0$	0.367	0.369	0.373	$_{15}q_0$	−0.273	−0.268	−0.259
$_5q_{15}$	0.039	0.035	0.038	$_{20}q_0$	0.391	0.392	0.397	$_{20}q_0$	−0.221	−0.220	−0.209
$_5q_{20}$	0.048	0.049	0.049	$_{25}q_0$	0.421	0.421	0.426	$_{25}q_0$	−0.160	−0.158	−0.148
$_5q_{25}$	0.055	0.053	0.054	$_{30}q_0$	0.453	0.452	0.457	$_{30}q_0$	−0.095	−0.096	−0.086
$_5q_{30}$	0.062	0.058	0.060	$_{35}q_0$	0.486	0.484	0.490	$_{35}q_0$	−0.027	−0.032	−0.020
$_5q_{35}$	0.067	0.066	0.068	$_{40}q_0$	0.521	0.518	0.524	$_{40}q_0$	0.042	0.036	0.049
$_5q_{40}$	0.075	0.082	0.079	$_{45}q_0$	0.557	0.557	0.562	$_{45}q_0$	0.114	0.115	0.124
$_5q_{45}$	0.084	0.102	0.093	$_{50}q_0$	0.594	0.603	0.603	$_{50}q_0$	0.190	0.208	0.208
$_5q_{50}$	0.105	0.127	0.115	$_{55}q_0$	0.637	0.653	0.648	$_{55}q_0$	0.280	0.316	0.306
$_5q_{55}$	0.139	0.165	0.152	$_{60}q_0$	0.687	0.710	0.702	$_{60}q_0$	0.394	0.448	0.428
$_5q_{60}$	0.191	0.217	0.202	$_{65}q_0$	0.747	0.773	0.762	$_{65}q_0$	0.541	0.613	0.582
$_5q_{65}$	0.266	0.291	0.275	$_{70}q_0$	0.814	0.839	0.827	$_{70}q_0$	0.738	0.826	0.784
$_5q_{70}$	0.369	0.394	0.381	$_{75}q_0$	0.883	0.903	0.893	$_{75}q_0$	1.010	1.113	1.062
$_5q_{75}$	0.484	0.516	0.492	$_{80}q_0$	0.940	0.953	0.946	$_{80}q_0$	1.372	1.503	1.429
$_5q_{80}$	0.655	0.664	0.657	$_{85}q_0$	0.979	0.984	0.981	$_{85}q_0$	1.924	2.064	1.982

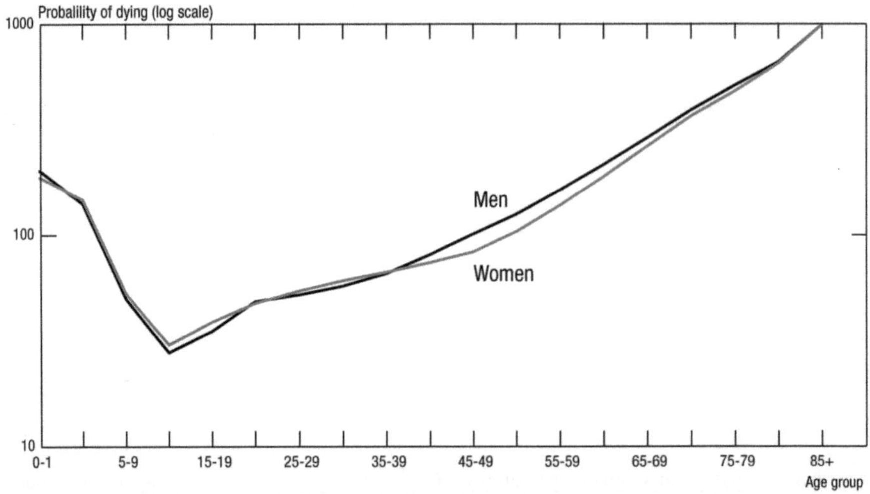

Fig. 9.4 Pre-industrial standard: comparison of probabilities of dying by age and sex

of mortality variations, both in level and in age-at-death structure was poorly represented by a linear regression on logits. Consequently some researchers propose "distorting" the mortality regime in the lowest and highest age groups by introducing two further parameters in order to better reflect current changes in mortality (Zaba 1979; Mitra 1983; Ewbank et al. 1983; Murray et al. 2003).

Douglas Ewbank et al. (1983) propose a model[4] in which the parameters κ and λ measure the gradient of the mortality curve and specify its impact on the lowest and highest age groups.[5]

Introducing these two parameters makes it possible to distinguish more clearly between populations whose levels of mortality are similar but where differences are particularly marked in the lowest and highest age groups. The Brass method becomes considerably more complex to apply, however. It is very hard to estimate the value of these parameters in the mortality regime of pre-industrial populations, and even more complicated to make the model work subsequently, whatever its theoretical interest, because we do not have enough reliable, precise data about those lowest and highest age groups.

In this initial application of the Brass method to pre-industrial populations, we see no need to operate a model that is over-complex, with additional components based on the least robust segments of the observed mortality curves. It is preferable to take as a standard a mortality regime as close as possible to the conditions of life (and death) of archaeological populations, thereby significantly reducing any disparities between estimation and observation.

The work presented below therefore uses the linear progression proposed by Brass.

9.4 Establishing Model Tables with the Brass Method

For reasons linked to the nature of our sources, we cannot use the Brass method to fit and smooth a mortality curve taken from osteological data by means of our reference table. But we can establish a series of model tables whose parameters are the various palaeodemographic indicators previously chosen: the juvenility index, mean age at death of the over-20s and the proportion of 5–14-year-olds in the population. Compared with the model tables proposed in the previous chapter, the estimates of deaths at key ages (0–5; 15–19 and 20–24; after 70) will follow from the properties of the logits of the cumulative probabilities.

[4] The equation they propose is
logit $_xq_0 = \alpha + \beta * T(\kappa, \lambda)$, with T such that

if $_xq_0 \geq 0.5$, $T(\kappa, \lambda) = \dfrac{\left(\dfrac{_xq'_0}{1 - {_xq'_0}}\right)^k - 1}{2\lambda}$, and if $_xq_0 < 0.5$, $T(\kappa, \lambda) = \dfrac{\left(\dfrac{_xq'_0}{1 - {_xq'_0}}\right)^{-k} - 1}{-2\lambda}$.

[5] Parameter κ measures mortality at ages 0–5 relative to mortality at around age 35. If $\kappa > 0$, the number of survivors falls more sharply in the youngest age groups in the observed table than in the reference table; and vice versa, if $\kappa < 0$. In the youngest age groups, therefore, the survival curve is determined by β and κ. Parameter λ measures the mortality of the 70–74 age group compared with the 60–64 age group. If $\lambda > 0$, the number of survivors in the oldest age groups declines more gradually in the observed table; and vice versa, if $\lambda < 0$.

9.4.1 Construction Method

The following life tables were constructed from the pre-industrial standard as defined above, by varying the values of the model's two parameters (α, mortality level and β, mortality gradient). It is not possible to assume just any value of α or any value of β, and even less to combine them indiscriminately. The range of possible values is relatively narrow and needs to be determined.

For Brass (1971), the European Standard can be taken as a summary of the mortality patterns observed in Europe (consequently close to the model tables of the UN and Coale and Demeny's West model), as long as the parameters of the equation lie within the following limits: $0.6 \leq \beta \leq 1.6$ and $-2 \leq \alpha \leq 0.5$. He sets slightly different limits for the African Standard, adapted to the populations of developing countries where child mortality is still fairly high; Brass (1975) proposes adjusting the tables by varying parameters α and β between +0.8 and −0.8, and +0.7 and +1.4, respectively.

A more recent study (Sergei Vassin 1994) tends to place European mortality before the epidemiological transition[6] within narrower limits, both for level of mortality and its structure (shape of curve). He starts from the idea that, in a stable population, mortality must never be so high as to prevent population replacement. This exercises a double constraint: infant mortality must not exceed certain limits (which he sets at a maximum of 450 per 1,000 and an almost contemporary minimum of 3 per 1,000); and mortality in the oldest age groups must also fall within a narrow interval (100 per 1,000 and 650 per 1,000). These constraints are associated with fairly wide thresholds,[7] which are reduced if one assumes a life expectancy at birth of between 25 and 65 years ($0.75 \leq \beta \leq 1.25$ and $-1.5 \leq \alpha \leq 0.1$), and even more so if one takes a value below 30 years for life expectancy at birth. The thresholds of the two parameters are then $-1.5 \leq \alpha \leq -1$ and $0.75 \leq \beta \leq 1.1$ (Vassin 1994, Fig. 11, p. 63).

These thresholds, based on a contemporary standard (European Standard), need to be adapted for pre-transitional populations. The fact that our standard comes close to the object of our study, together with the observation of a life expectancy at birth more often between 30 and 40 years in our tables, is a strong reason to look directly for the thresholds within which our standard can plausibly vary.

To define these, we began by setting $\alpha = 0$ to examine the effect of parameter β alone on the mortality curve (Figs. 9.5 and 9.7), and then, conversely, we set $\beta = 1$

[6] As defined by Abdel Omran (1971), the epidemiological transition comprises three successive sequences: the first, which he calls the "age of pestilence and famine", is characterized by high mortality and a life expectancy at birth of 20–40 years. The "age of receding pandemics", is marked by a regular decline in mortality and a gain in length of life (life expectancy at birth close to 50 years) enabling the population to enter a growth phase. The current third phase of "degenerative and man-made diseases" sees the emergence of new diseases and of pathologies related to population ageing. As child mortality falls and length of life increases, the shape of the mortality curve changes from one phase to the next.

[7] $0.5 \leq \beta \leq 1.5$ and $-1.5 \leq \alpha \leq 1$.

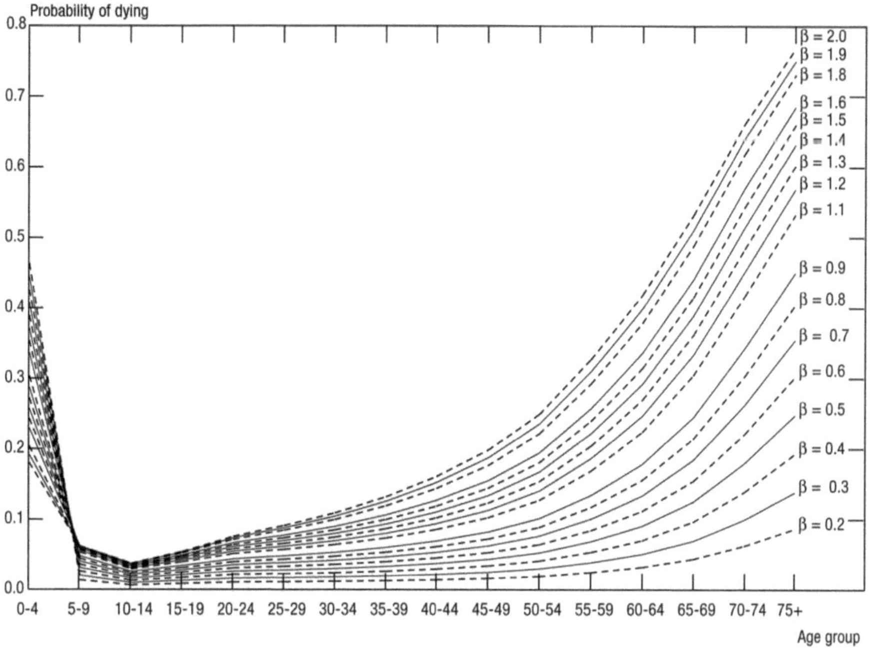

Fig. 9.5 Variation of the mortality curve under the influence of β alone (mortality before and after median age: 38.4 years)

to see the impact of α (Figs. 9.6 and 9.8), varying their values by increments of 0.1. We defined the interval of variation for parameters α and β on the basis of the extreme values that two probabilities of dying could take, reflecting their respective influence: ${}_5q_0$ and ${}_{20}q_{40}$. These values are those observed in our source tables,[8] some of which are summarized in Table 9.4.

The analysis shows that for values of ${}_5q_0$ and ${}_{20}q_{40}$ between the upper and lower limits defined in Table 9.4, the values of α must lie between −0.3 and +0.5, and of β between 0.8 and 1.5.

The next stage was to combine the various values of α and β, within the intervals previously defined, to construct a set of tables within the fairly wide spectrum of pre-industrial mortality. However, some combinations of values must be excluded if all the probabilities obtained in this way are to remain within the upper and lower limits observed in our source tables (Table 9.4), notably the value 0.8, or even 0.9, for β when parameter α is negative; and the value 1.6, or even 1.5, for β when parameter α is positive. Ultimately, the only values chosen for β lie within a

[8] The minimum and maximum values are sometimes outliers, particularly for ${}_5q_0$, for which we preferred to use the thresholds of 0.175 and 0.574.

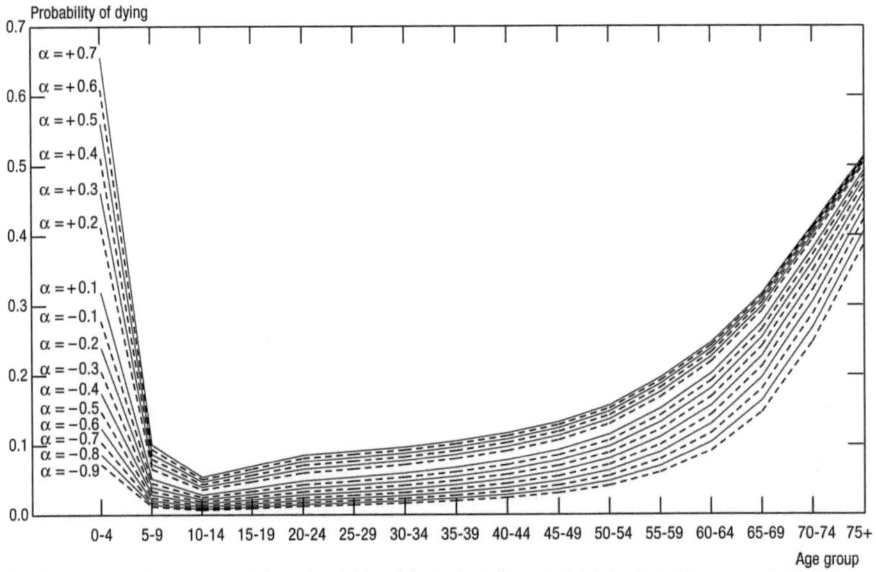

Fig. 9.6 Variation of the mortality curve under the influence of α alone (infant and child mortality)

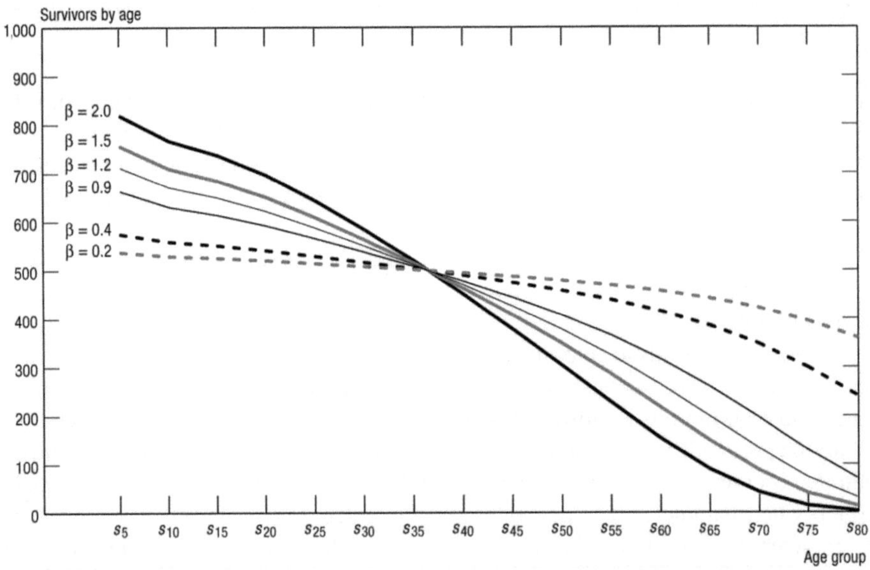

Fig. 9.7 Variation in survivors by age under the influence of β alone (with $S_0 = 1{,}000$)

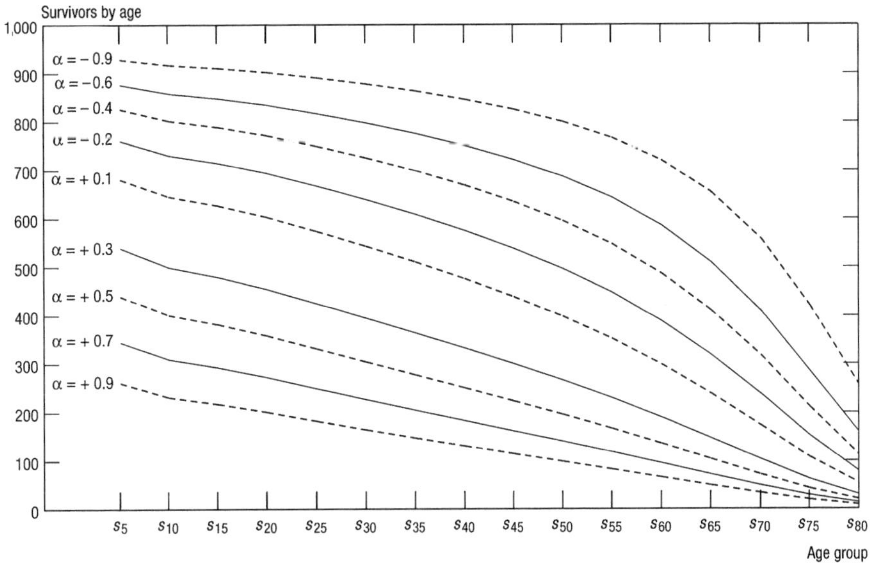

Fig. 9.8 Variation in survivors by age under the influence of α alone (with $S_0 = 1{,}000$)

Table 9.4 Statistical observations of child ($_5q_0$) and adult ($_{20}q_{40}$) probabilities of dying

		Both sexes	Females	Males
$_5q_0$	Minimum	0.1280	0.1199	0.1138
	Mean	0.3169	0.3084	0.3147
	Maximum	0.6072	0.4628	0.4900
	Median	0.2979	0.3024	0.3147
	Standard deviation	0.0866	0.0716	0.0808
$_{20}q_{40}$	Minimum	0.2183	0.2154	0.2730
	Mean	0.3689	0.3432	0.3886
	Maximum	0.5903	0.5813	0.6348
	Median	0.3496	0.3208	0.3673
	Standard deviation	0.0063	0.0814	0.0876

Key: The mean value corresponds to the pre-industrial standard

narrower interval than before (between 0.8–0.9 and 1.4–1.5), for values of α between −0.3 and +0.5. The life tables associated with this combination of values (Tables 9.2 and 9.3 at Springer Extra) all vary in parallel to the reference table. The levels of mortality differ but the pattern of mortality by age remains the same. Figure 9.9 illustrates this principle via the number of survivors by age.

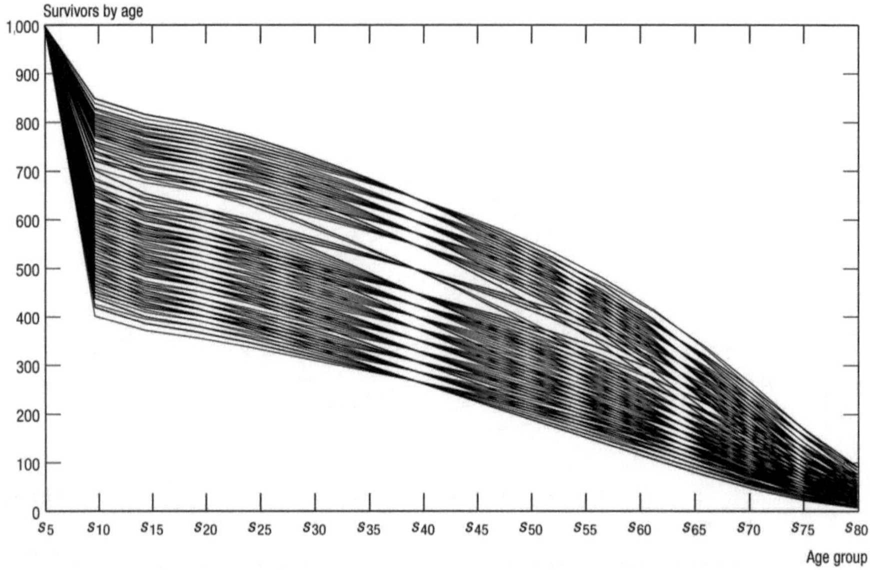

Fig. 9.9 Curves of survivors by age associated with various combinations of values defined for model parameters α and β (−0.3 < α < + 0.5 and 0.8 < β <1.5). *Source*: Tables 9.4 and 9.5 at Springer Extra

9.4.2 Panels of Proposed Life Tables (Both Sexes Combined)

A set of 95 tables for both sexes combined[9] was created by applying the linear regression between the logits of the ${}_xq_0$ of the reference table (pre-industrial standard) and those of the tables generated by introducing for each cumulative probability a given value of parameter α and of parameter β, within the intervals described above, varying by increments of 0.1 for β and 0.05 for α.

$$\text{logit}\,{}_xq_0 = \alpha + \beta \times \text{logit}\,{}_xq_0{}'$$

with ${}_xq_0$ as the calculated probability and ${}_xq_0{}'$ the probability from the reference table.

To return to ${}_xq_0$ (x) values from the logit ${}_xq_0$ (y) estimates, the formula is

$${}_xq_0 = \frac{1}{1 + e^{(-2y)}}$$

[9] It will be recalled that the palaeodemographic indicators constructed on the juvenile segment can only be used with the "both sexes" models, because of the extreme difficulty of determining a child's sex with current techniques (see Chap. 4). "Male" and "female" models may, however, be of use with the input "mean adult age at death". These will be provided later.

Table 9.5 Values of palaeodemographic indicators as estimated in the 96 life tables and observed in the 167 source tables

	Values estimated in model tables		Values observed in source tables	
	Minimum	Maximum	Minimum	Maximum
IJ	0.059	0.197	0.034	0.245
$P_{5\text{-}19}$	0.08	0.218	0.052	0.238
a_{20}	48.16	61.28	48.14	63.03

Finally, to reconstruct the entire life table, the ${}_aq_x$ values corresponding to the ${}_xq_0$ had to be recalculated using the formula

$${}_5q_x = \frac{{}_{x+5}q_0 - {}_xq_0}{1 - {}_xq_0}$$

This yields the succession of probabilities ${}_5q_0$, ${}_5q_5$, ${}_5q_{10}$, ${}_5q_{15}$, … up to ${}_5q_{80}$.

It only remains to calculate the various parameters associated with each of the tables (survivors by age, deaths by age) and the values of the juvenility index, indicator P_{5-19} and mean adult age at death (Tables 10.1 and 10.2 at Springer Extra).

We obtain a fairly large set of plausible life tables for the mortality regime of pre-industrial populations. The small size of the chosen increments tends to increase the similarity between tables. We could have kept only half of them, or even one-third, but we preferred to maintain a finely grained distribution to optimise entry into these model tables from palaeodemographic indicators.

We have only kept for these models the tables whose probabilities lie strictly within the interval of values observed in the source tables (Table 9.5). The palaeodemographic indicators associated with this set of tables also vary within the interval of observed values (Table 10.2 at Springer Extra).

9.4.2.1 Remarks and Limitations on Use

The regular variation of the α and β coefficients makes it possible to provide a finely grained set of model tables. However, the palaeodemographic indicators cannot necessarily provide greater certainty, because the data upon which they are based are themselves only probable.

In most cases, two combinations of α and β do not give two identical values for the various parameters associated with the table (probabilities of dying, survivors or deaths by age, juvenility index, indicator P, mean adult age at death). However, caution is advised where the quantitative similarity of certain indicators may conceal notable differences in the mortality patterns they reflect. When the user has only one entry parameter, assumptions need to be made concerning the mortality of the under-5s and of adults as measured by minute variations in parameters α and β (it will be recalled that the greater α, the higher child mortality). When there

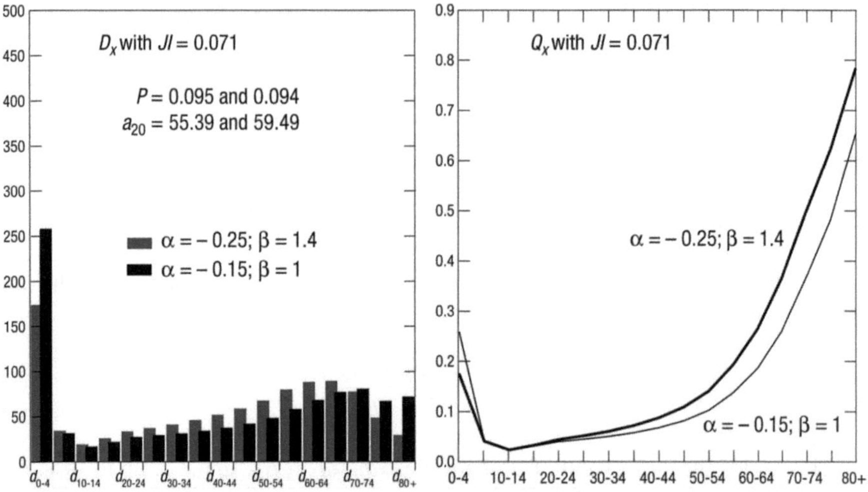

Fig. 9.10 Age distributions of deaths (*left*) and mortality curves (*right*) for two values close to…

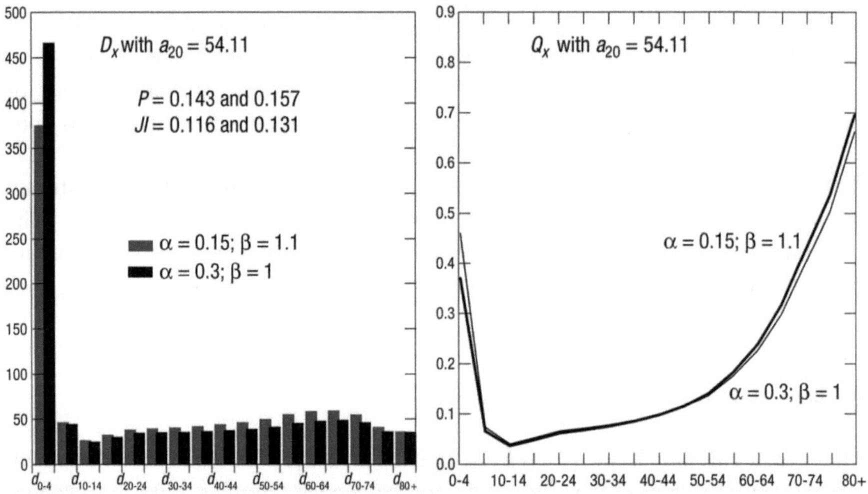

Fig. 9.11 Age distributions of deaths (*left*) and mortality curves (*right*) for two values close to…

are two palaeodemographic indicators, the value of the second one should guide the choice of life table.

We provide a graphical illustration below for each of the chosen palaeodemographic indicators (Figs. 9.10, 9.11, and 9.12). For example, for a juvenility index of 0.071 (Fig. 9.10), the mean adult age at death differs noticeably from one level of mortality to another (similarly for indicator P, see Fig. 9.12). With "mean adult age at death" (Fig. 9.11), an identical value is also counterbalanced by a different value of the juvenility index or P. An example of archaeological application is given for the Frénouville site (fourth century AD) in Sect. 11.1.

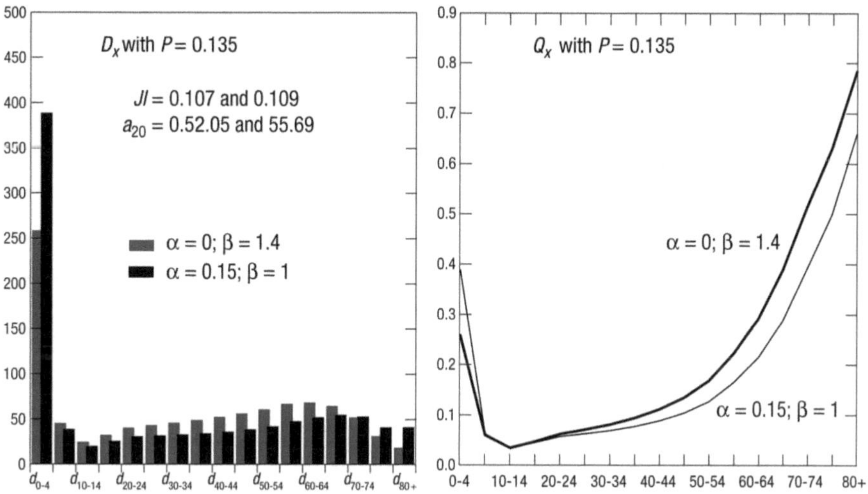

Fig. 9.12 Age distributions of deaths (*left*) and mortality curves (*right*) for two values close to...

9.4.3 *Atypical Life Tables*

The palaeodemographic indicators provided by the life tables calculated above may be quite far removed from those obtained from osteological data. The juvenility index (*JI*) or indicator *P* may reach values that the model as defined above does not cover. Consequently we have varied parameters α and β without considering the constraints imposed by the observed life tables, so that the values of JI and P are higher than the thresholds previously obtained (Tables 10.3 and 10.4 at Springer Extra). We have done the same with mean adult age at death (Table 10.5 at Springer Extra).

The modelling procedure shows that high values for the juvenility index and indicator *P* are associated with much higher mortality levels than those observed in our tables, at all ages (Figs. 9.13 and 9.14). Use of these atypical models is therefore based on the assumption of a mortality structure that has not as yet been observed. This anomaly should suggest a bias in the osteological data (such as a shortfall in adult skeletons). Otherwise, it would have to be admitted that the mortality of archaeological populations differs substantially from that of pre-industrial populations observed at various times and on various continents. This mortality structure, with extremely high infant and early childhood mortality (50–60 % dying before the age of 5) and extremely high adult mortality, makes population replacement almost impossible. Such situations can only have occurred over an extremely short period of excess mortality or high immigration. If there is no archaeological evidence for these exceptions and the site under study presents a long and uninterrupted chronological sequence, then the possibility of bias in burial practices needs to be considered.

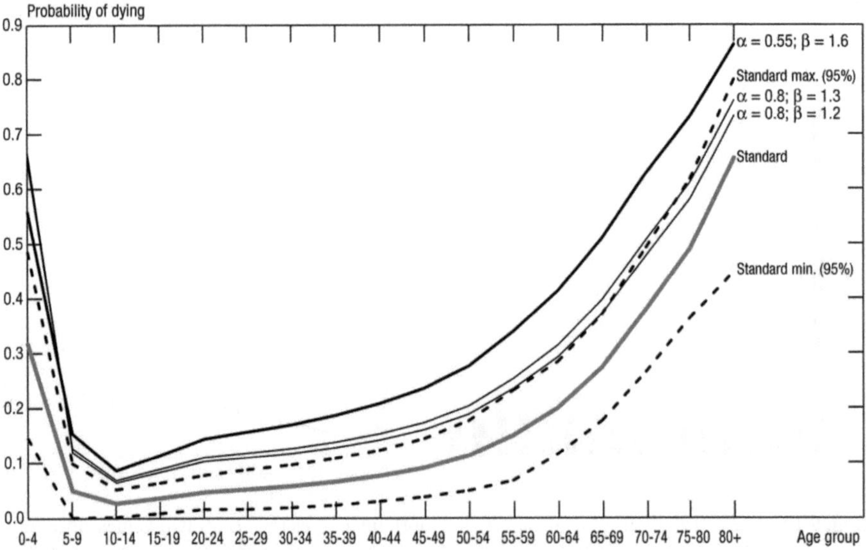

Fig. 9.13 Probabilities of dying by age associated with various combinations of model parameters α and β, such that the juvenility index exceeds 0.245 *Key:* The series of probabilities all lie outside the limits of the pre-industrial standard, with a 95 % confidence interval

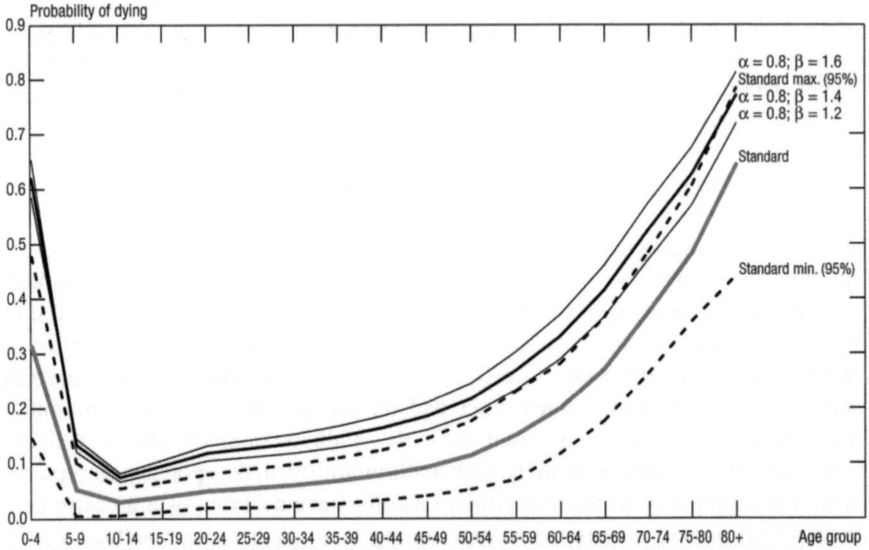

Fig. 9.14 Probabilities of dying by age associated with various combinations of model parameters α and β, such that indicator *P* exceeds 0.240 *Key:* The series of probabilities all lie outside the limits of the pre-industrial standard, with a 95 % confidence interval

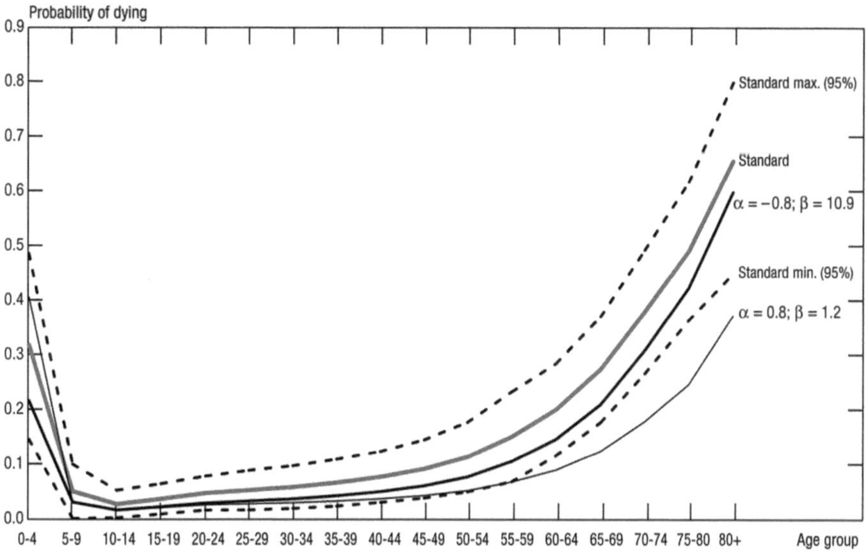

Fig. 9.15 Probabilities of dying by age associated with various combinations of model parameters α and β, such that mean adult age at death exceeds 63 years *Key*: Probabilities of dying by age associated with various combinations of model parameters α and β, such that mean adult age at death exceeds 63 years

The problem is less clearly marked for high values of mean adult age at death (Fig. 9.15), which tend to imply rather a much lower mortality at all ages than is observed. Some caution would appear to be required nevertheless in any conclusions.

9.5 Conclusions

The estimation of probabilities of dying in the youngest age groups varies considerably because of the wide distribution of observed values. Previous models, based on a linear regression between the logarithms of the probabilities and the palaeodemographic indicators, were of poor quality for the early ages. Life tables constructed by the logit method, albeit based on the same sample of tables, are much more reliable because the chosen value is the mean of the probabilities, which excludes the extreme cases that impaired the quality of the others. This relational model is one of the best ways of studying infant and child mortality. While using the mortality regime predefined for pre-industrial populations, a vast number of virtually possible situations can be observed with a clear awareness of the underlying assumptions.

Nevertheless, we must not overlook the unstable nature, in time and space, of mortality between 0 and 5 years. A number of patterns are possible, and probability ratios (in particular) may be reversed. In this sense, the infant and child mortality of pre-industrial populations is as erratic as that observed in developing countries and would probably require that further parameters be defined. Indeed, the consequences of these variations, however minute, should not be underestimated. A slight improvement in living conditions for children aged 0–5 may have major demographic consequences (particularly on population growth). Similarly and conversely, the infant diseases that regularly wiped out one-quarter or one-third of this age group had a profound impact on population dynamics.

Part III
Developing a Study Protocol

Chapter 10
Final Overview

An archaeological excavation has delivered a set of skeletons. What conditions must be met if we wish to use this set of skeletons in a demographic study? And if these condition are met, what options should be chosen and what are the basic stages in the process?[1]

10.1 Assessing the Usefulness of the Study

Before beginning any palaeodemographic study, the first stage is to define the research topics, considering the state of both the biological sources (conservation of the skeletons) and of the archaeological sources (quality of written evidence about the set). This choice can only be made after long, close cooperation between the various researchers involved and after examining the archaeological, anthropological and palaeodemographic arguments, not forgetting the historical dimension.

To make up for the biases and gaps inherent in osteological data, palaeodemographers must address numerous difficulties relating to the excavation (quality of recovery), choice of sample (what group is it representative of?), state of skeletons (differential conservation), inaccuracies in anthropological methods (particularly in estimating age and sex) and choice of a demographic model (to fit the characteristics of the set under study).

The larger the volume of documentary evidence for the archaeological and historical background of the skeletons, the greater the chances of carrying out a comprehensive and precise demographic study. This truism can be stated more bluntly: don't expect exceptional results if the historical and archaeological

[1] The purpose of the methods proposed in this handbook is to reveal the major demographic characteristics of a buried population on the basis of a single source: the exhumed remains of human bones. Studies of the size of populations in a given territory using other types of data, such as buildings, pottery and food remains are quite another matter, and they are not covered here.

I. Séguy and L. Buchet, *Handbook of Palaeodemography*, INED Population Studies 2, 165
DOI 10.1007/978-3-319-01553-8_10, © Springer International Publishing Switzerland 2013

background is poorly understood. If over-wide chronological margins are chosen, with intervals of a century, for example, much of the information will remain undetectable. The study of a population taken from a cemetery used for two centuries where the graves are not accurately dated can only give an "average" image of that population's demographic characteristics over the entire period, with the risk that the average is an artificial one that does not reflect any historical reality. Short phases of migration or population decline, for example, will be impossible to identify.

Things are quite different where the topo-chronology established by archaeologists allows the sample to be divided into clearly identified phases, down to the quarter-century. In that case it is possible to reconstruct the demography of these populations in dynamic terms, tracking its changes in response to births and deaths, and even migrations, and to identify any growth phases.

10.2 Sample Representativeness

The existence of easy-to-use demographic tools should not obscure the fact that a palaeodemographic estimate is only of value for the sample under study (does it represent the population it is taken from, the buried population, or does it only reflect itself?).

10.2.1 Importance of Sample Size

The reliability of the final study results closely depends on the representativeness of the sample. Not every set can be used for a palaeodemographic study. Remember that even in exceptional archaeological circumstances one can never be sure that a set of exhumed skeletons is significantly representative of the buried population, let alone the living one. Sample representativeness must be considered on several levels prior to any study.

10.2.1.1 On the Scale of the Burial Ground

Because of the variability of individual characteristics, it is difficult to extract useful information from a very small number of skeletons, be this due to poor bone conservation or to a small dig area. There is no rule for setting a threshold and the archaeological background must also be taken into account; the fact that skeletons are few in number will not have the same import if they are found in an isolated enclosed space or if they come from a small sector of a large burial ground. Common sense should prevail, together, of course, with discussion among colleagues. When conclusions are later drawn from the study, it is important to consider the limitations imposed by the sample size.

10.2.1.2 On the Scale of the Region

For an accurate understanding of the demographic history of a region, one would need to study all the cemeteries used by the inhabitants of that region (including outside their territory: emigration, war, etc.). This is clearly impossible; there will be less than a score of cemeteries at best, and their excavation will rarely have been exhaustive.

The conclusions reached in that case form just part of a larger picture and their value partly depends on the representativeness of the sites included. Failure to bear that in mind can lead to considerable errors of interpretation. Comparing mortality curves calculated from different series might cause an incautious palaeodemographer to deduce major variations in mortality from one site to another, or one period to another, when the observed differences are merely the effect of random variations in samples comprising too few skeletons.

The plausibility of assumptions thus closely depends not only on the method used but also on the numbers analysed. This is why it is important to specify the statistical limitations of the results obtained. Variance in estimates of mortality rates can be calculated to take account of random variations affecting samples comprising too few skeletons.

10.2.2 Population Structure

In addition to the number of skeletons, the structure of the "burying" population, the people who used the cemetery to bury members of their group, has a strong influence on the age distribution of deaths. In the standard example of a partially excavated cemetery, the exhumed skeletons are only a sub-set of the buried population, not necessarily representative, because the age, sex or social background of an individual may have played a part in the choice of their grave. Selective burial practices may apply to one component of the population (children, foreigners, outsiders, etc.) or to all the deceased in one group, as in the case of military cemeteries or those of religious communities. However, selection occurs not only at the time of burial; it also operates beforehand, in the formation of the group from which those who did not survive can be observed. In this particular group, this "burying population", will necessarily have an age-sex distribution of deaths different from that of a parish graveyard. The cemeteries belonging to hospices and hospitals are a good illustration: the buried population taken from a group of sick people is necessarily different from that of a village graveyard.

Before any palaeodemographic analysis, it is therefore important to determine whether all components of the population appear to be correctly represented (sex ratio, age structure, grave goods, etc.) or, failing that, to identify from historical and archaeological documentation the rules of selection before or at the time of burial.

However, a study should not be abandoned on the grounds that the sample is the result of a selection and is not "natural". The information that may emerge can be of great historical value (as is the case for "catastrophe cemeteries") and one should resist the temptation to "standardise" the available sample to make it a "natural" population, when its specific features may be its most interesting characteristic.

10.3 Basic Anthropological Data: A Pragmatic Choice of Methods

Once the historical and qualitative outlines of the population available for study have been specified, anthropologists/palaeodemographers must collect from each skeleton the biometric data required to estimate its sex and age at death, two fundamental parameters for palaeodemography.

10.3.1 Determining Sex

Sex determination does not present any real methodological problems in the case of adult subjects, and only a desire for the smallest margin of error justifies preference for one method over another. The hip bone (*os coxae*) is favoured, with some justification, by many anthropologists. However, because of its fragility, it is often seriously damaged in archaeological excavations, and in the absence of the hip bone, it is possible to use cranial or post-cranial features to determine the sex of skeletons with a respectable success rate, sometimes as good as for the hip bone.

On the other hand, determining the sex of a child presents problems that are as yet only partially resolved. When it is impossible to distinguish the proportions of males and females among juveniles with an acceptable margin of error, it is preferable to work on data for both sexes combined (except, naturally, for the special cases of cemeteries of nuns or soldiers).

For the choice of methods for determining the sex of adults, see Sect. 3.1.1.

10.3.2 Estimating the Age at Death of a Skeleton

This naturally poses a number of technical problems for which there is an abundant literature and a variety of solutions, some more useful than others. However, age determination is not an end in itself for the palaeodemographic approach. The information collected from the skeleton is not merely intended to complete the columns of bio-archaeological factsheets, but rather to enable palaeodemographers to discern the behaviours of a set of individuals. For that purpose, the data, skeleton by skeleton, must be analysed using a comprehensive approach.

10.3.2.1 The Difficult Task of Choosing a Biological Age Indicator

Despite a century of scientific debate, anthropologists and palaeodemographers still fail to agree on the best indicator. In fact, there is no satisfactory osteological age indicator and, despite affirmations to the contrary, none of the currently known indicators display a sufficient statistical correlation with chronological age to be considered as expressing a subject's true age. Whatever option is chosen, the ages obtained can never be directly used for a palaeodemographic approach. This observation applies to both juveniles and adults, even though for children, who are growing, the criteria used are more stable than for adults, whose skeletons bear the marks of biological ageing in ways that differ from one individual to another.

For adults, after reviewing the various indicators proposed and critically analysing the associated reference populations, we chose to take the cranial suture closure? as indicator, despite the many criticisms levelled at it. Its failings are known – and clearly identified in the many studies devoted to it – but also its advantages. The very process of identifying these failings has made it possible to propose correctives and adaptations that have made suture closure into a clearly understood tool, unlike other indicators, most of which have not been sufficiently tested and require improvement. Indeed, some indicators will no doubt fall out of fashion once exposed to the arguments of critical analysis.

Consequently, although cranial suture closure does not give good results as an indicator of individual age, it is of great interest for a "collective" approach to age at death. However, this choice is not final, and merely expresses the state of palaeodemographic research at a particular point in time. Tooth cementum analysis (TCA) is promising, but still requires substantial refinement.

The methods presented in this handbook were designed to adapt easily to the emergence of new indicators.

Adult subjects – Recommended age indicator: cranial sutures closure (see Sect. 3.2.2).

For juveniles, we propose a new method for interpreting mineralisation that is both easy to use and statistically rigorous. Among existing methods, those that are statistically satisfactory are generally complex to apply, while the simplest to use are lacking in statistical rigour.

Juveniles – Recommended age indicator: dental mineralisation (see Sect. 3.2.1).

10.3.2.2 A Single-Criterion Approach to Age

For adults and for children, we have preferred an approach to age based on a single biological criterion, clearly defined from a high-quality reference collection. The value of the reference population (reliable age data, and quantifiable and reproducible anthropological observations) is a determining factor for a palaeodemographic study.

In the current state of research, we have chosen to use a single biological criterion, whether osteological or dental, taken from a comparison population whose ages are clearly established from civil records and whose biological characteristics have been measured using well-defined protocols. While multi-criterion approaches produce better correlations with individual ages, they are subject to major statistical biases because they use collections of varying size in various places that have seldom been critically examined (see Sect. 3.2.2).

In general terms, care must be taken to avoid using age indicators for a single study that have not been developed from a single reference population, because the estimates obtained may well diverge considerably. The challenge for the years ahead will be to establish comparison collections that are properly documented and meet the highest statistical standards.

10.4 Estimating Age at Death for a Buried Population

10.4.1 Why Shift from Individual Ages to Collective Age?

Whichever the indicator or method chosen, no estimate of age, for adults or juveniles, can be expressed without a margin of error, so it is extremely difficult, indeed hazardous, to envisage a demographic approach based on a series of individual age estimates subject to this inconvenient uncertainty. Nor should we give in to the temptation (as so many palaeodemographers have done!) of "turning a blind eye" to this margin of error, and using the observed averages to propose reconstituted patterns that may well be attractive but are always controversial.

Nor can the solution be to determine wider age categories ("young", "mature", etc.). The risk of allocating an individual to one category rather than another is just as high; furthermore, the use of wide categories causes a loss of information and the subjective nature of the observation make it irreproducible and thus limits any comparison between sites (see Sect. 3.3).

The solutions available to palaeodemographers[2] all consist in adopting a comprehensive, probabilistic approach to cover all the observations made on the skeletons so as to obtain a "probable distribution" by age group (1-year, 5-year, 10-year or other), while taking care to separate adults from juveniles.

By using a "reasoned" reference collection to estimate children's age at death (see Sect. 4.2) from a single age indicator – the extent of tooth mineralisation – the same methodology can be adopted for non-adults as for adults (see Sect. 4.1).

[2] Unlike forensic scientists, who seek to estimate the age of an isolated individual.

10.4.2 *Maintaining a Probabilistic Approach to Age*

Given the poor correlation between chronological age and any biological age indicator, it is important to maintain a probabilistic approach to estimating the age at death of a skeleton, whether for the group as a whole or for each individual parameter (see Sect. 3.3). Using probabilistic methods requires a change of perspective. The results obtained are no longer unique but one plausible solution among many.

10.4.3 *Use of a Pre-industrial Reference Standard*

To minimise the risks of a possible biological drift in age indicators, especially those for adults, the solution adopted in this handbook is to create reference populations – one for juveniles, one for adults – that come as close as possible to the standards for pre-industrial populations (see Chap. 3).

And, rather than depend on a uniform age-sex structure that would not correspond to "normal" situations, we decided to start from the age distribution at death observed in largely rural populations with little or no access to modern medicine, before, or shortly after, the start of their demographic transition.

These reference populations are the single and constant reference for determination of age at death. Once the biological characteristics have been properly established (as far as they can be), they remain invariable and only the age structure may be modified by the characteristics of each site.

10.5 New Palaeodemographic Tools

The basis of our approach needs to be sufficiently broad and stable for our future research to be built on generally accepted components. Our purpose is also to harmonise palaeodemographic techniques, whether for adults or juveniles. Even if the biological indicators differ, the analysis methods will use the same concepts: probable distribution of deaths by age group and calculation of the mean age at death associated with that distribution.

The existence of visibly differing biological processes (as far as our current evidence shows) between adult men and women is a reason for preferring separate analyses by sex. Similarly, where sample size allows, it is better to work with 5-year age groups, which minimises information loss, provided that statistical precision is not compromised; it is always easier to aggregate data back into 10-year age groups than the reverse.

These principles cannot always be applied to anthropological series and anthropologists/palaeodemographers must assess each situation individually.

10.5.1 The Revised "PFP" Method (Probability Vectors)

This method[3] was proposed by Claude Masset in 1982 to calculate the probable distribution by age-at-death classes of a buried population on the basis of the relationships between biological indicators and chronological age observed in the reference population. Since the baseline is constant, as long as care is taken with the statistics, this method can be used to compare cemetery populations separated by time and space. It also provides limited but reliable evidence concerning the demography of buried populations.

This approach is adopted separately for juveniles and adults.

- The new vectors for adults: see Sect. 5.2.2.
- The new vectors for juveniles: see Sect. 5.2.3.

In some cases, this method can reflect demographic features specific to the site under study. For that purpose, a frequency (probability vector) matrix is needed that fits the demographic context of the population whose skeletons have been exhumed. Since early experiments gave useful results (see Sects. 11.2.1 and 11.2.2), it is possible to use "palaeodemographic models" (i.e., matrices integrating these specific factors, such as more comfortable living conditions and standards, mass immigration, major epidemics) that may be preferred where the documentation associated with the skeletons so permits.

- Adapted vectors (palaeodemographic models): see Sect. 5.2.4.

The PFP method has its limitations, however. In Chap. 12 of the prospective section, Daniel Courgeau discusses and challenges it, and then in Chap. 13, with Henri Caussinus, proposes a new approach based on the principles defined in the Rostock Manifesto. Since these principles, stated by Robert D. Hoppa and James W. Vaupel in 2002, have yet to be demonstrated, this proposal remains of prospective value only.

10.5.2 New Life Tables for Palaeodemographers

Where possible, a palaeodemographic study aims to go beyond the mere production of inventory factsheets or the calculation of probable distributions by age-at-death groups. Its purpose is to reconstruct the population using a dynamic approach, taking account of natural balance (births and deaths) and even migration. This exercise is difficult, however, because most of the indicators habitually used in demography are unavailable to palaeodemographers, who are obliged to construct their own!

[3] The IPFP (Iterative Proportional Fitting Procedure) method as used in palaeodemography does not actually require any iteration. It could therefore be called the PFP method, but it is better known as "probability vector method" (see Sect. 5.2).

Here too, the approach should be guided by a few basic principles. For example, the relationship established, via one or more palaeodemographic indicators, between osteological data and demographic tools must be as direct as possible and not require a succession of estimates.

The mortality model used must be compatible with the demography of pre-industrial populations. Like the life tables for contemporary populations, it must be able to reliably reconstruct the complete mortality curve of populations for which we have only fragmentary information. Using a set of mathematical properties that link the demographic parameters, it then becomes easy to access the demographic characteristics of the theoretical population associated with the calculated table (life expectancy at age x, probability of dying, number of deceased and number of survivors by age). Although more complex to use, we have opted for the theoretical model of stable populations, of which stationary populations are only a special case.

All our work is thus based on life tables specially developed for pre-industrial populations, with inputs directly accessible from osteological data. These inputs, also called "estimators", are mainly those proposed by Bocquet-Appel and Masset: mean adult (over 20) age at death, juvenility index (*JI*) and proportion aged 5–19 in the over-5 population (P_{5-19}).

The two mortality models presented in the handbook are not entirely equivalent. The first – the "Logarithmic" model – can be used to fit mortality to observed data, by specifying the quality of the regression. It can also take account of possible (positive or negative) variations in population growth.

The other – the "Logit" model – is better suited for measuring the variations caused by changes in the mortality of children or adults. It is more flexible, but does require some care to avoid overstepping the bounds of the demographically possible. Since it does not capture the effects of population growth rate on palaeodemographic indicators, we recommend that it be used in cases where growth is slow (according to archaeological or historical evidence), or for very short chronological sequences.

- "Logarithmic" model: see Chap. 7.
- "Logit" model: see Chap. 8.

Naturally, the use of these mortality models is not essential for all palaeodemographic studies. These tools are only of value if they correspond to the osteological data and the archaeological purpose, and are used appropriately (Box 10.1).

- Examples of application to juveniles: see Sect. 11.1.1.
- Examples of application to adult subjects: see Sect. 11.1.2.

This does not mean that palaeodemographic research is complete, for a new method for estimating age at death, currently being developed (see Chap. 12) will soon propose a different approach. For the time being, the examples of archaeological applications presented in the next chapter are based on the principles we have outlined here.

Box 10.1 How to Use Mortality Models

Like Ledermann's or Coale and Demeny's tables, our mortality models have been developed to measure and correct the anomalies observed in certain poorly known population structures. Their value for palaeodemographers arises from the fact that no adjustment of variables is required for the transition from osteological data to the demographic tool, and that they most accurately reflect mortality by age and sex of populations that have not begun their demographic transition. They are consequently the perfect tool for palaeodemographers, on the strict condition that they are used properly.

Estimating Bias in Osteological Series

Our mortality models are essential tools for detecting any anomalies in the age-sex distribution of deaths of a buried population.[1] To apply them properly, it is important to use not of all the observed age classes, but solely those that archaeologists and anthropologists consider to be the most reliable on the basis of their separate observations. As soon as a bias is suspected, it must be excluded from the observation so that it can be measured. Including underestimated data in the calculation distorts the entire mortality curve and invalidates any comparison with a model curve.

So, starting from incomplete or highly specific data, it is possible to answer questions about how the deceased were "recruited" (Masset 1987) for a given burial ground, and to estimate the resulting biases or anomalies, particularly the shortfall of children under 5, on condition that the above methodological error, only too frequent, is avoided.

Testing for Crisis Mortality

In general, the mortality models are only applicable to normal mortality patterns. This is because they reflect the mortality by age habitually observed in a given population. This "normality" is the essential precondition for their use in prospective studies: if the study population behaves like the one defined by the model, then one or other vital parameter can be estimated by reference to the defined model. Consequently these models cannot capture the

[1]This research area, the detection of anomalies in age and sex parameters and their interpretation in biocultural terms, is mainly pursued by Pascal Sellier (1996) and Dominique Castex (2005, 2007) and their doctoral students.

(continued)

Box 10.1 (continued)

impact of mortality crises (due to epidemics, war or famine) that regularly afflicted populations under the old demographic regime. For that purpose, it is better to turn to models of population dynamics.

Can our models taken from the "normal" mortality of pre-industrial populations be used to test the more "selective" mortality occurring during the demographic crises well known to historical demographers? Perhaps they can. But on the understanding, once again, that two static factors are involved in the constitution of osteological series: the age-sex composition of the population exposed to the mortality law at that point in time,[2] and the risk of dying between two ages (normal mortality distribution or crisis mortality). There is also a dynamic factor: the growth or decline of the study population. The combined influence of these three elements makes it extremely difficult, in our view, to detect mortality crises from osteological data alone. A surplus of infant skeletons might mean either burial practices or conservation conditions that reflect the "normal death toll" for this age group, or population growth (by family immigration) and/or a higher birth rate,[3] or an excess of deaths in this age group.

Mortality crises occur in regular cycles, and the appearance of childhood diseases, primarily the killer smallpox, follows well-attested patterns. Unless the chronological sequence under study is highly targeted and focuses on the crisis years themselves, it will be hard to obtain anything other than an image of "average" mortality, i.e. the equilibrium between years of excess mortality and years of recovery in life expectancy.

Naturally, these mortality models are not useful for all palaeodemographic studies. Such tools are only of value if they correspond to the osteological data and the archaeological purpose involved.

---

[2] This is an important point. It is not the living population as a whole that serves as a basis, but the sub-population directly exposed to the epidemic or the atypical mortality distribution. Depending on context, the two notions may or may not overlap.

[3] Under health conditions that vary little, any increase in the birth rate leads inevitably to an increase in infant and child mortality.

Chapter 11
Examples of Archaeological Applications

The new "tools" proposed in this handbook make it possible to re-examine the demographic approach to buried populations and to address some of the recurrent questions of palaeodemography, such as the problem of estimating the proportion of under-20s, and not least, of the 0–4 age group, generally under-represented or non-existent in historic burial grounds. Similarly, intra-site demographic developments can be explored, since hypotheses of population growth or decline can now be integrated into the choice of models.

New research topics can also be addressed: dynamic approaches, modelling of demographic behaviours, regional variations, impact of migration and demographic crises. To illustrate these various approaches, we present four case studies. These mainly involve sites where the research question has required us to adjust our tools, to propose innovative approaches, and refine the proposed study protocol. The analysis of these sites is not necessarily complete nor is the discussion closed (see Chap. 13): In this chapter the aim is primarily to illustrate some current thinking in palaeodemography that may be of interest to historians, archaeologists and even demographers.

The chosen examples are classic cases, and have been divided into two groups. The first comprises sites where only biological data are available, namely two burial grounds from Late Antiquity, one rural and one urban, that have been comprehensively excavated and covered by thorough historical, archaeological and anthropological studies (Frénouville and Lisieux); the second comprises two sites with extensive written documentation that provide information on the demographic behaviour of the exhumed populations. Each belongs to a specific context, one a burial ground of the Modern period for an exclusively female population (Royal Abbey of Maubuisson); and the other, an urban cemetery of the contemporary period (Antibes). All these sites are comparable because they have been studied using the same methods[1]: same

[1] We use the PFP method, under its first definition (see Chap. 12), starting from the age distribution within a given stage. The fitting process therefore requires no iteration. The reference population used is fixed in geographical and chronological terms (nineteenth century Portugal, but representative of pre-industrial populations).

I. Séguy and L. Buchet, *Handbook of Palaeodemography*, INED Population Studies 2, DOI 10.1007/978-3-319-01553-8_11,

biological age indicators, same reference populations, and same methods for estimating age at death.

11.1 Sites for Which Only Biological Sources Are Available

In these first two examples, the demographic parameters are estimated for populations of skeletons for which there are no documents that could be used to give an alternative assessment of demographic behaviours. Since each population has its own biases, the research questions are different, as are the tools used. For archaeological reasons, the smallest feasible dating unit may be the century, whereas in other rare cases the models take account of variations in growth rate that permit accurate dating of the archaeological context and thus a relative chronology by quarter-century, i.e. the approximate span of a generation.

11.1.1 Frénouville Rural Cemetery (Calvados, N.W. France, Fourth Century AD Sector)

11.1.1.1 Research Question: Estimating the Proportion of Subjects in the Under-20 and 0–4 Age Groups

Archaeologists and anthropologists generally observe that the skeletons of young children are under-represented in burial grounds, and thus conclude that their palaeodemographic sample is severely skewed. However, the extent of this under-representation deserves some discussion, because it depends so heavily on the demographic assumptions made.

The model life tables proposed in this book, particularly those based on the logit method, can be used to estimate the expected proportion of the population aged under 20 or under 5, given the conditions of mortality and growth previously defined as those of the population under study.

11.1.1.2 Archaeological and Osteological Sources

The Frénouville burial ground[2] (site name "Le Drouly", Calvados) lies in open fields and covers roughly one hectare. Some 650 tombs have been found, dating from the end of the third century to the end of the seventh century. This long use of the site is marked by a radical change in tomb orientation in the fifth century: the Late Empire tombs are oriented north–south, while those of the Merovingian period

[2] Excavations directed by Christian Pilet (CRAHAM, Caen, France), (Pilet 1980); anthropological study by Luc Buchet (CEPAM, Nice, France) (Buchet 1978, 1998).

Table 11.1 Distribution by age group of the skeletons exhumed at Frénouville (fourth century AD)

Frénouville : archaeological data	130 adults and adolescents (over 15 years)
	7 adolescents (10–14 years)
	26 children aged below 10 years of which 50 % below one year (i.e. 12 or 13 children)

lie east–west. In both areas, the general layout is a cemetery in rows, sometimes broken by removal of stones for reuse. The present analysis concerns the first phase of cemetery use, namely the 163 graves of the Gallo-Roman period (late third to late fifth centuries). Of these, 130 contained skeletons of older adolescents (ages 15–17 years) and adults (18 years and above), of which 58 were of known age and sex, and 11 of known age. A further 33 graves contained children whose skeletons were not preserved; from the size of the empty graves, they were probably children aged under 10 years in 26 cases[3] and adolescents aged 10–17 in 7 cases (Table 11.1).

The Frénouville population was heterogeneous. Roughly two-thirds were of local origin while one-third were outsiders who had been living in the region from the third century AD, some of them from the Roman army (Buchet 1998). This military presence may well have attracted migrants up until the fifth century (Buchet 1998), and the local populations gradually mixed with the new arrivals during the next three centuries. In order to calculate the growth rate over two centuries and then deduce an annual value, it may be assumed that immigration continued at a steady rate from the late third to the late fifth centuries.

11.1.1.3 Preliminary Calculations

Calculating the Annual Growth Rate

The increase in the number of deaths observed from one period to another may be measured by the formula

$$\ln \frac{D(t_n)}{D(t_0)}$$

Where $D(t_0)$ represents the deaths occurring during the first phase of site occupation (from the late third to fifth centuries), and $D(t_n)$, the deaths occurring during the second phase (sixth to seventh centuries), assuming that all the deaths observed during a given period are distributed uniformly throughout that period.

[3] The grave lengths varied from 0.8 to 1.5 m; since the body occupied on average 65–70 % of this length (allowing for coffins), the height of the bodies buried there must have ranged from 0.50 to 1.40 m, corresponding to children under 10 years old.

From the known number of skeletons corresponding to the first site phase (N = 163) and the second phase (N = 487 for the sixth to seventh centuries), the population growth throughout the period is

$$\ln\frac{487}{163} = 1.095$$

which converts to an annual rate of

$$\frac{1.107}{200} = 0.0055.$$

Since we know that the Frénouville population has a high rate of immigration, we can develop $D(t_n)$ as

$$D(t_n) = D(p_{0,n}) + D(i_{0,n})$$

where $D(p_{0,\,n})$ is the proportion of deaths of local people between the late third and late fifth centuries, and $D(i_{0,\,n})$ the proportion of deaths among the population that immigrated over the same period.

According to the archaeo-osteological observations, we may estimate that $D(i_n) = 2\times$ D($p_{0,\,n}$), as above, and deduce an increase by the formula

$$\ln\frac{3 \times D(p_0)}{D(p_0)} = \ln(3) = 1.098$$

Which, converted to an annual rate, provides a value identical to that calculated above

$$\frac{1.098}{200} = 0.0055$$

Illustrating the importance of migration increase and the marginal effect of natural increase (based on the rate of natural increase observed in the early modern period in France: 0.0033).

Since the cemetery has been fully excavated and was used over a fairly long period, one may expect to find all the individuals who died during these two centuries. However, only 26 tombs were those of children aged under 5, representing just 15 % of the entire buried population. This is a surprising result given the high infant and child mortality among pre-industrial populations.

Can the "missing" proportion of the population be estimated from osteological data, given that we cannot base our calculations on either the juvenility index or indicator *P*, because it is impossible to obtain an accurate count of juveniles whose skeletons are not preserved?

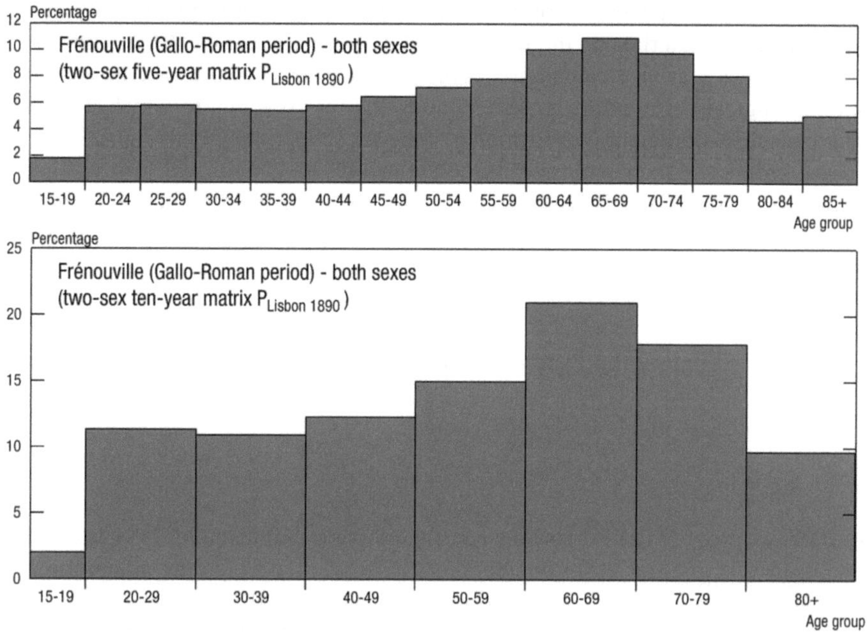

Fig. 11.1 Probable distribution of deaths by 5-year and 10-year age groups. *Key*. Probable distribution of deaths by age group associated with the mean adult age at death calculated for third to fifth century Frénouville (frequency matrix $P_{Lisbon1889}$). *Top*, 5-year age groups; *bottom*, 10-year age groups (except for first)

Estimating the Probable Distribution of Adult Deaths and the Associated Mean Age at Death

Provided that the Gallo-Roman population that used this cemetery had a demographic structure similar to the pre-industrial standard as defined by us, and that the distribution by stages of suture closure observed on the 69 "adults" is representative of all the exhumed adults (bearing in mind that for half of the adults and adolescents ages could not be estimated because of poor bone preservation), we can estimate the probable distribution of adult deaths by age (Fig. 11.1) and calculate the mean age at death associated with that distribution.

The probable distribution of deaths by 5-year age groups has the advantage of enabling a visual comparison with the theoretical distribution of deaths in a life table. This requires a sample large enough to be distributed across all the age groups ($N > 60$), however. If the adult numbers are too low, it is preferable to allocate the deaths by 10-year age groups (and aggregate the theoretical deaths in the life table similarly to enable comparison).

The automatic calculation of mean adult age at death (see macro utility in the supplementary materials on the INED website) gives a figure of 56.07 years.[4]

However, a reasoned calculation is preferable because of the specific archaeo-anthropological features of the site. Among the 69 skeletons to which a cranial suture closure coefficient was attributed, there are surely individuals from the 15 to 19 age group,[5] so their inclusion in the automatic calculation will lead to under-estimation of mean age at death. Starting from the probable distribution by 5-year age group and removing this bias, we obtain a value for mean adult (over 20) age at death of 56.78 years (and 57.09 years using 10-year groups).

11.1.1.4 Estimating the Proportion of Missing Juveniles

First Method: Use of Model Tables

- First hypothesis

In 2008, we proposed estimating the missing proportion of juveniles from life tables established for pre-industrial populations ("Power" model), on the assumption of constant growth in all age groups (Séguy and Buchet 2008). Mean adult age at death, taken from the probable distribution of deaths by age group (5-year, $a_{20} = 56.8$ years), is used as the input for the network of life tables corresponding to an annual growth rate of 0.005, the value closest to that estimated for Frénouville (0.0055).

In addition to the series of probabilities of dying by age (95 % confidence interval), Table 11.2 provides the age structure of the theoretical population and the allocation (proportional) of deaths by age group. The birth rate corresponding to this stable population is 26.22 per 1,000; the mortality rate is thus 21.22 per 1,000.

If we consider the period of occupation of the cemetery as a single point of observation, we can calculate the number of deaths occurring at Frénouville before age 10 from the number of deaths occurring after age 10 (N = 137). Under the demographic conditions defined above, the number of children dying before their tenth birthday should be 88 (or 39 % of all deaths), of whom 80 before their fifth birthday (Table 11.3 and Fig. 11.2 illustrate the observed and expected distributions of deaths by age at Frénouville).

However, if a constant rate of increase is applied to all ages, this does not take account of the particular context at Frénouville, where immigration was high throughout the period under consideration.

---

[4] With a distribution by 5-year group, the central age-point is defined more precisely ($a = x + 2.5$) than with 10-year groups ($a = x + 5$), leading to a slight variation in the calculated mean age at death (56.07 and 55.92 years, respectively). The sample size here is sufficient to justify calculation from 5-year groups.

[5] Some skeletons were not sufficiently well preserved to observe spheno-occipital fusion, which is a good indicator of entry into adulthood.

Table 11.2 Frénouville, Gallo-Roman period, estimated life table

Annual growth rate = 0.005				Associated stable population		
Age group	Adjusted R^2	Mean probability of dying (Q_x)	Probability in a 95 % CI	Age structure	Deceased of the table (D_x)	Life expectancy at age x (E_x)
00–04	*0.287*	*293.4*	*[285–301]*	*167*	*354*	*38.1*
05–09	*0.319*	*45.0*	*[42–47]*	*144*	*37*	*46.7*
10–14	*0.520*	*25.7*	*[24–26]*	*126*	*20*	*42.8*
15–19	*0.721*	*34.3*	*[33–35]*	*110*	*25*	*38.2*
20–24	0.841	43.5	[42–44]	94	30	34.0
25–29	0.852	49.3	[48–50]	80	32	30.1
30–34	0.903	55.0	[54–55]	67	33	26.6
35–39	0.926	62.0	[61–62]	55	34	23.2
40–44	0.913	72.8	[71–73]	45	36	19.9
45–49	0.905	86.3	[85–87]	35	39	16.9
50–54	0.868	107.9	[106–109]	27	43	14.1
55–59	0.836	142.2	[140–144]	19	49	11.5
60–64	0.828	193.8	[191–196]	13	56	9.2
65–69	0.804	260.4	[257–263]	8	59	7.2
70–74	*0.571*	*371.3*	*[366–376]*	*5*	*60*	*5.5*
75–79	*0.327*	*493.0*	*[486–499]*	*2*	*49*	*4.3*
80–84	*0.189*	*621.9*	*[612–631]*	*1*	*31*	*3.5*
85–89		750.0		0	14	3.5

Key: Input "estimated a_{20}" (56.8 years), both sexes combined, annual r = 0.005. The figures in *italics* and *smaller characters* are estimates with a low correlation coefficient ($R^2 \leq 0.78$)

- Second hypothesis

Here we start from the assumption that the population growth observed during the Gallo-Roman period is due to both migration and natural increase (excess of births over deaths). We accept that immigrants probably represented one-third of the population at the end of the fifth century (see above) and assume that it consists mainly of individuals rather than families. We also make the assumption of constant immigration throughout the two centuries observed. This amounts to ascribing to the young and mature adult age groups (15–59) of the Frénouville population an immigration rate that is higher than the growth rate for the other age groups.

We assume that the age 0–14 and over 60 age groups had a growth rate of some 0.001 and that the immigration rate for the 15–59 age group was 0.003. Knowing the life table, we can then calculate the theoretical population associated with these parameters (Table 11.4). When this is applied to the conditions observed at Frénouville, where there are 137 deaths over the age of 10, the expected number of child deaths before age 10 is 73 (35.2 % of all deaths), including 66 under age 5 (Table 11.5 and Fig. 11.3).

On this assumption of young adult immigration, the proportion of children is lower than when a constant rate of increase is considered (identical for all age groups). The estimated values remain within the ranges observed in historical demography.

Table 11.3 Frénouville, Gallo-Roman period, estimated distribution of deaths

	Deaths by age (Dx) proportionate to number of over-10s (N = 137)			Archaeological and anthropological estimate	
Age group	N	%	Age group	N	
00	52	23.0	00	13	26
01–04	28	12.3	01–04	13	
05–09	8	3.7	05–09		
10–14	4	2.0	10–17	7	7
15–19	6	2.5	15–19	4	130
20–29	14	6.1	20–29	126	
30–39	15	6.6	30–39		
40–49	17	7.5	40–49		
50–59	21	9.2	50–59		
60–69	26	11.5	60–69		
70–79	25	11.0	70–79		
80+	10	4.4	80+		
Total/of which	226	100.0	Total	163	
Over 10	138	61.0			
Under 10	88	39.0			
Under 5	80	35.0			

Key: The distribution of deaths associated with the life table in Table 11.2 has been calculated proportionately for the set of skeletons over 10 years of age

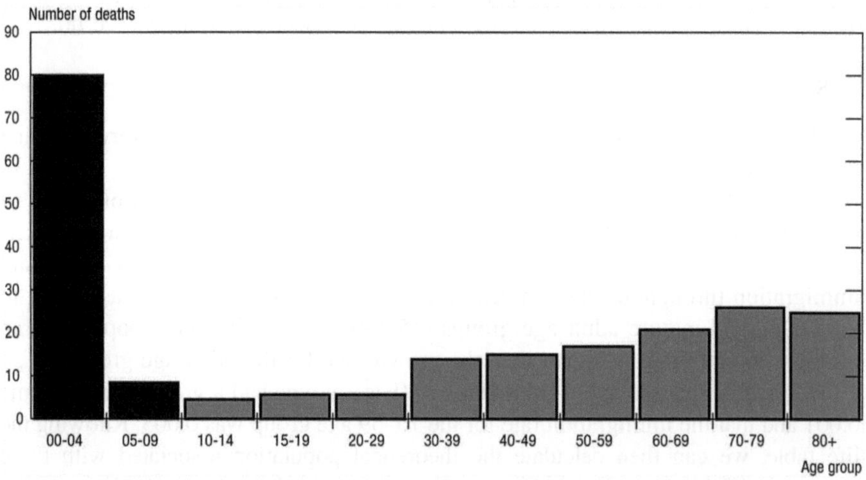

Fig. 11.2 Frénouville, Gallo-Roman period, estimated distribution of deaths by age group. *Key:* The expected distribution of deaths by age, according to the mortality distribution given in Tables 9.5 and 11.1, and the number of skeletons over 10 years of age (N = 137). *Black*: The two age groups with the greatest shortfall (26 skeletons are under 10 years of age)

Table 11.4 Frénouville, Gallo-Roman period, estimated life table

Annual rate of increase unevenly allocated across age groups				Associated theoretical population		
Age group	Adjusted R^2	Mean probability of dying (Q_x)	Probability in a 95 % CI	Age structure	Deceased of the table (D_x)	Life expectancy at age x (E_x)
00–04	*0.287*	*293.4*	*[285–301]*	*158*	*314*	*38.6*
05–09	*0.319*	*45.0*	*[42–47]*	*140*	*34*	*48.3*
10–14	*0.520*	*25.7*	*[24–26]*	*125*	*18*	*45.3*
15–19	*0.721*	*34.3*	*[33–35]*	*107*	*23*	*39.9*
20–24	0.841	43.5	[42–44]	93	28	35.9
25–29	0.852	49.3	[48–50]	80	30	32.3
30–34	0.903	55.0	[54–55]	68	31	28.8
35–39	0.926	62.0	[61–62]	57	32	25.4
40–44	0.913	72.8	[71–73]	46	35	22.1
45–49	0.905	86.3	[85–87]	37	38	19.0
50–54	0.868	107.9	[106–109]	28	43	16.0
55–59	0.836	142.2	[140–144]	21	50	13.2
60–64	0.828	193.8	[191–196]	16	65	12.1
65–69	0.804	260.4	[257–263]	11	70	9.7
70–74	*0.571*	*371.3*	*[366–376]*	*6*	*73*	*7.6*
75–79	*0.327*	*493.0*	*[486–499]*	*3*	*61*	*6.1*
80–84	*0.189*	*621.9*	*[612–631]*	*1*	*39*	*5.1*
85–89		750.0		1	18	5.2

Key: Entry parameter "estimated a_{20}" (56.8 years), both sexes combined, r variable among age groups: $r = 0.001$ for ages 0–14 and over 60, and $r = 0.003$ for ages 15–59. The figures in italics and smaller characters are estimates with low correlation coefficients ($R^2 \leq 0.78$)

It is therefore possible to propose plausible reconstructions of the juvenile population at Frénouville, even if it left few archaeological traces. Two hypotheses that are consistent with the archaeo- and osteological data have been proposed. However, the weak statistical correlation between the chosen estimator (mean adult age at death) and the probability of dying in the youngest age groups does appear to undermine these results somewhat, so we looked for another approach.

1. The first, $\alpha = 0.2$; $\beta = 0.9$, is not far from the pre-industrial mortality standard we have defined (where $\alpha = 0$ and $\beta = 1$). However, it does show a higher infant and child mortality and a slightly lower adult mortality (Fig. 11.4). The main demographic parameters associated with this mortality distribution are $_5q_0 = 430.2$ per 1,000; life expectancy at birth = 30.0 years; life expectancy at age 20 = 36.8 years. The expected number of individuals under age 10 is 119 (46.5 % of all deaths), including 110 under age 5 (43.0 %).
2. The second, $\alpha = 0.4$; $\beta = 0.8$, ascribes a predominant weight to child mortality: it is much higher than that of the pre-industrial standard and strongly affects the distribution of deaths by age (Fig. 11.5). The main demographic parameters associated with this mortality distribution are $_5q_0 = 548.52$ per 1,000; life expectancy at birth = 24.0 years; life expectancy at age 20 = 36.7 years. The

Table 11.5 Frénouville, Gallo-Roman period, distribution of deaths

	Deaths by age (Dx) proportionate to number of over-10s (N = 137)			Archaeological and anthropological estimates	
Age group	N	%	Age group	N	
00	43	20.6	00	13	26
01–04	23	11.0	01–04	13	
05–09	7	3.4	05–09		
10–14	4	1.8	10–17	7	7
15–19	5	2.3	15–19	4	130
20–29	12	5.7	20–29	126	
30–39	13	6.3	30–39		
40–49	15	7.3	40–49		
50–59	19	9.2	50–59		
60–69	28	13.5	60–69		
70–79	28	13.4	70–79		
80+	13	5.6	80+		
Total/of which	210	100.0	Total	163	
Over 10	137	65.2			
Under 10	74	34.8			
Under 5	66	31.6			

Key: The distribution of deaths associated with the life table in Table 11.4 has been calculated proportionately from the number of skeletons over 10 years of age

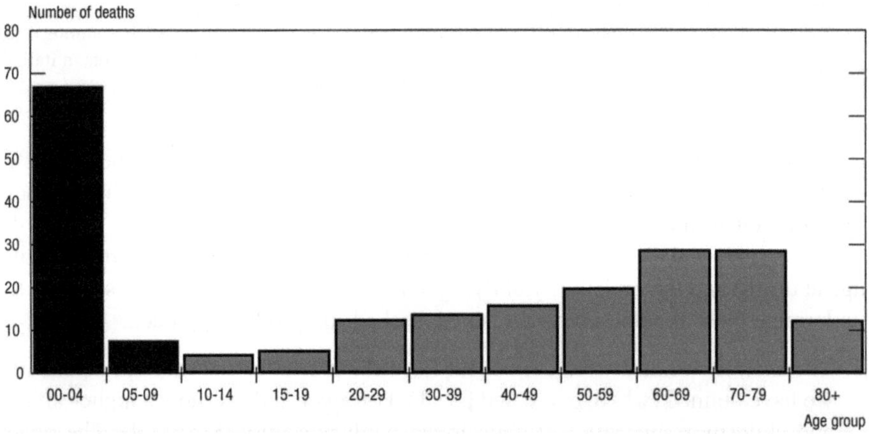

Fig. 11.3 Estimated number of deaths by age group. *Key:* The expected distribution of deaths by age, according to the mortality distribution given in Tables 11.2 and 11.3, and the number of skeletons over 10 years of age (N = 137). *Black*: The two age groups with the greatest shortfall

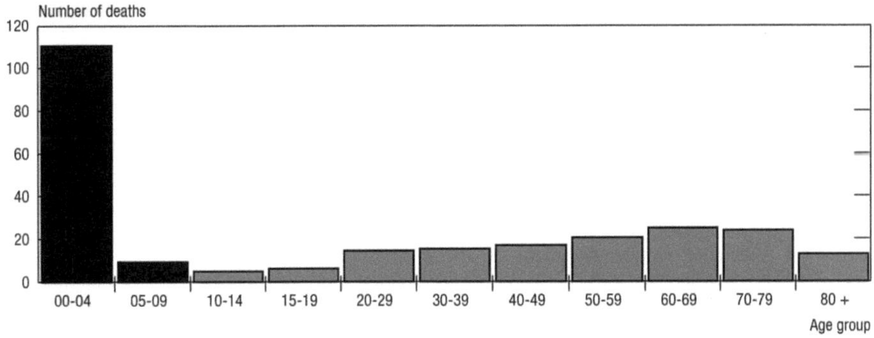

Fig. 11.4 Estimated number of deaths by age group according to various selected parameters. *Key*: Expected distribution of deaths by age, according to a mortality regime defined by parameters $\alpha = 0.2$ and $\beta = 0.9$, and number of skeletons over 10 years of age (N = 137). *Black*: The two age groups with the greatest shortfall

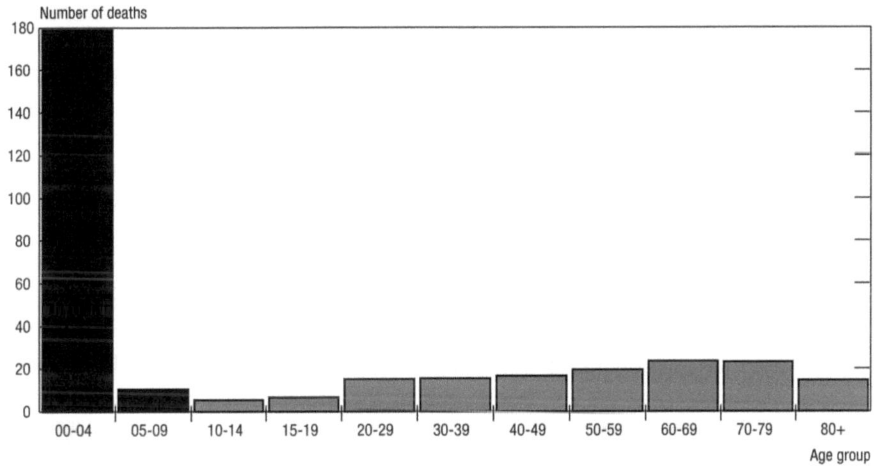

Fig. 11.5 Estimated number of deaths by age group according to various selected parameters. *Key*: Expected distribution of deaths by age, according to a mortality regime defined by parameters $\alpha = 0.4$ and $\beta = 0.8$, and number of skeletons over 10 years of age (N = 137). *Black*: The two age groups with the greatest shortfall

expected number of individuals under age 10 is 189 (57.9 % of all deaths), including 179 under 5 (54.85 %).

3. The third, $\alpha = -0.2$; $\beta = 1.2$, is the reverse of the other two. Child mortality is lower than the pre-industrial standard, and it is adults close to the median age who contribute most to the distribution of deaths by age group (Fig. 11.6).

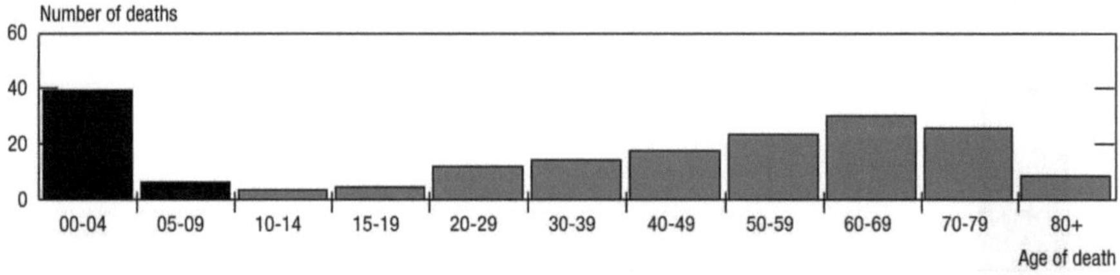

Fig. 11.6 Estimated number of deaths by age group according to various selected parameters. *Key*: Expected distribution of deaths by age, according to a mortality regime defined by parameters $\alpha = -0.2$ and $\beta = 1.2$, and number of skeletons over 10 years of age (N = 137). *Black*: The two age groups with the greatest shortfall

Second Method: Use of the Logit Method to Estimate the Missing Proportion of Juveniles

We can estimate the proportion of juveniles in another way, by reasoning from the structure and level of mortality, i.e. by using the logit method.

For values close to those obtained for this site – between 56.8 and 57.1 years – the values of the regression parameters vary in the proportions given in Table 11.6. These three proposed solutions correspond to three mortality patterns that are both possible and plausible, albeit differing widely from each other.

The main demographic parameters associated with this mortality distribution are ${}_5q_0 = 212.82$ per 1,000; life expectancy at birth = 42.0 years; life expectancy at age 20 = 37 years. The expected number of individuals under age 10 is 45 (17.5 % of all deaths), including 39 under age 5 (15.1 %).

Of the three, this is the estimate[6] that comes closest to the number archaeologically observed (N = 26).

The missing osteological data for the under-10s makes it impossible to use a second indicator (*P* or *JI*) to choose one of the models. However, we can take account of the archaeological evidence we have, namely a veterans' settlement site for the Roman army, to prefer the second model. During the first decades of the site's occupation, most of the population was adult. The small number of children exhumed may consequently reflect demographic reality.

11.1.1.5 Discussion and Conclusion

Through careful use of archaeological sources alongside the osteological data, the solution that comes closest to the living conditions described can be chosen from among the range of plausible options.

[6] Note that similar values of α and β will naturally provide an age distribution at death close to this one, but mean adult age at death can vary (for example, with $\alpha = -0.25$ and $\beta = 1.3$, the age distribution of deaths is virtually identical but the mean age at death is 56.68 years). It is not so much the value of the palaeodemographic indicator that must be considered first, but rather the mortality trends in the two age groups that modify the model.

Table 11.6 Values of parameters α and β that come close to the value of "estimated a_{20}" for this site

Parameter values	Estimated a_{20}
$\alpha = 0.2$; $\beta = 0.9$	56.91
$\alpha = 0.4$; $\beta = 0.8$	56.98
$\alpha = -0.2$; $\beta = 1.2$	57.16

Table 11.7 Summary of hypotheses used and results obtained

	Logit method			Power method	
		Infant-child mortality regime		Life tables for pre-industrial populations. Input: estimated mean adult age at death: 56.8 years	
For an estimated mean adult age at death of 57.33 years and an observed population size over age 10 of 137	Mortality regime close to pre-industrial standard $\alpha = 0.2$ and $\beta = 0.9$ (e_{20} : 36.8 years)	Higher than that of the pre-industrial standard $\alpha = 0.4$ and $\beta = 0.8$ (e_{20} : 36.7 years)	Lower than that of the pre-industrial standard $\alpha = -0.2$ and $\beta = 1.2$ (e_{20} : 37.0 years)	Annual growth rate = 0.005	Annual growth rate = variable by age group
${}_5q_0$	430.2	548.5	212.8	289.4	289.4
Estimated share of under-5s	110	189	39	80	66
Estimated share of under-10s	119	176	45	88	74
Estimated total population	256	326	182	225	210

The power model has at least two advantages: (1) there is no need to choose one hypothesis over another, which is appreciable when archaeological documentation is imprecise; (2) the population growth rate can be varied. On the other hand, this model lacks flexibility, and does not allow for possible variations in mortality by age group.

Conversely, the logit model provides considerable latitude for observing the consequences of minimal variations, not on the curve as a whole, but on the two segments we have isolated (α, representing the mortality of the youngest group, 0–5, and β, representing the mortality of mature adults around the mean age at death of our pre-industrial standard, 38.4 years). So the unknown proportion of juveniles we are seeking to estimate may vary considerably (Table 11.7).

Depending on the various assumptions made (mortality regime, population growth rate), the 130 skeletons of adolescents and adults exhumed at Frénouville may correspond to a total population of 182–326 individuals, of whom 17.5–57.9 % died before age 10, and 15.1–54.8 % before age 5.

The situations described are all compatible with the demographic behaviour of pre-industrial populations. The choice of one over another must be based on archaeological or historical evidence. For Frénouville, therefore, we prefer the hypothesis of infant and child mortality well below the generally accepted standards, given the weight of male, and to a lesser extent female, immigration of "mature" individuals. A reconstruction with 40–50 % juveniles would therefore be inappropriate in this context.

11.1.2 Lisieux Urban Cemetery (Calvados, N.W. France, Fourth Century AD Sector)

11.1.2.1 Research Question: Estimating the Population Living in Lisieux in the Fourth Century AD

Only in exceptional cases is historical, archaeological and anthropological evidence about housing and burial practices available for a single site. The "Lisieux-Michelet" cemetery has provided several hundred skeletons from the Late Empire and Early Middle Ages. Some 793 graves date from the fourth century alone. Two methods for estimating the settlement pattern were tested and compared. One is based on archaeological data – the urban structure and its environment – and the other on a palaeodemographic approach using as source data the skeletons from the cemetery.

11.1.2.2 Archaeological and Osteological Sources

The town of Noviomagus Lexoviorum (Lisieux) was founded in the first century AD in the heart of the territory of a Gaulish tribe, the Lexovii. The Roman town's expansion peaked in the second half of the second century. Following disturbances at the end of the third century, the town was fortified and became part of the continental side of the Saxon Shore defensive system. In around 280, a castrum was erected in the centre of the settlement. It was rectangular in shape (400 × 200 m) and protected a small area of 8 ha. All the archaeological evidence supports the thesis that the districts outside the walls were deliberately demolished and that the urban fabric withdrew to the interior of the castrum. This radical rearrangement of residential space, together with economic and religious influences, must have deeply marked the mentalities of the time. The changes can be seen even among the dead. There was a shift from cremation to burial, some burial areas were moved and new cemeteries appeared in the countryside, as at Frénouville (see above) and Saint-Martin-de-Fontenay.[7]

A major burial area, discovered by chance in the courtyard of the former Michelet school, 250 m from the castrum, was exhaustively excavated.[8] More

[7] Saint-Martin-de-Fontenay (Calvados, France): excavations directed by Christian Pilet (CRAHAM, Caen, France); anthropological study: Armelle Alduc-Le Bagousse (CRAHAM, Caen, France) and Luc Buchet (CEPAM, Nice, France). Publication: Pilet et al. 1994.

[8] The burial ground was excavated in four consecutive field seasons from 1990 to 1993. Excavations directed by Didier Paillard. Anthropological study: Armelle Alduc-Le Bagousse and Luc Buchet.

than 1,150 skeletons, in two distinct groups dating from the Late Empire and Early Middle Ages – with no intervening use – were inventoried. In size (0.8 ha) and burial density (up to six clearly identified levels of graves), the Late Empire phase is the larger. The earliest graves, oriented north–south, were dug in the early fourth century and the latest at the very beginning of the fifth century.

The extensive size and general layout of the burial area suggest that this was the main, or even sole, cemetery of the town. The large quantity of burial goods has made it possible to date the burials by quarter-century periods. It is this series of 793 individuals that is used for the present study.

11.1.2.3 Estimating the Population Size of Lisieux from Archaeological Data

The archaeological data suggest that Lisieux occupied a maximum area of around 60 ha at its height in the second half of the second century. This area consisted of spaces for living, working and recreation, making it difficult, indeed hazardous, to estimate the population of Noviomagus on the basis of surface area alone. Furthermore, one would need to know the actual floor area of the major public buildings (forum, baths, etc.) and premises used by craftsmen.

Starting from Christian Goudineau's hypothesis (1980, pp. 261 and 309–310) that the populations of most major towns did not exceed 5,000–6,000, we may estimate that for an equal area, Early Empire Lisieux had a maximum density of 80–100 residents per hectare. The reduction in urban area from the late third century may well have involved a reduction in density, but less marked than that observed within the walls. This is why Didier Paillard et al. (2006) took a density of 50 residents per hectare as their working hypothesis for their estimated fourth-century population of Noviomagus.

If we suppose that population density is related to urban area (8 ha), the estimated fourth-century population of Lisieux living within the town walls must have been around 400. In view of the remarks made above (areas of public buildings and workshops, etc.), this estimate can be considered as a maximum.

11.1.2.4 Estimating the Population Size of Lisieux from Anthropological Data

Estimating population size from anthropological data is no easy task.[9] First, one must critically examine the quality of the sample available for analysis. With

[9] An initial estimate of the fourth-century population of Lisieux was proposed in 2006 (Paillard et al. 2006). Since then, more refined methods have been used to calculate new estimates, and the results are presented here.

748 skeletons under study, the sample can be considered as large enough, but the distribution by sex and age at death requires some commentary.

Observed Distribution by Sex and Age

Unlike what is usually observed in rural cemeteries of that period in what is now Lower Normandy, there is a relatively high number of juveniles (208 have been identified, 26.2 % of all buried subjects) and all age classes are represented, from newborns to older adolescents.

The adult/juvenile ratio did not remain constant throughout the cemetery's period of use; juveniles were proportionately fewer between 325 and 375 AD[10] (Table 11.8).

Although the number of buried children is higher than in other cemeteries, it is difficult, in order to explain the variation between periods (Fig. 11.7), to estimate the relative importance of demographic and taphonomic processes. The age distribution at death of juveniles[11] in the first and third quarter-centuries corresponds to the expected distribution among pre-industrial populations, but in the second and fourth quarter-centuries, the number of infants and the ratio $\frac{D_{5-9}}{D_{10-14}}$ seem very low, but these figures should not be dismissed out of hand. The previous example showed the impact of high population growth on the juvenility index.

The sex distribution of adults also varies. Throughout the fourth century, males outnumber females. Only in the early years of the fifth century does the balance shift in favour of females.[12] This male excess is more marked during the earliest phase (62 %) and the last quarter of the fourth century (61 %), where burial goods and evidence of traumatic injuries suggest the presence of soldiers (Paillard et al. 2006).[13] The inversion of the sex ratio (56 % women) at the start of the fifth century may not be due to chance; military activities may explain the absence of some of the male population.

---

[10] This is confirmed by a chi-square test. The observed value of χ^2 is 11.414, and since the p-value (0.022) is below the chosen threshold (0.05), the null hypothesis (that the proportions are constant) cannot be confirmed. At least one proportion differs from the others.

[11] Probable age distribution at death estimated from stages of tooth mineralisation.

[12] The low number of subjects of indeterminate sex gives these variations in the sex ratio a high level of significance.

[13] The probabilities calculated with the Z-test display a significant difference at the 0.05 threshold between the numbers for the first quarter of the fourth century and those of the early fifth century. There is only a weakly significant difference (0.08 level) between the last quarter of the fourth century and the start of the fifth century.

Table 11.8 Distribution of adults and juveniles (under 18) by chronological phase

Phase	Total number of individuals	Adults (M/F and sex undetermined)	Juveniles	% JU/AD	N males	N females	Total both sexes	Sex ratio	Sex undetermined (%)
300–325	169	118	51	30.2	69	42	111	0.62	5.9
325–350	195	155	40	20.5	80	61	141	0.57	9.0
350–375	183	147	36	19.7	74	65	139	0.53	5.4
375–400	141	97	44	31.2	53	34	87	0.61	10.3
+/–400	60	41	19	31.7	17	22	39	0.44	4.9

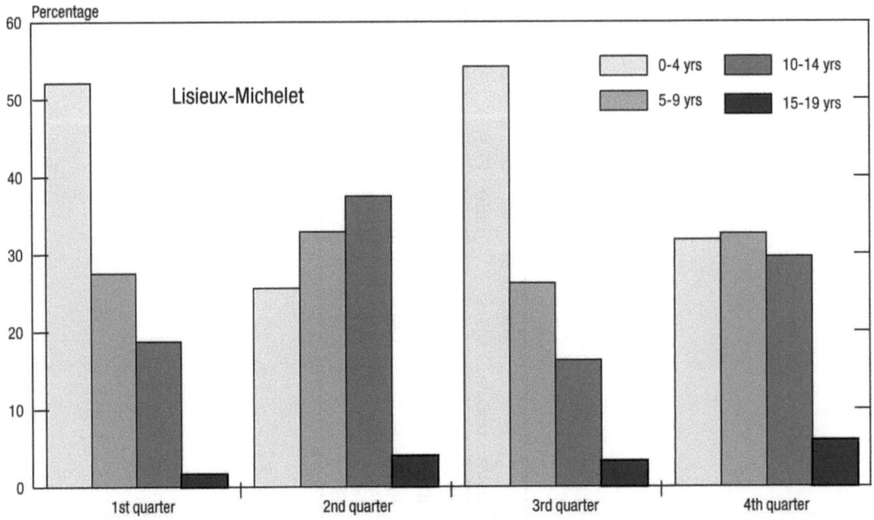

Fig. 11.7 Probable distribution of age at death of juveniles by quarter-century

Estimating the Proportion of Missing Juveniles

For this purpose we used the mortality models that link anthropological information to the expected age distribution of deaths. The chosen input is either the estimated mean adult (over 20) age at death[14] (a_{20}) or the juvenility index (*JI*),[15] so that only regressions that provide the best results are used (called "best model"). We assume, therefore, that all persons who died over the age of 5 were buried in the cemetery, for every quarter-century (although for some periods the representativeness of juveniles may be questionable, see above).

The growth of the Lisieux population throughout the fourth century must also be taken into account. Thanks to the precise nature of the topo-chronological evidence, it is possible to single out four phases of uneven length. The skeleton count reveals significantly different population numbers for the four chronological phases thus defined (see Table 11.8). Population growth (positive or negative, by natural increase or migration) can thus be introduced by means of the following formula:

$$r = \frac{1}{a} \times \ln \frac{D_{(t+a)}}{D_t}$$

where D is the number of exhumed adult skeletons, t the observation phase, and a its length.

[14] The mean adult age at death was calculated by the method proposed in this book, using the utility available on the INED website (reference population Lisbon 1889, distribution by 5-year age groups).

[15] The presence at this site of a large number of children makes it possible to calculate a juvenility index for each quarter-century.

Table 11.9 Lisieux-Michelet, fourth century, both sexes combined

Lisieux fourth century	c. First quarter 310–325	Second quarter 325–350	Third quarter 350–375	c. Fourth quarter 375– c. 400
Number of years (*a*)	15	25	25	30
Annual growth rate	0	0.0109	−0.0021	−0.0021
N adults	118	155	147	138
Value of "estimated a_{20}" (5-year groups)	54.35	53.48	55	55.3
Juvenility index (*JI*)	0.203	0.11	0.075	0.138

Key: Demographic parameters by quarter-century: number of adults, estimated mean adult age at death (a_{20}), and juvenility index (*JI*)

Table 11.10 Lisieux-Michelet, first quarter of fourth century, life table associated with "best model"

Age group	Calculated probability of dying	Standard deviation of Qx (95 % CI)	Survivors according to table (S_x)	Deaths according to table (D_x)	Life expectancy at age *x* (E_x)
00–01	275.0	[273.0–276.8]	1,000	275	26.4
01–04	215.4	[213.5–217.1]	725	156	35.2
05–09	117.9	[117.1–118.5]	569	67	40.4
10–14	53.2	[52.8–53.6]	502	27	40.4
15–19	44.8	[44.2–45.2]	475	21	37.6
20–24	58.1	[57.6–58.6]	454	26	34.2
25–29	66.5	[65.8–67.0]	427	28	31.2
30–34	74.0	[73.4–74.5]	399	30	28.2
35–39	83.7	[83.2–84.2]	370	31	25.3
40–44	97.6	[97.0–98.2]	339	33	22.3
45–49	114.4	[113.7–115.0]	306	33	19.3
50–54	141.3	[140.4–142.1]	271	38	16.7
55–59	182.1	[180.8–183.3]	232	42	14.0
60–64	233.2	[232.0–234.3]	190	44	11.6
65–69	302.0	[300.5–303.4]	146	44	9.3
70–74	406.2	[346.2–476.5]	102	41	7.3
75–79	522.3	[445.2–612.7]	60	32	5.6
80+	750.0		29	29	4.0

Key: Entry parameters: "estimated a_{20}" (54.35 years) and *JI* both sexes combined (0.203), r = 0

We calculated for each phase the annual and instantaneous growth rate using the method explained earlier (Frénouville example, Sect. 11.1 and Chap. 8, Box 8.1). Table 11.9 summarises the parameters to be included in the choice of mortality models: both sexes combined, growth rate, input and *JI*, number of adult deaths.

The mortality models provide the demographic parameters associated with the values given as inputs, including the theoretical age distribution at death, and life expectancy at birth associated with that distribution (Table 11.10, for the first quarter of the fourth century). This distribution can then be weighted according to the observations made at Lisieux by linking the number of exhumed adult skeletons for the first quarter of the fourth century (118) with the number of deaths over age

Table 11.11 Estimated population of Lisieux, by quarter-century, as a function of the anthropological data and demographic assumptions

Lisieux-Michelet- fourth century	First quarter 310–325	Second quarter 325–350	Third quarter 350–375	c. Fourth quarter 375–c.400	
Number of years (a)	15	25	25	30	Anthropological data
Annual growth rate	0	0.0109	−0.0021	−0.0021	
N adults	118	155	147	138	
Value of "estimated a_{20}" (5-year groups)	54.35	53.48	55	55.3	
Juvenility index (IJ)	0.203	0.11	0.075	0.138	
Life expectancy at birth (e_0) in associated stable population	26.4	38.2	34.9	29.5	Estimated demographic parameters
Mean age at death in stable population	26.4	29.2	36.7	31.2	
Estimated number aged 0–17 in Lisieux	137	137	77	114	
Estimated total number of deaths (0-ω years)	255	292	224	252	
Estimated total population of Lisieux	449	490	314	249	

18 given by the life table, i.e. 655 (assuming a linear distribution of deaths in the 15–19 age group and hence a two-fifths contribution from the 18–19 age group).

In this way we obtain the expected number of juvenile deaths at Lisieux, according to the parameters and hypotheses formulated for the first quarter of the fourth century. By adding the known number of adults and the estimated number of juveniles, we can calculate the total population, i.e., without migration or changes in the fertility or mortality regimes, ***the minimum number of individuals who must have been born to result in the observed number of adult deaths.***

The same operation is repeated for the other three quarter-centuries, using each time the relevant parameters as shown in Table 11.9. The introduction of a growth rate, whether positive or negative, requires using the mathematical concept of semi-stable populations, in other words, assuming that the population immediately achieves a stable state. On a first hypothesis, we assume no migration and therefore that population growth, positive or negative, is due solely to variations in fertility and mortality. Given the extremely short period over which this change had to occur, we assume that it immediately produced the expected effects (semi-stable population model), particularly on the structure of age at death.

Proceeding as before, but from the *deaths of the stable population* associated with the life table, we obtain the expected number of juvenile deaths at Lisieux in each quarter-century. By adding the number of adults and estimated number of juveniles, we estimate the minimum number of individuals that must have been born to result in the number of adult deaths observed in each quarter-century (Table 11.11).

The calculation of the number of people living at Lisieux in each period is based on the equation that demographers apply to stationary populations (Pressat 1983), adapted to cemetery populations (in a stationary population, the number of births and deaths balances out and life expectancy at birth equals mean age at death):

$$P = \frac{D \times e_0}{a}$$

where P is the mean annual population, D the total number of deaths observed (i.e. all the exhumed skeletons plus the "missing" skeletons from certain age groups, as estimated from the life table), e_0 life expectancy at birth and a the number of years the cemetery was in use.

However, we must now include the growth, positive then negative, of the population of Lisieux over the fourth century and weight the terms of the equation by the consequences of this growth on the population structure.

The approximate formula[16] we use is

$$P \approx \frac{D \times e_0[1 + \rho(e_0 - a_0)]}{a}$$

where a_0 is mean age at death (estimated, like e_0, from the life table) and ρ the instantaneous growth rate.

The population estimates obtained (Table 11.11) therefore differ slightly from those calculated under the stationary population hypothesis (Paillard et al. 2006), except for the first quarter-century, which has a zero growth rate.

11.1.2.5 Discussion and Conclusion

The results obtained from the anthropological data, under assumptions of varying growth rates during the fourth century (supported by archaeological observation), come close to the estimate initially made from population density per hectare (roughly 400, see above).

The advantage of the palaeodemographic estimate is that it offers a dynamic view of population change that is more informative than the mean population size calculated over a century. It places the start of demographic decline at the end of the first half of the fourth century, whereas archaeo-anthropological observations only note a gradual slowdown in burial rate in the last quarter of the fourth century. It also measures that decline, since the town lost nearly 45 % of its population in 50 years.

[16] We are grateful to Daniel Courgeau for developing Alfred J. Lotka's formula (1939, p. 20): $P(a) = N * e^{-\rho a} * p(a)$, where $P(a)$ is the population of age (a), N the number of births in that year; ρ the instantaneous population growth rate; $p(a)$ the probability of survival from birth to age (a), and adapting it to a cemetery population under the hypothesis of a Malthusian population with a constant growth rate (here, the number of deaths does not equal the number of births and life expectancy at birth does not equal mean age at death).

The palaeodemographic estimate does not appear, on first glance, to agree with the archaeological and palaeopathological observations that identify harsh living conditions for the residents of the Lisieux castrum throughout the fourth century, particularly during its second quarter, which was a critical phase in the history of the settlement, with serious disturbances that caused economic disruption leading to poor living conditions. These events do not emerge so clearly in our demographic reconstruction, where some indicators, such as a rapid growth rate and a relatively high life expectancy at birth, may appear to signal improvements in living conditions. However, this high growth rate may well have been due, not to a significant increase in fertility or a sharp drop in mortality (particularly among young children), but rather to a large inflow of men, women and, to a lesser extent, children, possibly fleeing the disorder and growing poverty in the countryside during the second quarter of the fourth century. In this way, the town's apparent demographic boom more likely reflects the flight of country-dwellers to a protected area (the castrum) and is merely a precursor of the misfortunes soon to follow.

Indeed, during the following period, the third quarter-century, the population of Lisieux declined and the general indicator of estimated mortality (life expectancy at birth) fell. Mean adult age at death, meanwhile, rose slightly. Together with the previous indicators, this supports the hypothesis of a fall in the birth rate or, more probably, the emigration of the youngest individuals, accompanied by their mothers, as suggested by the sharp rise in the sex ratio at that time (see Table 11.8).

The final quarter of the fourth century is, in every way, the most critical period. The extent of traumatic injuries observed (including a number of violent deaths) and the sudden decrease in the proportion of male graves just before the cemetery was abandoned, possibly reflect a renewal of military activity in the region. Far-reaching socioeconomic changes occurred at this point and the hypothesis of a gradual abandonment of the town as early as the second half of the fourth century is supported by the archaeological evidence: the cemetery remained out of use from the start of the fifth century until the seventh.

11.2 Examples from Sites for Which Biological and Statistical Sources Are Available

For the end of the early modern period (particularly second half of the seventeenth and the eighteenth centuries) and the contemporary period, written sources, such as parish records and then civil records, reveal the demographic features of past populations.[17] In some exceptional cases, there are both historical archives (written data) and biological archives (exhumed skeletons). This is true for the two examples we have chosen, where we have both the population exhumed from the

[17] An earlier work (Signoli et al. 2005) contains a study of these sites using a different calculation method, based on Masset's (1982) proposed probability vectors.

excavated sector and the historical sources referring to the population from which this sample was taken. Furthermore, these sources are sufficiently detailed to enable us to reconstruct the age-sex structure of the population and to establish the mortality distribution corresponding to the palaeodemographic sample.

Such exceptional situations enable us to compare the demographic data established from historical documents contemporary with the excavated cemetery against the results obtained by palaeodemographic methods from the skeletons, and not least, to identify certain biases in these methods and open up new avenues for research. These extremely rare sites give us the opportunity to test our tools, devise models applicable to similar contexts and advance the debate concerning a fundamental question, namely the representativeness of the palaeodemographic sample.

11.2.1 Monastic Cemetery at Maubuisson, Val d'Oise (Paris Region) (Seventeenth and Eighteenth Century Convent Population)

11.2.1.1 Research Question: What Palaeodemographic Standards Should Be Applied to a Population Whose Specific Features Are Known?

To answer this question, the chosen example is taken from the study of skeletons in a convent of Cistercian nuns not far from Paris: the Royal Abbey of Maubuisson (Saint-Ouen-l'Aumône, Val-d'Oise, seventeenth–eighteenth centuries), for which the historical sources can be used to reconstruct the population structure and establish the life table specific to this community.[18]

The site has another feature of particular interest: it is one of the rare cases where there is almost perfect identity between the burying population (the living population that uses this burial ground) and the buried population. Once a nun had entered the convent she was rarely authorised to leave it; the only exceptions were the abbesses and nuns still alive when the abbey was closed down in 1791. From 1677 to 1791, our observation period, 162 nuns were buried at Maubuisson and 37 of them have been exhumed. The statistical representativeness of the palaeodemographic sample can, in this case, be quantified (since the skeletons represent just over one-fifth of the burials) and studied qualitatively: does the age distribution at death, estimated from osteological data, correctly reflect the demographic behaviour of the population under study? If not, can it be improved?

[18] Excavation: Philippe Soulier, Christian Toupet and Jean-Yves Langlois; anthropological analyses by Christine Dumont, Bertille Danion and Jean-Yves Langlois, reviewed by Véronique Gallien; documentary research: Monique Wabont.

11.2.1.2 Archaeological and Osteological Sources

Two burial areas at the abbey have been excavated: the chapterhouse, reserved primarily for senior nuns and those of aristocratic birth, and the west gallery of the cloister, intended for lay sisters and novices. Our study uses only the 37 adult skeletons dating from the early modern period (29 from the cloister west gallery and 8 from the chapterhouse).

The skeletons are fairly well preserved and their age at death has been determined anthropologically from stages of cranial suture closure.

11.2.1.3 Contribution of Written Sources

These 37 nuns were recorded in the "Register of professed nuns of Our Lady Royal known as Maubuisson, deceased since the 6th of November 1652", which from 1677 until the convent's closure in 1791 lists the date of death of all the nuns, their age and the number of years since taking their vows. From this information, it was possible to construct the life table for the Maubuisson nuns during the final years of the French Ancien Régime (Table 11.12 and Fig. 11.8) but not to identify the 37 exhumed nuns, since their graves are now unnamed.

One of the major points to be considered is the specific age structure of the mortality of the Maubuisson nuns. These women, mainly from aristocratic backgrounds, enjoyed privileged living conditions during their childhood and adolescence so were in excellent health when they entered the convent (indeed this was a requirement for taking vows in this highly selective royal abbey). Monastic life, albeit harsh in some respects, also sheltered them from the risks facing contemporary laywomen during their reproductive years. Thanks to these various factors, their rate of survival at each age was higher than that of other women.

The age distribution of deaths at Maubuisson, according to the mortality distribution[19] (Table 11.12) necessarily differs from our reference for pre-industrial populations, as can be seen in Fig. 11.8.

11.2.1.4 Choosing the Right Method

Using frequency matrices (or the probability vector method) has the advantage of flexibility; in particular, we can vary the reference population so that it matches as closely as possible the demographic characteristics attributed to the study population.

---

[19] The age distribution of deaths results from a combination of two demographic phenomena: the age structure of the population and its mortality distribution. It is preferable to use the distribution of deceased women taken from the life table (D_x) rather than the age distribution of deaths observed in parish registers (d_x). This makes it possible to measure the impact of mortality independently from the age structure of the population.

Table 11.12 Life table of nuns at Maubuisson (Val d'Oise)

Age group	Observed probability of dying	Survivors from the table (S_x)	Deceased from the table (D_x)	Life expectancy at age$_x$	Female probability of dying in France (1770–1779)	Female life expectancy in France (1770–1779)
15–19	0	1,000	0	51.9	33	41.4
20–24	0	1,000	0	46.9	44	37.7
25–29	12	1,000	12	41.9	54	34.3
30–34	9	988	9	37.3	62	31.1
35–39	16	979	15	32.7	70	28.0
40–44	50	963	48	28.1	77	24.9
45–49	43	915	39	24.5	83	21.8
50–54	79	876	70	20.5	105	18.6
55–59	125	807	100	17.0	140	15.4
60–64	195	706	137	14.1	203	12.5
65–69	266	569	151	11.9	280	10.1
70–74	211	417	88	10.3	370	8.1
75–79	372	329	122	7.4	490	6.3
80–84	600	207	124	5.2	620	5.0
85–89	692	83	57	4.3	730	4.1
90+	1,000	25	25	3.5	1,000	3.5

Key: The life table of the 179 Maubuisson nuns (1677–1791) is compared with that of French women, 1770–1779 (based on Yves Blayo 1975)

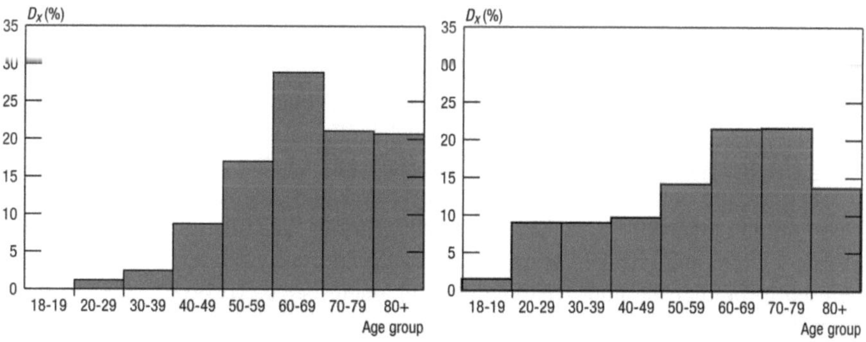

Fig. 11.8 Theoretical distribution of deaths by age. (**a**) For a population exposed to the same mortality distribution as the Maubuisson nuns (1677–1791). (**b**) For a population exposed to the Lisbon (1889) mortality distribution

In the Maubuisson example, the demographic parameters of the population are known (age-sex structure, mortality distribution). We have therefore constructed a fictitious reference population that has the biological characteristics (in this case, the rate of closure of cranial sutures) of our $P_{\text{Reference}}$ population (see Chap. 4) but

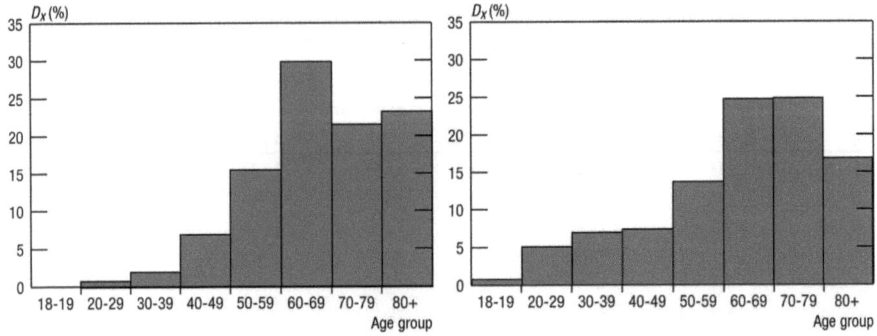

Fig. 11.9 Distribution of deaths by age obtained with $P_{Maubuisson1677\text{-}1791}$ *(left-hand figure)*. Distribution of deaths by age obtained with $P_{Lisbon1889}$ *(right-hand figure)*

the population structure (age distribution of deaths taken from the life table) of the study population. Given the small number of skeletons, it is preferable to use 10-year age groups for the various reference matrices.

Using a reference matrix that matches the demography of the Maubuisson nuns makes it possible to obtain a distribution of deaths by age group for the buried population very close to that observed from the archive documents (Fig. 11.8a). Statistical tests for comparing distributions (Wilcoxon, Mann–Whitney U) confirm this impression, although care must be taken with the statistical validation of the results. The distributions obtained using "$P_{Maubuisson1677\text{-}1791}$" (Fig. 11.9) or "$P_{Lisbon1889}$" (Fig. 11.9) do not differ significantly, at the 0.05 level, from the observed mortality structure.[20]

However, the results obtained with frequency matrices adapted to the features of pre-industrial populations are indubitably better than those obtained from a "standardised" reference population (see Buchet et al. 2003, for the results obtained using the probability vectors proposed by Masset).

We took the comparative study a little further and calculated the mean age at death of the over-20s associated with the various age distributions of deaths (Table 11.13), and checked to see whether the mortality models proposed could be used to predict the nuns' demographic characteristics. To save space, we have restricted the analysis to input $a_{20} = 68.62$ years, "female" model, with zero growth rate.

The estimated mortality curve matches quite well with the successive probabilities of dying established from the written sources (Fig. 11.10), and does show levels of mortality lower than those observed nationally.

[20] For $P_{Maubuisson1677\text{-}1791}$, the risk of rejecting a true null hypothesis is 94.53 %, exactly the same as for $P_{Lisbon1889}$ (94.53 %).

Table 11.13 Mean age at death (years) of the Maubuisson nuns, based on various sources and hypotheses

	Written sources based on		Osteological sources[a] Reference population	
	Registers of deaths	Maubuisson distribution of deaths	Lisbon (1890)	Maubuisson (1670–1791)
Age at death	67.81	66.88	63.72	68.62

[a]Whereas the osteological data establish a mean age at death for individuals aged over 20, the historical data show that this is in fact a mean age at death for those aged over 25, since no nun is recorded as dying between the two ages. This partly explains the variation between a_{20} (68.6 years) and the value calculated from the "Register of nuns" ($a_{20} = 67.8$ years). Similarly the value of a_{20} associated with the life table (66.9 years) includes deaths from age 20

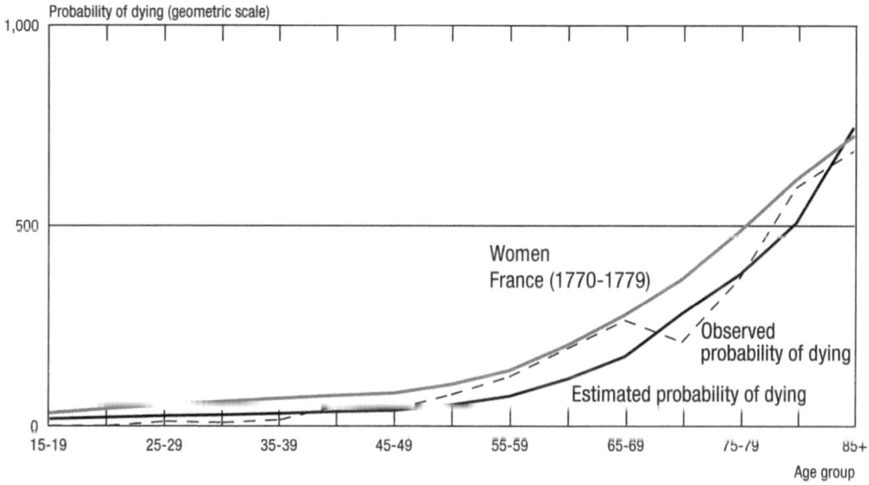

Fig. 11.10 Mortality by age of Maubuisson nuns (seventeenth–late eighteenth century) compared with all French women (late eighteenth century). *Key*: Mortality curve estimated from the female "estimated a_{20}" model and mortality curve calculated from written sources. Comparison of mortality in the Royal Abbey of Maubuisson (seventeenth–late eighteenth century) with that of the female French population in the eighteenth century

11.2.1.5 Discussion and Conclusions

In the case of a cemetery for a socially advantaged female population, we suggest using the "Maubuisson" palaeodemographic model rather than the standard model constructed from the characteristics of the population of late-nineteenth-century Lisbon, because the structure of age at death (reference population) significantly affects the estimate obtained. This drawback may turn out to be a useful advantage.

The particular features of this site were also used by Henri Caussinus and Daniel Courgeau to design the statistical model they proposed Chap. 13.

11.2.2 The Urban Cemetery of Antibes, Alpes-Maritimes (Small Port and Garrison Town, End Nineteenth Century)

11.2.2.1 Research Question: The Representativeness of the Anthropological Sample in Terms of Palaeodemographic Analysis

For our purposes, the study of the Antibes cemetery is of exceptional interest, because, as in the case of Maubuisson, we have both statistical documents providing information on the population of Antibes, and a set of skeletons representing a sample of the population buried in this cemetery.[21] Moreover, this example represents a "natural" population, in the sense that both sexes and all age groups are theoretically present, and concerns an urban site, with problems of reuse of graves, so there is a risk that the sample may not be as exhaustive as assumed.

Using a palaeodemographic model that reflects the specific features of the Antibes population, and the same approach as described above for the Maubuisson site, we discuss how far the palaeodemographic sample is representative of both the buried population and the "burying" population, i.e., all the people who may potentially use the cemetery.

11.2.2.2 Archaeological and Osteological Sources

A planned building development (*îlôt Terminus*) near Antibes railway station gave rise to an archaeological assessment, carried out in 1996, which confirmed the presence of graves on the site, known to have been part of the town's former cemetery. The graves belonged to the last cemetery extension in 1877–1897. When the Antibes town council decided to transfer the human remains to the new cemetery at Rabiac, just over a kilometre from the old one (this took place from 1897 to 1902), some skeletons were not claimed by family members and these are the ones found in the excavation. In fact, it turned out that none of the remains buried in the excavated area had been claimed and transferred to the new cemetery (which would have left the outline of an empty grave).

A rescue excavation was carried out in 1998 by a team from AFAN.[22] It produced the skeletons of 165 adults and 17 juveniles which were in very different states of preservation and generally highly fragmented (only 73 adult skeletons were sufficiently well preserved for age and sex to be identified). The

---

[21] An initial approach to this question and a discussion of palaeodemographic models were published in 2003 (Buchet et al. 2003).

[22] Excavation led by Philippe Vidal, AFAN-Mediterranean unit (AFAN is now INRAP: *Institut National de Recherches Archéologiques Préventives*).

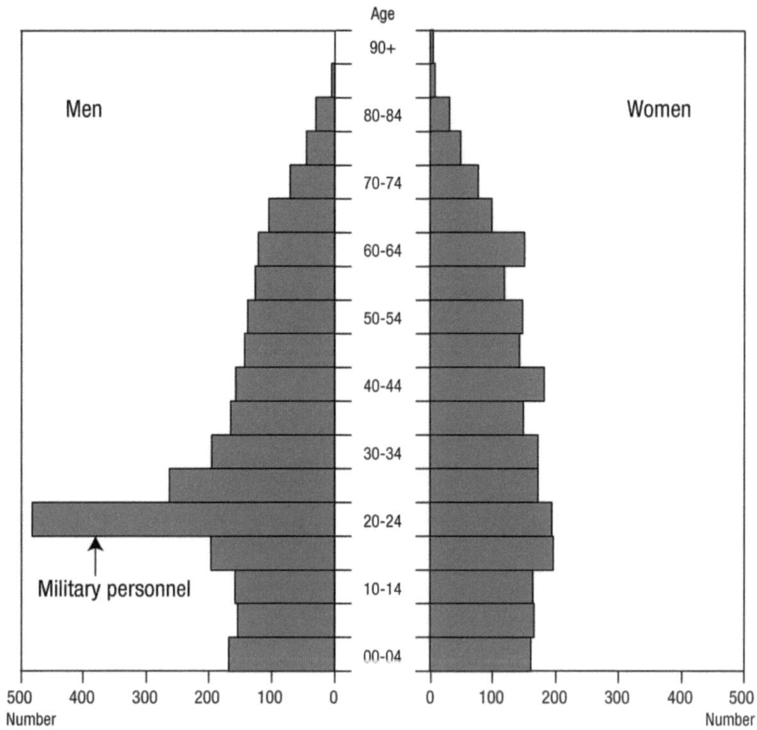

Fig. 11.11 Population pyramid of Antibes residents in 1881

anthropological analysis and the demographic study were performed at Sophia Antipolis (Buchet and Séguy 2003; Buchet et al. 2003).

11.2.2.3 Demographic Sources

Antibes presents certain marked demographic features revealed in censuses that must be taken into account: rapid growth, mainly by immigration, from the end of the nineteenth century; a large number of immigrants, mainly Italians employed as housemaids and seasonal labourers in farming, construction and crafts; military garrisons accounting for 10 % to 20 % of the total population, depending on the year (Fig. 11.11).

Census records include the sex and age of all residents in the town at given dates. Combined with the deaths recorded from 1877 to 1897, these data can be used to establish the mortality by sex and age of Antibes residents in various periods. Using the census records from 1881 and the deaths in the years just before and after (1879–1883), we calculated, with the method recommended by Louis Henry and Alain Blum (1988, pp. 158–160), the mortality distribution for the early part of the period of cemetery occupation (Fig. 11.12).

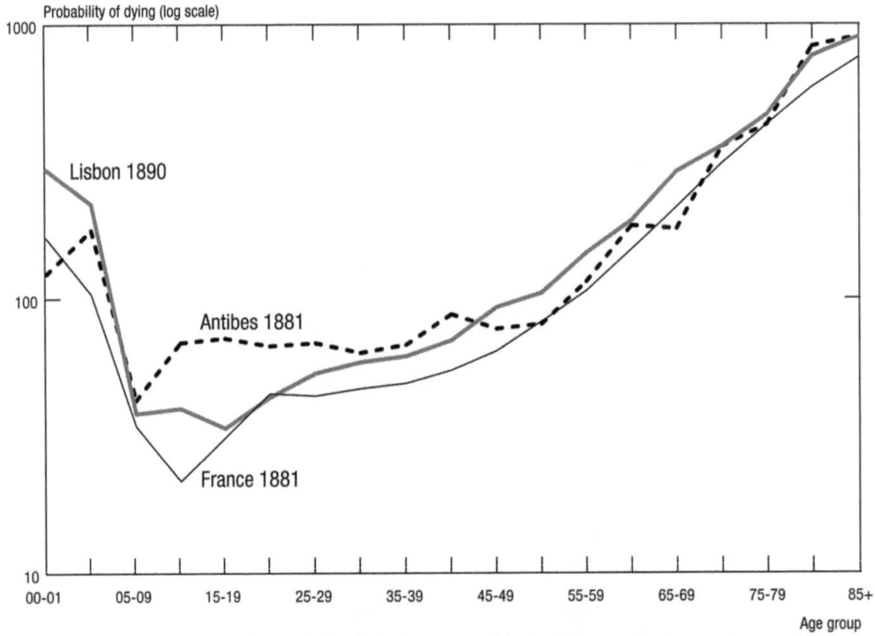

Fig. 11.12 Mortality by age of Antibes residents (both sexes combined) in the late nineteenth century compared with that of the French population as a whole and of Lisbon in 1889

Taking both sexes together, the age-specific probabilities of dying reveal for Antibes in 1881 high infant and child mortality, characteristic of pre-industrial populations, and mortality of young people and adults well above the national average (France, 1881, based on France, Meslé and Vallin 2001). This excess mortality partly reflects the specific features of this port and garrison town and also, probably, the deficiencies in the 1881 census (under-recording of certain categories, particularly housemaids aged 18–25, causing an over-estimation of the risk of dying at those ages). After age 50, the Antibes mortality curve joins that of France as a whole.

Note that despite certain similarities (same period, ports), Lisbon's mortality characteristics differ noticeably from those of Antibes (Fig. 11.12). While infant mortality is higher in Lisbon, the risk of dying between the ages of 5 and 45 is lower than in Antibes. This leads to an age distribution of deaths rather different from the one we have adopted as a standard ($P_{Lisbon1889}$).

11.2.2.4 Palaeodemographic Approach

Distribution by Age Group of Adult Deaths and Associated Mean Age

As for the Maubuisson site, knowledge of the demographic context for the archaeo-anthropological sample (exhumed skeletons) makes it possible to use a specially

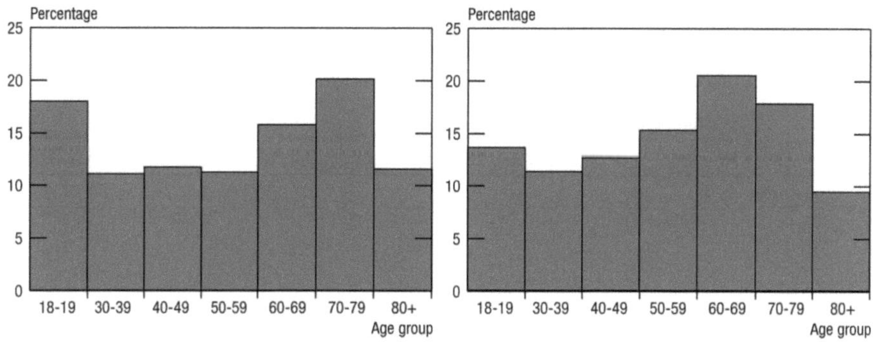

Fig. 11.13 Age distribution of deaths obtained with $P_{Antibes1881}$ *(right-hand figure)*. Age distribution of deaths obtained with $P_{Lisbon1889}$ *(left-hand figure)*

adapted frequency matrix, or palaeodemographic model. We consequently constructed a reference population from the distribution of deaths by age observed in Antibes in 1881 (see utility in supplementary materials on INED website). The results obtained using this reference population, "Antibes 1881", were preferred to those provided by the "Lisbon 1889" reference population (Fig. 11.13), although the two distributions do not differ significantly at the 0.05 level (Kolmogorov-Smirnov test). The rationale adopted here was the one that underpins the development and use of "palaeodemographic models", the idea being to include demographic behaviours specific to the population under study, where these are available.

Mean age at death of the over-20s was calculated from the probable distribution of deaths associated with the two palaeodemographic models ("Antibes 1881" and "Lisbon 1889"). These values are lower than those provided by the historical sources[23] (Table 11.14) and the observed difference between the two estimates of a_{20} is due to the influence of the structure of the reference population.

Distribution by Age Group of Juvenile Deaths

In view of the small number of juveniles (N = 17) taken from the partial excavation of the cemetery, it is not possible to use the PFP method of tooth mineralisation stages. They were therefore divided into age groups according to their individual ages, using for the 13 cases where the estimated age covered two or more age classes, the probability of belonging to one or other group.

[23] The difference between values based on death records and those based on the mortality distribution is due to large migrant inflows to Antibes, notably in the late nineteenth century, and to the emergence of a new category of migrants, namely retirees.

Table 11.14 Mean age at death of Antibes residents, calculated from various sources and hypotheses

	Written sources according to...		Osteological sources	
			Reference population	
	Registers of deaths	Antibes distribution of deaths	Lisbon (1890)	Antibes (1881)
Age at death	59.89	57.31	55.89	56.84

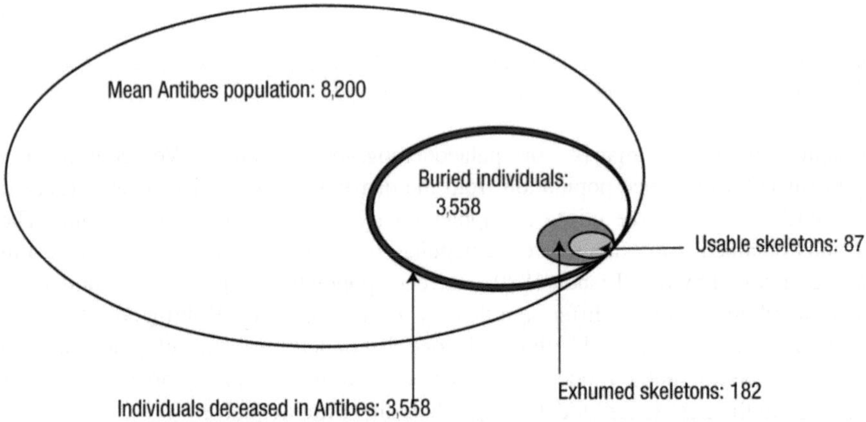

Fig. 11.14 Representativeness of an excavated sector of Antibes cemetery (1877–1897)

11.2.2.5 Sample Representativeness

Excavation of this late cemetery was necessary for specific methodological purposes, primarily to measure the representativeness of the palaeodemographic sample. For the modern and contemporary periods, demographic data sources are a useful tool for complementing the anthropological data (exhumed skeletons); they can be used for direct quantitative and/or qualitative comparison with the set of deceased subjects they concern, and with the "burying" population that used the cemetery (Table 11.15 and Fig. 11.14).

From the Exhumed Sample to the Analysed Sample

Of the 165 adult skeletons, sex and age could be determined for only 73, and for the 17 juvenile skeletons a probable age was attributed.

The representativeness of the palaeodemographic sample (individuals whose age and/or sex could be determined) with respect to the exhumed population is no worse at Antibes (47.80 %) than at most archaeological sites (see the mean calculated for selected Late Antiquity and Early Middle Ages sites, Table 11.16).

Table 11.15 Representativeness of the various samples

	Number of subjects		Representativeness/Analysed Population (%)		Representativeness/Exhumed Population (%)
Mean Antibes population	8,200	Antibes population	1.05	Exhumed population/Antibes population	2.22
Buried population	3,558	buried population	2.45	Exhumed population/buried population	5.12
Exhumed population	182	exhumed population	47.80		
Analysed population	87				

Table 11.16 Representativeness of the number of skeletons analysed with palaeodemographic methods

Sites	Number of exhumed skeletons	Number of analysed skeletons	Representativeness (%)
Les Rues-des-Vignes (Nord)[a]	208	134	64.4
Frénouville (Calvados)	697	299	42.9
Verson (Calvados)	289	173	59.9
Lyon-Saint-Irénée (Rhône)	170	97	57.1
Lyon-Saint-Laurent (Rhône)	178	61	21.1
Sézegnin (Switzerland)	644	320	49.7
Mean			49.2

[a]From Luc Buchet 1998

From the Analysed Sample to the Buried Population

The availability of death records for the site's entire period of use makes it possible, on certain assumptions,[24] to know the total number of burials carried out in the Antibes cemetery (3,558). The population exhumed during the excavations (182) represents 5.12 % of all deceased, and the analysed sample (87) just 2.45 %.

This poor result is not surprising. In urban areas, cemetery graves are regularly cleared and reused and there no chance of finding the entire buried population. Furthermore, for obvious land use reasons, it is rarely possible to excavate the entire area of a former cemetery, but only a part. In a number of ways, therefore, the Antibes cemetery "yields" as much evidence as a normal urban cemetery.

From the Analysed Sample to the Population of Antibes

The town's population is known in detail from the censuses held every 5 years: 6,752 (including 935 military personnel) in 1876; 9,329 (including 1,840 military personnel and 2,039 non-nationals) in 1896,[25] or a mean Antibes population of 8,040 for the period 1876–1896. In fact, allowing for variations in the growth rate,[26] the mean population corresponding to the cemetery's period of use (1877–1897) is closer to 8,200 (see Fig. 11.14).

Since the town had only one cemetery, the "burying" population from which the buried persons were taken coincides in this case with the mean population of the

[24] In particular, assuming the following to be negligible quantities: transported remains (Antibes residents deceased elsewhere and buried in Antibes), those lost at sea, non-residents deceased in Antibes and buried in their place of origin (whose death is recorded in the civil records but whose place of burial is unknown).

[25] Source: Antibes municipal archives – 1F2 827.

[26] The annual growth rate was fairly low between 1876 and 1881, roughly 0.63 %, taking the population to 6,795. By the end of the century, however, Antibes saw rapid population growth (some 3 % per year) and its 1897 population can be estimated at approximately 9,610.

Table 11.17 Age distribution of deaths recorded in Antibes and of the exhumed skeletons

	Recorded deaths				Distribution of exhumed skeletons			
Age groups	(*N*)	(%)		Expected deaths (D_x) (%)	(*N*)		(%) minus those of undetermined age	(%) with three of undetermined age
0–4	732	21	27	28	2	+3	1.12	9.34
5–9	69	2		3	3		1.68	
10–14	73	2		5	2		1.12	
15–17	62	2		3	7		3.91	
18–19	41	1	73	2	165		92.18	90.66
20–29	338	10		8				
30–39	222	6		7				
40–49	270	8		7				
50–59	311	9		7				
60–69	403	11		10				
70–79	594	17		13				
80+	443	11		7				
Total	3,558	100		100	179	182	100	100

town of Antibes. At this level of analysis, representativeness dwindles even further, since the exhumed sample now only amounts to 2.22 % of deaths of Antibes residents, and this may become critical for the palaeodemographic sample (Table 11.15): can these 87 analysed skeletons, 1.06 % of the population then living in Antibes, really reflect the town's demographic characteristics?

Representativeness in Terms of Population Structure

This question is even more acute when the age-sex structure of the buried population is examined. The palaeodemographic sample is not homogeneous: the adult population is over-represented with respect to the age distribution of deaths recorded in Antibes from 1877 to 1891. Table 11.17 shows the distribution of deaths by age group as observed and as expected from the mortality distribution specific to late-nineteenth-century Antibes. Juveniles, especially the under-5 s, are clearly under-represented, a very common state of affairs in palaeodemography (Fig. 11.15).

Representativeness of the Child Population

Because of poor preservation (child skeletons), only 14 were used in the analysis. The distribution of these 14 skeletons into age classes at death is nothing like the distribution expected from the demographic data. There are a number of reasons for this, and primarily the small sample size (Fig. 11.15 and Table 11.17). Furthermore, this partial excavation of the cemetery may have omitted the sector traditionally reserved for children, and the planned closure of the cemetery may have caused most families to transfer the remains of their prematurely deceased children to the new cemetery.

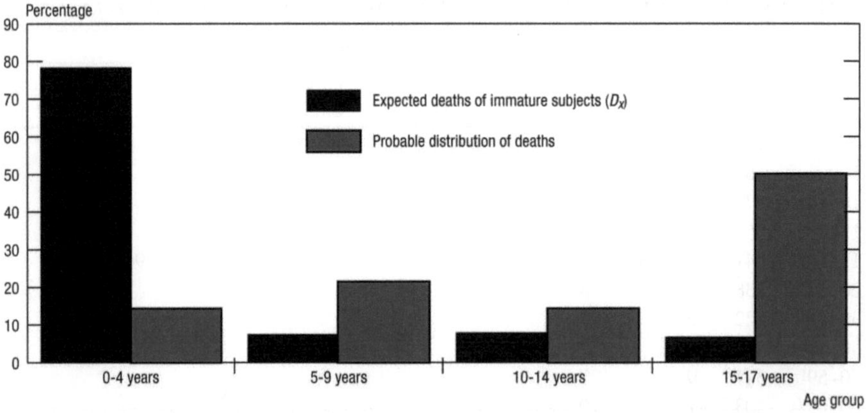

Fig. 11.15 Probable distribution of deaths of under-18s, in the palaeodemographic sample and according to the mortality distribution of Antibes1881

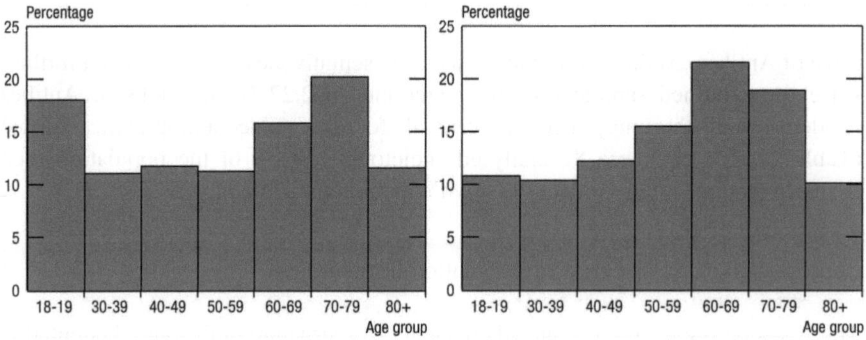

Fig. 11.16 Probable distribution of deaths of over-18s, in the palaeodemographic sample and based on the mortality distribution of Antibes1881

Representativeness of the Adult Population

The use of a frequency matrix adapted to the demographic situation of the Antibes population increases the reliability of the analysis. The probable distribution of deaths by age group of the palaeodemographic sample comes close to that observed from the archive documents. The two distributions were compared in a statistical test (Kolmogorov-Smirnov), which concluded that the series do not differ significantly at the 0.05 level. We may say, therefore, that the two distributions (Fig. 11.16 and Table 11.18) are not significantly different and that the sample of analysed skeletons is representative of the distribution of deaths by age group in the Antibes population.

Note, however, the over-representation of the 18–19 age group, similar to the excess in the 15-17 age group. We are tempted to explain this anomaly, not by higher mortality in this age group, but by the effect of migration. Since the graves of young immigrants and young conscripts who happened to die in Antibes were less

Table 11.18 Antibes – nineteenth century. Distribution of deaths by age group

Age group	(A) Recorded deaths	(B) Expected deaths (D_x)	Expected adult deaths (D_x) based on (A)	(C) Probable distribution of skeletons (10-year groups Antibes 1881), both sexes combined	(D) Antibes nineteenth century Model a_{20}; $r = 0.015$	Expected juvenile deaths (D_x) based on (A)	(E) Probable distribution of juvenile skeletons (dental analysis)
0–4	20.6	27.8			85.1	78.2	14.3
5–9	1.9	3.1			7.9	7.4	21.4
10–14	2.1	4.8			3.9	7.8	14.3
15–17	1.7	2.8			2.8	6.6	50.0
Sub-total children	26.3	38.4			100.0	100.0	100.0
18–19	1.2	1.8	1.6	3.6	2.0		
20–29	9.5	7.8	12.9	14.5	11.6		
30–39	6.2	6.6	8.5	11.1	11.6		
40–49	7.6	7.2	10.3	11.8	12.5		
50–59	8.7	7.1	11.9	11.3	14.9		
60–69	11.3	10.2	15.4	15.8	19.4		
70–79	16.7	13.2	22.7	20.2	19.3		
80+	12.5	7.6	16.9	11.6	8.8		
Sub-total adults	73.7	61.6	100.0	100.0	100.0		

Key: Proportional distribution of deaths by age group based on various sources: the first (A), observed in death records; the second (B), as expected from the Antibes mortality distribution (late nineteenth century), the third and fourth as reconstructed from the adult palaeodemographic samples using the probability vector method (C) and model tables (D), and the fifth (E) based on the juvenile palaeodemographic sample (tooth stages) often transferred to the new cemetery, their skeletons were more likely to emerge from the excavation.

As a test, we introduced into Table 11.18 the distribution of deaths estimated from our model tables (input a_{20}; instantaneous annual growth rate $= 0.015$). The results are mostly good, including for those ages where correlation with input a_{20} is weak (0–14 years and over-70s).

11.2.2.6 Conclusion: The Palaeodemographic Sample May Indeed Reflect the Buried Population Under Certain Conditions

The *îlôt Terminus* site in Antibes proves that it is possible to make good use of anthropological data even where numerical representativeness is poor. A poor yield is not a valid reason for rejecting a sample, provided that qualitative representativeness can be achieved by means of suitable palaeodemographic models (specific reference population).

Knowledge of the mortality distribution of the population makes it possible to estimate the age-sex structure of the exhumed population. Where, in addition, the "sampling fraction" of the palaeodemographic sample is known, it is possible to go further and attempt to reconstruct the buried population.

Some sites from the modern and contemporary periods already lend themselves to this approach (Signoli et al. 2005). Others will emerge, making it possible to identify general trends and to propose models applicable to periods for which no written data are available, but which can be considered as fairly similar in socio-demographic terms.

Part IV
Further Analysis

Part IV
Further Topics

Chapter 12
Critique of Current Methods

Daniel Courgeau

To estimate the age structure of a population of skeletons, palaeodemographers usually only have the structure by the stage of evolution of certain biological indicators. As we shall see, these indicators alone are generally not sufficient for this estimation. So another source of information is needed to link the indicators to individuals' actual ages. It may be a *reference population* for which researchers have both types of measurement. The combination of the two should provide an age structure for the *observed population*. However, there are various possible solutions to the problem, and these have been extensively discussed among palaeodemographers. In this chapter we examine these solutions and that discussion.

First, the names given to the methods used in palaeodemography are often incorrect and contradictory. For example, what Masset, Bocquet-Appel, Jackes and others call the *vector method*, namely IPFP and IBFP,[1] are referred to by Konigsberg, Frankenberg and others by such terms as the ALK and IALK[2] methods. Here they will be designated not by initials but names that clearly indicate their underlying hypotheses, because these methods have long been used in other disciplines and have a specific meaning that must be clearly defined. Obviously, the links between the more general name and the specific methods used by palaeodemographers will be spelt out.

Second, we present in detail the aim of each method, independent of the discipline that uses it, in order to show its purpose. Each one will be applied to precisely identical examples so as to demonstrate its usefulness for palaeodemography.

The author wishes to express his warmest thanks to Henri Caussinus for his highly relevant comments on the draft of this chapter, leading to a number of improvements. Thanks also to Isabelle Séguy and Luc Buchet for their many comments on the same draft. The author is, of course, entirely responsible for the content.

[1] Iterative Proportional Fitting Procedure, Iterative Bayesian Proportional Fitting Procedure.

[2] Age Length Key, Iterated Age Length Key.

D. Courgeau (✉)
INED, Paris, France
e-mail: daniel.courgeau@wanadoo.fr

I. Séguy and L. Buchet, *Handbook of Palaeodemography*, INED Population Studies 2, DOI 10.1007/978-3-319-01553-8_12,

We shall attempt to see which is the best suited to answering the questions asked by palaeodemographers.

We shall start from the established hypotheses underpinning these estimates (biological uniformity), which are crucial for solving the problem posed. We shall, in particular, assume that the two sources mentioned above provide sufficiently reliable and fully usable information to estimate the age structure of buried populations. These hypotheses have been fully discussed throughout the book (see also Kemkes-Grottenthaler 2002; Usher 2002).

We set aside earlier estimates of less value to modern research: an excellent critical presentation of them can be found in Masset (1973b). We shall only present the main two approaches currently used by most palaeodemographers.

12.1 Tables of Minimum Distance Between Each Cell

The principle of this method is to reconstitute the cells of a matrix when only its marginal values (the row totals and column totals) are known and then to use an earlier reference matrix to improve the estimate. The criterion used is the greatest proximity between each cell in the reconstituted matrix and those of the earlier matrix. In palaeodemography, this means starting from the observed biological stages of a population under study and a reference population for whose members both age and biological stage are known, and reconstituting the breakdown by age and stage of the population under study. This makes it possible to deduce that population age structure.

12.1.1 Historical Background

This approach was first used by Kruithof (1937), who was working on telephone networks. Starting from a full phone traffic flowchart taken from a reference population, he sought to estimate a new matrix of traffic flows for an observed population where only the marginal values were known (row and column totals).

This work was taken further in the 1940s (Deming and Stephan 1940; Stephan[3] 1942) to estimate the cells of a contingency table subjected to a number of constraints on one or more of its marginal values, where all the cells of an initial table are known and are to be approximated as closely as possible. This method was used for census data, which were often incompletely tabulated.

It was later developed by Leontief (1949), Stone, Bates and Bacharach[4] (1963) in economics, Friedlander (1961), Thionet (1963, 1964), Caussinus (1965), Fienberg (1968, 1970) and Bishop et al. (1975) in statistics, Tugault (1970),

[3] In the second article, Stephan recognised that the results of the earlier one did not coincide with those of the least squares method, as Deming and Stephan (1940) had wrongly stated, but they did provide an approximate solution.

[4] These authors called it the RAS method, from the symbols they used in the input–output matrices. This name was common in the 1960s.

Table 12.1 Example of a simple matrix

		Age		
		Group 1	Group 2	Stage totals
Stage	Stage 1	m_{11}	m_{12}	3
	Stage 2	m_{21}	m_{22}	2
Age totals		4	1	5

Willekens (1977) and Willekens et al. (1981) in demography, etc., to be applied to increasingly complex cases, especially those where there was not even an initial matrix taken from a reference population. A detailed presentation is given in the chapter on estimates from incomplete data in Courgeau (1988), because these methods are widely used to estimate matrices of migration flows between zones, where information is incomplete (only the two marginal values known; earlier matrix known with, for example, only one or two marginal values for the period under study).

12.1.2 Table Subject to Constraints

This is a brief presentation of the simplest case where only the marginal values of the table are available and these are the constraints. The aim is to estimate all the cells without having a reference population, and then to do so with one.

Assume a two-entry matrix where only the marginal values are known (Table 12.1).

It can easily be seen that if only the age and stage totals are known, it is not possible to choose each intersecting age-stage cell arbitrarily. Here only two matrices are possible:

$$M_1 = \begin{pmatrix} 3 & 0 \\ 1 & 1 \end{pmatrix} \text{ and } M_2 = \begin{pmatrix} 2 & 1 \\ 2 & 0 \end{pmatrix}.$$

To choose between these two matrices, a further hypothesis must be made. Let us assume that each matrix corresponds to a macrostate generated by microstates induced by the deceased individuals themselves. Since we have five individuals i_1, i_2, i_3, i_4, i_5, it can be seen that the three individuals comprising m_{11} in M_1 can be chosen in a large number of ways:

$$(i_1, i_2, i_3) \text{ or } (i_1, i_2, i_4) \text{ or } (i_3, i_4, i_5) \text{ etc.}$$

If we then assume that each of these microstates is equally probable, the number of possible choices of three individuals for m_{11} is

$$\frac{5!}{3!(5-3)!} = \frac{5!}{3!2!} = \frac{120}{10} = 10$$

where $5! = 1 \times 2 \times 3 \times 4 \times 5$ (factorial).

This leaves just two individuals, for whom the number of choices is simply 2. The last individual is necessarily counted in the last non-zero cell. There are therefore $\frac{5!}{3!2!} \times 2 = 20$ possibilities for five individuals to form matrix M_1. This is known as the entropy of the macrostate.

Similarly, we show that there are

$$\frac{5!}{2!3!} \times \frac{3!}{2!1!} = \frac{5!}{2!2!1!} = \frac{120}{4} = 30$$

possibilities of forming matrix M_2 from five individuals. So this is the matrix we shall take as the most likely estimate for the matrix for which only the marginal values are known. In other words, we optimise the entropy.

It may be supposed that this method, where the two sets of marginal values are known, is also applicable in cases where, as in palaeodemography, we only have the row marginal values (stage totals). In such cases there is more than one possible solution. For the example used here, we find the same maximum number of possibilities for two different matrices:

$$M'_1 = \begin{pmatrix} 2 & 1 \\ 1 & 1 \end{pmatrix} \text{ and } M'_2 = \begin{pmatrix} 1 & 2 \\ 1 & 1 \end{pmatrix}$$

this number being $\frac{5!}{2!3!} \times \frac{3!}{2!1!} \times \frac{2!}{1!} = 60$ for the first matrix and $\frac{5!}{1!4!} \times \frac{4!}{2!2!} \times \frac{2!}{1!} = 60$ for the second. And yet the two matrices are quite different, since their column marginal values (age totals) are (3 2) and (2 3), respectively.

This shows clearly that, as stated above, this method cannot provide an estimate of the age structure when we only have the stage distribution of the observed population, except, of course, for the obvious case where it is known that a single age group corresponds to a single stage. In other cases, it is generally necessary to have a reference population whose age and stage structure is known. This brings us to the second case, in which we seek to come as close as possible to the reference.

12.1.3 Table Subject to Constraints and Coming as Close as Possible to an Initial Table

Where the two sets of marginal values of an observed table are known, along with a reference table, the problem has been addressed in detail by many of the authors cited above, and leads to an iterative method, now universally known as the IPFP method. Where only one set of marginal values for the estimated table is known, the problem has received less attention, and we examine it here with some simple examples.

12.1.3.1 Method Used When Only One Set of Marginal Values Is Known

Example 1

This example starts from a matrix taken from a reference population which gives the joint distribution of two characteristics:

$$M^0 = \begin{pmatrix} 40 & 10 \\ 20 & 30 \end{pmatrix}.$$

If the observed population still has the same age totals as before, $\begin{pmatrix} 3 \\ 2 \end{pmatrix}$, a new estimation of age structure may be obtained that includes this extra information. The principle is once again to estimate a matrix M whose terms come closest to those of matrix M^0 and which provides the age totals of the observed population.

The proximity between two matrices can be measured in different ways, however. The simplest, on the face of it, is a Euclidean distance, the sum of the squares of the differences between each corresponding term in the two matrices divided by two. But this distance will depend largely on terms with a high value, where the differences may be greater than for terms of a low value. So the difference may best be judged in relative terms. In this case, one can use a chi-square distance that will weight the squares of each difference by the numbers observed in the reference population. This distance,[5] proposed by Deming and Stephan (1940) and Friedlander (1961), can be expressed as

$$d\left(M, M^0\right) = \frac{1}{2}\sum_{ij} \frac{\left(m_{ij} - m_{ij}^0\right)^2}{m_{ij}^0}.$$

We shall seek to minimise it, subject to the constraints

$$m_{11} + m_{12} = 3 \,\text{and}\, m_{21} + m_{22} = 2.$$

Because of these constraints, we shall use the Lagrange multiplier method. This calculates the partial derivatives of the distance, subject to constraints, with respect to each variable m_{ij}, which must be zero to obtain the minimum distance. This may be expressed, for example, for m_{11} and m_{12}:

$$\frac{m_{11} - m_{11}^0}{m_{11}^0} = \lambda \,\text{or}\, m_{11} = m_{11}^0(\lambda + 1)$$

[5] Minimising a chi-square distance involves additive adjustments, whereas optimising an entropy involves multiplicative adjustments. The corresponding algorithms are very similar in approach and results, but are, strictly speaking, different.

$$\frac{m_{12} - m_{12}^0}{m_{12}^0} = \lambda \,\mathrm{or}\, m_{12} = m_{12}^0(\lambda + 1)$$

where $-\lambda$ is the Lagrange multiplier for the sum of the differentials $dm_{11} + dm_{12}$. Summing the two equations, the value of $\lambda + 1$ may be estimated as $\frac{m_{1.}}{m_{1.}^0} = \frac{3}{50}$. Applying this to the other two cells, we now obtain a single solution:

$$m''_1 = \begin{pmatrix} \frac{12}{5} & \frac{3}{5} \\ \frac{4}{5} & \frac{6}{5} \end{pmatrix}.$$

This does give us the stage totals $\begin{pmatrix} 3 \\ 2 \end{pmatrix}$, but the age totals are no longer integers (3.2 1.8). This is therefore a theoretical solution not verified in practice, since the numbers of individuals must be integers, but it may be considered as the age structure of a larger population, namely (0.64 0.36). The matrix obtained is the closest to the initial matrix in terms of chi-square distance. The use of other distances[6] would clearly lead to different estimates, which are generally not far from this one.

It can also be seen that this method requires only one iteration and that consequently the name Iterative Proportional Fitting Procedure is no longer really appropriate. We propose calling it simply Proportional Fitting Procedure (PFP).

Example 2

This example uses the palaeodemographic data from Bocquet-Appel (2005, p. 297). It provides first a matrix of initial data where both the individuals' ages and the stages in which their femurs are classified are known (Table 12.2).

We wish to estimate from this matrix the age structure of a new population (Loisy-en-Brie, Late Neolithic) for which we know only the femur stage distribution (Table 12.3). This is a clear case for a PFP procedure, with a reference matrix, Table 12.2, and only one set of marginal values for the observed population. Although the aim here is not to estimate the cells of the matrix, which are of no

[6] Such as $d(M, M^0) = \sum_{ij} \frac{\left(m_{ij} - m_{ij}^0\right)^2}{s_{ij}}$, where $\frac{1}{s_{ij}}$ is the weight attached to m_{ij}^0 (see Stephan 1942), which may be measured, for example, by the variance of m_{ij}^0; $d(M, M^0) = \sum_{i,j} m_{ij} \ln \frac{m_{ij}}{m_{ij}^0}$ which is a measurement known as Kullback–Leibler divergence (also information divergence, information gain, relative entropy, or KLIC).

Table 12.2 Population classified by age and femur stage (reference population)

Stages\ages	23–34	35–46	47–58	59–70	71–82	83–94	Total
I	8	1	0	0	0	0	9
II	22	10	3	0	0	0	35
III	47	35	26	6	5	0	119
IV	13	29	35	30	25	5	137
V	1	4	10	10	9	4	38
VI	0	0	1	0	5	3	9
Total	91	79	75	46	44	12	347

Table 12.3 Numbers of femurs observed at Loisy-en-Brie, classified by stage[a]

Stages	Numbers
I	2.0
II	8.0
III	31.5
IV	40.5
V	12.0
VI	2.0
Total	96.0

[a]The numbers given by Bocquet-Appel for stages III and IV are in decimal form, probably to allow for some approximate determinations

Table 12.4 Transformation of Table 12.2 so that the probabilities in each row sum to unity

	23–34	35–46	47–58	59–70	71–82	83–94	Row sum
I	0.889	0.111	0.000	0.000	0.000	0.000	1
II	0.628	0.286	0.086	0.000	0.000	0.000	1
III	0.395	0.294	0.219	0.050	0.042	0.000	1
IV	0.095	0.212	0.255	0.219	0.182	0.037	1
V	0.027	0.105	0.263	0.263	0.237	0.105	1
VI	0.000	0.000	0.111	0.000	0.556	0.333	1

interest to palaeodemographers, but rather to estimate one set of marginal values from another, the method will still seek to find the closest matrix to Table 12.2, term by term, that provides stage totals equal to the numbers in Table 12.3.

The solution to this problem corresponds exactly to that presented above, minimising the chi-square distance when one set of marginal values for the table to be estimated is known and another matrix is available, estimated from a larger number of individuals (Deming and Stephan 1940). In this case, we assume that the Loisy-en-Brie population is distributed across the osteological stages in the same way as the reference population: we work on row probabilities (the probabilities in each row sum to unity). This gives Table 12.4.

Applying these probabilities to each stage observed at Loisy-en-Brie, we obtain the following age at death structure:

$${}_{12}P_{23} = 0.244,\ {}_{12}P_{35} = 0.225,\ {}_{12}P_{47} = 0.222,\ {}_{12}P_{59} = 0.142,$$

$${}_{12}P_{71} = 0.132,\ {}_{12}P_{83} = 0.035^{7}$$

Once more it can be seen that where we have only one set of marginal values for the matrix to be estimated, a simple calculation with no iteration is sufficient, whereas with the standard IPFP method, where both sets of marginal values for the matrix to be estimated are known, more than one iteration is required.

General Case

Let n_{ij} be the number of individuals at stage i (out of a total of l stages) in age group j (out of a total of c groups), and $n_{i.}$ the various stage totals in the reference population. For the observed population, let π_i be its stage structure, the only known element. It can be shown that its age structure $\hat{p}_j$ is expressed by the formula

$$\hat{p}_j = \sum_{i=1}^{l} \pi_i \frac{n_{ij}}{n_{i.}}. \tag{12.1}$$

This corresponds to the matrix with the lowest possible chi-square distance from the reference matrix, subject to the numbers observed for each stage.

The results are identical to those obtained with the traditional method of age estimation in palaeodemography, usually known as the *probability vector method.* This method goes back to Masset (1971) and was officially proposed by that author in 1973. Since then it has been used by many authors (Simon 1982; Blondiaux 1989; Pilet et al. 1990; Danion et al. 1994; Buchet 1998). It is clearly described in Masset (1973, 1982, 1995) as the sum of the l vectors corresponding to the various probabilities for each individual in the observed population located at stage i of belonging to each age group j, based on probabilities calculated from the reference population. This sum must naturally be divided by the total number of individuals observed to obtain the age structure given by the above formula. So it is indeed the same principle that underlies both this method and the PFP method.

Now we shall show that this estimator $\hat{p}_j$ is also the same as that proposed by another approach.

[7] This notation may appear unusual but corresponds to the results previously presented in tabular form as below, or as a histogram.

Age group	23–34	35–46	47–58	59–70	71–82	83–94	Row total
Proportion	0.244	0.225	0.222	0.142	0.132	0.035	1

12.1.3.2 The ALK Method

The ALK (Age Length Key) method was initially proposed by Friðriksson (1934) for determining the age of all the caught fish of a given species via a random sample taken from a single catch. Although it is easy to classify the total population by length group, measuring the age of each fish by close observation of their otoliths[8] is too expensive. So a small random number of fish are taken from each length group, the reference population, and their ages are accurately measured. The problem then is to estimate the age of fish taken from the total population on the basis of their length alone. The basic assumption of this method is that the fish in each length class in the reference population are a random sample of the observed population (Kimura and Chikuni 1987). If so, then it is possible to estimate the age distribution of the entire observed population. This method is still used (Holden and Raitt 1974; Farley and Basson 2005) to estimate the age of caught fish. It has also been used in other fields, notably in palaeodemography.

It was the first method proposed by Konigsberg and Frankenberg (1992) under the same name, Age Length Key, to estimate the age distribution of past populations from a cross-distribution. But in this case we note that the basic assumption is unlikely to hold because we are working with two populations whose age structures have no particular reason to be the same.

More specifically, estimation by the ALK method involves first calculating a matrix of age group frequencies for each of the stages in the reference population. The numbers in each stage in the observed population are then attributed to the various age groups according to the frequencies calculated in the reference matrix for each of the observed stages. It only remains to sum the values in this matrix for each age group and divide each of these numbers by the total observed population to obtain the age distribution of the observed population. It can be seen that this amounts to calculating for the age group j

$$\hat{p}_j = \sum_{i=1}^{I} \pi_i \frac{n_{ij}}{n_i}.$$

This is the same estimator (12.1) as for the PFP method with chi-square distance, although it is based on apparently different principles. Naturally, if another distance is used, the two estimators will differ, but generally only slightly.

[8] Otoliths, literally "ear stones", are crystal structures found in the internal ears of fish and other vertebrates that are sensitive to gravity and linear acceleration. Their growth rings are used to estimate the fish's age.

12.1.4 Critique

Note that the distribution calculated in this manner necessarily depends on the age distribution of the reference sample and it is "flattened under the influence of the reference sample", in the words of Masset (1995). This is a direct result of the assumption that each cell of the estimated matrix must be as close as possible to each cell of the reference matrix. Naturally, the greater the correlation between age and stage, the more satisfactory the estimate. Unfortunately, these correlations are rather weak in palaeodemography, usually around 0.5 (Bocquet-Appel and Masset 1982; Table 3.1, Sect. 3.2 above), so the reference population has a major impact on the age structure of the observed population.

Similarly, the assumption of the ALK method that the reference population is drawn from the observed population no longer holds, even for two populations of the same fish species taken from different catches. This problem has been raised by a number of researchers (Kimura 1977) and is a crucial one in palaeodemography, where the two populations are necessarily different, as mentioned above.

Note too that the two methods are used to estimate the theoretical observed matrix that is closest, term for term, to the initial matrix, which explains the dependency between the two matrices and ignores the invariance hypothesis (Müller et al. 2002), also known as biological uniformity hypothesis (see Sect. 2.3) whereby for any human remains of a given age at death, the likelihood of being classified in a given stage only depends on that age, whatever the population to which the bones or teeth belong. Similarly, for fish, when the age structures of the reference population and the observed population are not necessarily the same, the more general hypothesis is made that the length structure for each age must be identical for the two populations. This hypothesis is similar to the previous one, introducing a dissymmetry into the tables under consideration. It therefore becomes necessary to look for a more satisfactory method that takes this hypothesis fully into account.

12.2 Tables of Minimum Distance Between Each Column

Here it is not each cell in the reference matrix which must come as close as possible to the corresponding cell in the matrix of observed stages but each column with respect to its marginal value. In palaeodemography, this means starting from the distribution by stage within each age group in the reference population. Then the weightings are found that, after multiplication by the various previously estimated distributions, give the numbers per stage in the observed population. The weightings will then correspond to the numbers per age in the observed population. In this case, the invariance hypothesis is perfectly verified. This problem, which differs from the previous one, must be solved by different methods.

12.2.1 Historical Background

Here too the original task was to determine the age of an observed fish population for which only the length distribution is known. In this case, the reference population does not come from the same observed population but only a population of the same species, for which both length and age are known, as measured once again from the otoliths. Hasselblad (1966) provided an iterative method for an estimate of this type, followed by Orchard and Woodbury (1972), then Chikuni (1975). It was statistically developed by Kimura and Chikuni (1987), who proposed to call the method IALK, showing that it involved iterations. Unlike the ALK method, it only assumes that the length distributions for each age in the reference population are applicable to the observed population, which does not belong to the same total population (Kimura and Chikuni 1987) and may therefore have a quite different age structure. These authors use the same algorithms and most of them use the term "mixture" for the method. It is, in fact, a special case of the more general likelihood maximisation (or expectation maximisation, EM) algorithm proposed by Dempster et al. (1977).

In his unpublished 1977 thesis, Bocquet-Appel proposed to estimate the age structure of a population for which only the stage structure of its osteological remains are known by means of an iterative method starting from a uniform age structure. Masset (1982), in another unpublished thesis, proposed this *method of successive approximations* to avoid the over-flat result of the probability vector method. To that end, he and Bocquet-Appel wrote an iterative program, called *Approx*, which he appended to his thesis. We shall see below that, when starting from a uniform distribution, this program gives the same results as the IALK method. But Masset wrongly states that this method "only gives really satisfactory results if the subjects tested and those forming the basis of the sample are one and the same" (Masset 1982, p. 225), whereas in fact it avoids the need for that assumption, as we saw above. The application example for this method provided by Masset on pages 275–276 of his thesis, using a population with seven age groups, leads to results that are hard to accept. Although he starts from a reference population with seven age classes and seven stages, and the vector of the stages in an observed population of 60 individuals contains no zero term, he obtains an age structure for the observed population that is rather implausible:

$$(34.10\ \ 1.72\ \ 0\ \ 24.18\ \ 0\ \ 0\ \ 0),$$

because it contains four zero proportions. Faced with these disappointing results, he falls back on the less sophisticated method, thought to be more reliable, of probability vectors. In 1996, as we shall see, he adopted this method again with Bocquet-Appel.

Meanwhile, Konigsberg and Frankenberg (1992), looking for a more satisfactory method than ALK and wishing to avoid starting from a uniform distribution like Masset and Bocquet-Appel, realised that the IALK method avoided these biases.

First, while the ALK method gives good results when the reference population and the population whose age structure is to be estimated come from the same population, the IALK method avoids this constraint. Moreover, the estimation can be based not only on a uniform age distribution but on a distribution of any sort. Konigsberg and Frankenberg apply it to palaeodemography using the maximum likelihood method (see Box 12.1 for the principle).

Bocquet-Appel and Masset (1996) continued with their approximation method, which they now, wrongly, called IPFP, explicitly referring on page 572 to Deming and Stephan (1940). In the first part of this chapter, we showed that Bocquet-Appel and Masset sought to minimise the chi-square distance between each cell in the reference population matrix and the unknown matrix of the observed population, for which only one marginal value is known. IPFP is thus a misnomer, because the aim here is to minimise the distances between each of the columns corresponding to the same age: as we shall see, it is more like the IALK method. In order to distinguish it, we shall keep its original name of approximation method. In the same article, the authors also indicated the difficulties involved in getting this method to converge on acceptable results, and consequently proposed using it only to calculate the average age at death of individuals in the population. This restriction removes much of the method's usefulness.

These two approaches, which we shall call, for simplicity's sake, American and French, were the subject of considerable controversy between 1992 and 2002. But ultimately, they turn out to be practically identical (Konigsberg and Frankenberg 2002; Konigsberg and Herrmann 2002). Whereas the French insisted on the need to start from a uniform age distribution in order for their approach to yield the right result, the Americans understood that their approach always gives the same result, whatever the initial distribution. We return to this point below. However, the French realised that the IALK method could lead to solutions with a large number of zero age groups, as we showed when Masset's thesis was first presented. This explains the highly sceptical attitude of French palaeodemographers towards this method.

Box 12.1 Maximum likelihood

Henri Caussinus and Daniel Courgeau

Principle

In general, the statistical model states that data x are the observations of a random variable X, whose probability distribution depends on an unknown parameter θ. The probability density of X is a function of the data x and parameter θ, called likelihood function, often denoted L, and its value for x and θ, therefore, $L(x,\theta)$.

In frequentist (non-Bayesian) statistics, the most widely-used method is probably the so-called maximum likelihood method, which consists of estimating parameter θ by the value that maximises $L(x,\theta)$. The estimate is often denoted $\hat{\theta}(x)$, showing that it depends on observations x, or $\hat{\theta}$ for short.

(continued)

Box 12.1 (continued)

In practice:

- Parameter θ can be of any value, but is usually (and, in the cases studied here, invariably) a real number or a family of k real numbers, with possible restrictions; for example, variance is necessarily positive or a probability between zero and unity.
- The data can be of any value, but are usually a list (vector) of real numbers; this may be a sample of n values x_i, observations of n independent random values X_i from the same distribution. The likelihood here is the product of the individual densities $f(x_i,\theta)$ and it is shown that, on very general hypotheses, the method is efficient for high values of n (see Part 2); but it is of much wider application, and although its efficiency cannot be guaranteed, one may hope that it still provides reasonable estimates.
- Where the data are discrete, the density $L(x,\theta)$ is understood as the probability of observation x.
- It is clear that the same estimate $\hat{\theta}(x)$ is obtained if $L(x,\theta)$ is multiplied by a quantity that does not depend on θ (but may depend on x); this will simplify the expressions in many cases (in formal terms, the only change concerns the reference with respect to which the density is taken, an operation that must obviously be neutral for the estimation). Similarly, in finding the maximum, it is equivalent to replace $L(x,\theta)$ by an increasing function of this expression, its natural logarithm $\ln [L(x,\theta)]$, for example.
- It is usual practice to find the maximum of a function by setting the derivative to zero, in the case of one variable, or the partial derivatives in a multi-dimensional case. This practice, which is justified for large samples on fairly broad hypotheses, calls for certain precautions, however. First, the likelihood function may not be concave and the equations that set the derivatives to zero may have more than one solution, so a choice must be made; however, this case will not be encountered in the questions covered by this book. Second, where the set of parameter variations is bounded, the maximum likelihood may be achieved on the boundary without corresponding to a zero-value derivative; this case is of direct interest to us, as we shall see in the examples.

Properties

The most significant properties of the maximum likelihood method of estimation are asymptotic properties concerning samples of n independent observations, where n tends to infinity; in practice, where n is sufficiently high. In this case,

$$L(x,\theta) = \prod_i f(x_i,\theta) \text{ and } \ln[L(x,\theta)] = \sum_i \ln[f(x_i,\theta)]$$

(continued)

Box 12.1 (continued)

Moving on to the random variables, we have

$$\ln[L(X,\theta)] = \sum_i \ln[f(X_i,\theta)].$$

So $\ln[L(x,\theta)]$ is a sum of independent random variables to which, on very broad hypotheses, we may apply the general results of the probability calculus such as the law of large numbers and the central limit theorem (normal distribution). We can then show that for high values of n

- The maximum likelihood is "regular", i.e., is obtained by setting at zero the derivative or derivatives of likelihood with respect to θ
- The probability distribution of $\hat{\theta}$ is approximately normal, centred on θ (absence of asymptotic bias), with a variance close to

$$\left\{E\left[-\frac{d^2\ln[L(X,\theta)]}{d\theta^2}\right]\right\}^{-1} \tag{12.2}$$

in the case of a one-dimensional parameter (E stands for expected value); in the multi-dimensional case, the matrix of second-order partial derivatives replaces the second derivative above to produce the matrix of variances and covariances of $\hat{\theta}$. Replacing θ by its estimate in this expression, we obtain an estimate of the variance of $\hat{\theta}$. So it is possible to give approximate confidence intervals for θ or carry out tests. (Space does not permit demonstrations here, but note that the maximum of the likelihood function is less stable when it is poorly marked, i.e., the likelihood function is "flatter" near this maximum, its second derivative is smaller, making it intuitively likely that this derivative will appear in the variance of the estimator.)

Example

We may illustrate the point with a textbook example linked to the problems of estimating the age distribution where the stage distribution is known, with a reference distribution assumed to be error-free (Example 1 in this chapter with two age groups and two stages). The reference probabilities are

0.6667	0.25
0.3333	0.75

There is a single unknown parameter: let us take θ to be the probability p of the first age class. The stage probabilities are $0.6667p + 0.25(1-p)$ and $0.3333p + 0.75(1-p)$.

(continued)

Box 12.1 (continued)

For m independent observations we therefore have the likelihood

$$L(x,p) = \prod_i [0.6667p + 0.25(1-p)]^{x_i} [0.3333p + 0.75(1-p)]^{(1-x_i)}$$

or

$$\begin{aligned} L(x,p) &= [0.6667p + 0{,}25(1-p)]^{m_1} [0.3333p + 0.75(1-p)]^{m_2} \\ &= [0.25 + 0.4167p]^{m_1} [0.75 - 0.4167p]^{m_2}. \end{aligned}$$

(In the first expression above, x_i is the indicator that takes the value 1 if the observation is of stage 1 and 0 if it is of stage 2, and this formulation shows that it is a sample of size m; in the second expression, m_1 is the total number of stage 1 observations, and m_2 the total number of stage 2 observations, using the notation adopted in this book. One could also start from the binomial distribution of m_1, which would show in the likelihood a factor $m\,!/(m_1\,!\; m_2\,!)$, which, as we have seen, would have no effect.)

Figure 12.1 shows the variations of ln[L(x,p)] for two site data, respectively ($m_1 = 3$, $m_2 = 2$) and ($m_1 = 4$, $m_2 = 1$), the two cases considered in this chapter. It can be seen that if ($m_1 = 3$, $m_2 = 2$), there is a "regular" maximum; however, for ($m_1 = 4$, $m_2 = 1$) the regular maximum is "virtual", obtained for a value of p that is outside the range of possible values, since the "true maximum" for p is 1 and does not correspond to a zero value of the derivative.

Interpretation: although p on the x-axis can vary only between 0 and 1, the second curve has been extended to higher values (grey) to show the "virtual regular maximum".

For a sample of size m, we have

$$\ln[L(x,p)] = m_1 \ln(0.25 + 0.4167p) + m_2 \ln(0.75 - 0.4167p)$$

hence

$$\frac{d\ln[L(x,p)]}{dp} = \frac{0.4167m_1}{0.25 + 0.4167p} - \frac{0.4167m_2}{0.75 - 0.4167p}$$

which cancels to give $p = \frac{0.75m_1}{0.4167m} - \frac{0.25m_2}{0.4167m}$.

So we obtain for the two cases considered above $\hat{p} = 0.84$ and 1.32 respectively. Note that these values depend solely on the relative frequencies m_1/m and m_2/m.

(continued)

Box 12.1 (continued)

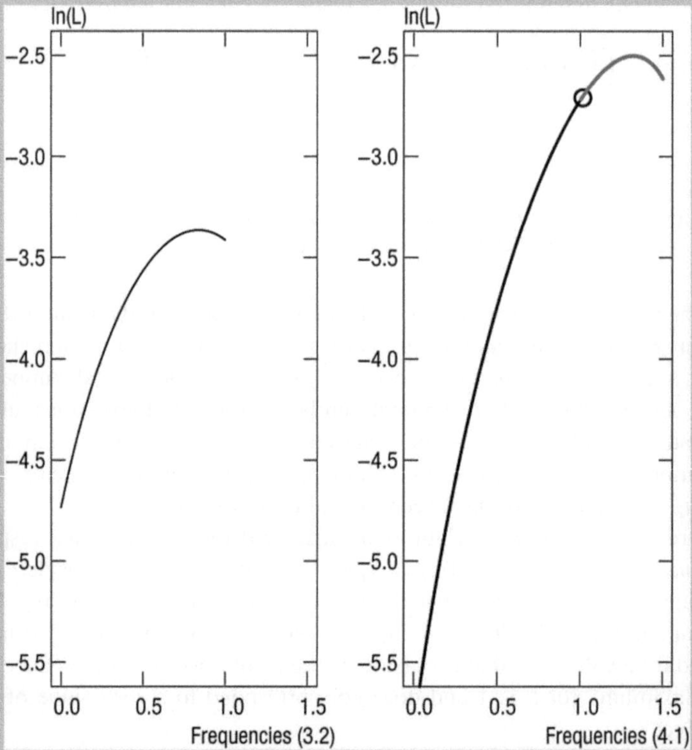

Fig. 12.1 Log-likelihood for $(m_1 = 3, m_2 = 2)$ *left*, and $(m_1 = 4, m_2 = 1)$, *right*

Let us also calculate expression (12.2). Here we need to take the second derivative of the logarithm of likelihood expressed for the random values M_1 and M_2, whose observed values are m_1 and m_2. This gives

$$\frac{d^2 \ln[L(X,p)]}{dp^2} = -(0.4167)^2 \left[\frac{M_1}{(0.25 + 0.4167p)^2} + \frac{M_2}{(0.75 - 0.4167p)^2} \right].$$

Following the standard result from the binomial distribution, we have $E(M_1) = m(0.25 + 0.4167p)$ and $E(M_2) = m(0.75 - 0.4167p)$, therefore:

$$E\left[-\frac{d^2 \ln[L(X,p)]}{dp^2} \right] = (0.4167)^2 m \left[(0.25 + 0.4167p)^{-1} + (0.75 - 0.4167p)^{-1} \right].$$

(continued)

Box 12.1 (continued)

We obtain an approximation to the variance of the estimator of p by inverting this expression and replacing p by its estimate. Thus, for $m = 5$, $m_1 = 3$, $m_2 = 2$ (therefore $\hat{p} = 0.84$), as in the example above, we obtain 0.2765 as the approximation to the variance, hence a standard deviation of 0.526; this is high because of the small sample size, but it is important to remember that this value is extremely unreliable, since it is based on the assumption that the sample is large. If, however, the sample were larger, for example $m = 100$ with $m_1 = 60$, $m_2 = 40$, the estimate of p would be the same but with an approximate variance divided by 20, namely 0.0138, with a standard deviation of 0.118. These are clearly more satisfactory and, not least, more reliable values.

We end by noting that if we had $m_1 = 80$, $m_2 = 20$ and a sample of 100, there would still be an "irregular" maximum likelihood for p equal to 1. However, with the values considered for the reference probabilities, to obtain $m_1 = 4$, $m_2 = 1$ for five observations is highly realistic, whereas $m_1 = 80$, $m_2 = 20$ for 100 observations is extremely unlikely because of the law of large numbers.

12.2.2 Maximum Likelihood Estimator

We will take the general case with the same notation as above to formulate and solve the problem. Let us look first at the IALK method.

We apply the frequencies of the distribution of the biological indicator, conditioned by the age groups in the reference population, namely $f_{ij} = \frac{n_{ij}}{n_{.j}}$. For the observed population, we have stage frequencies π_i. Applying the maximum likelihood method shows that the age structure $\hat{p}_j$ may be obtained by successive iterations from any initial structure, which is often taken to be $\frac{1}{c}$, i.e. uniform:

$$\hat{p}_j^{n+1} = \sum_{i=1}^{I} \pi_i \frac{\hat{p}_j^n f_{ij}}{\sum_{j=1}^{c} \hat{p}_j^n f_{ij}} \tag{12.3}$$

As many iterations as necessary are run to bring $\hat{p}_j^n$ as close as required to $\hat{p}_j^{n+1}$. An estimate of the age structure of the observed population is thus obtained. Konigsberg and Frankenberg (2002) recognise that the estimators they obtain in this way are estimators of maximum likelihood and not Bayesian estimators. Furthermore, this solution is only valid if the estimators are all positive: if some are zero, this may perhaps not be a maximum likelihood solution. It is also possible to estimate the variances of these estimators, as Cribari-Neto and Zarkos (1999) have done. We shall not develop these estimates here but shall use them below to estimate their standard deviation.

This method is easy to generalise to the case where there is more than one biological indicator or where these indicators are not discrete but continuous (Konigsberg and Frankenberg 1992).

It can also be used to introduce a continuous, rather than discretised, age without changing the principle. This is well laid out in Hoppa and Vaupel's edited volume (2002a), following a workshop on the topic at the Max Planck Institute for Demographic Research in Rostock, attended by a large number of English-speaking anthropologists, but with no French specialists invited. Konigsberg and Herrmann (2002) clearly indicate the similarity of the results obtained by IALK and these more sophisticated methods: "Our current methods fit fairly comfortably within the approaches taken during the Rostock workshop".[9]

First, the age distribution of a given stage in the reference population – with age now treated as a continuous variable – is provided by various types of non-parametric or parametric regression models. However, the volume's main originality is the use of a parametric event-history model (Courgeau and Lelièvre 1989) to model the probability density of the observed population's mortality. Provided the model does not include too many parameters (Gompertz two-parameter model, Gompertz-Makeham three-parameter model, Siler five-parameter model, etc.), we can estimate it using the maximum-likelihood method with the previously estimated age distribution of stages. Applying a notation similar to the previous one, we can summarise this formulation in the following form, where the age variable j is now continuous:

$$\pi_i = \int_j w_i(j) p(j, \theta) dj$$

and where $w_i(j)$ is the distribution of the stage i by age j, estimated in the reference population and $p(j,\theta)$ the age-specific probability density of the observed population, whose parameters θ we need to estimate using the l similar relations for each stage.

However, these methods introduce a number of additional hypotheses. They assume that the population is stationary or stable, so that the event-history model can apply to a population at a given moment, and that the age distribution of a given stage is continuous, so estimates differ according to the methods used. In principle, therefore, there is no reason why these hypotheses – which we have no way of verifying – should be fully satisfactory. For instance, a past population that has experienced an epidemic cannot be considered stationary or stable. Similarly, to impose on that population a parametric event-history model – ultimately rather simple and verified on current populations – may fail to capture situations where these models were not verified. Not least, these methods still assume that the reference population is

[9] The initial workshop was held in June 1999.

perfectly observed, although sampling errors in palaeodemography can be considerable. Not to allow for that fact, as with the IALK method, introduces a major risk of error into the estimate of the age structure of the observed population.

12.2.3 Approximation Method

The approximation method proposed by French-speaking authors is also iterative, but presented in an experimental rather than truly mathematical manner. Bocquet-Appel and Masset (1996) contains a detailed description of the procedure, where both p_j and f_{ij} depend on the iteration. We shall show that in fact it is not necessary for them to depend on the iteration, and in this case the method can, in some conditions, lead to the same result as IALK.

It starts from an initial value for the age breakdown which is assumed from the outset to be uniform, $\frac{m}{c}$, where the total number of members of the observed population m is distributed across the c age groups. The two important relationships for this algorithm are the following

$$\hat{p}_j^n - \sum_{i=1}^{l} \pi_i \frac{\hat{f}_{ij}^{n-1}}{\sum_{j=1}^{c} \hat{f}_{ij}^{n-1}} \text{ and } \hat{f}_{ij}^n - \hat{f}_{ij}^{n-1} \frac{\hat{p}_j^n}{\hat{p}_j^{n-1}}.$$

Starting from the initial value $\hat{p}_j^0 = \frac{m}{c}$ and the initial value $f_{ij}^0 = f_{ij}$, we deduce from the first relationship

$$\hat{p}_j^1 = \sum_{i=1}^{l} \pi_i \frac{f_{ij}}{\sum_{j=1}^{c} f_{ij}}. \tag{12.4}$$

Note that in this first iteration, the formulation (12.4) differs from the general formulation (12.3). The second relationship then gives

$$\hat{f}_{ij}^1 = \frac{c}{m} f_{ij} \hat{p}_j^1$$

and finally, by the first relationship

$$\hat{p}_j^2 = \sum_{i=1}^{l} \pi_i \frac{\hat{p}_j^1 f_{ij}}{\sum_{j=1}^{c} \hat{p}_j^1 f_{ij}}.$$

Now we are back at formula (12.3). It can also be seen that there is no point in considering $\hat{f}^{1}_{ij}$ or the number $\frac{m}{c}$, which cancels out from top and bottom. It only remains to see whether

$$\hat{f}^{n-1}_{j} = \frac{c}{m} f_{ij} \hat{p}^{n-1}_{j}$$

is verified, so

$$\hat{p}^{n}_{j} = \sum_{i=1}^{l} \pi_i \frac{\hat{p}^{n-1}_{j} f_{ij}}{\sum_{j=1}^{c} \hat{p}^{n-1}_{j} f_{ij}} \text{ and } \hat{f}^{n}_{ij} = \frac{c}{m} f_{ij} \hat{p}^{n}_{j}$$

which is easily shown by using the previous algorithm. As we have demonstrated that these relationships held for $n = 2$, they therefore hold for all values of n. Again we are back at the same formula (12.3) as for IALK, from the second iteration on. What Konigsberg and Frankenberg showed empirically has thus been proven mathematically in its most general form.

However, the main difference between the two methods is that the first can be used with any initial structure, as long as its age values sum to unity, whereas the second one requires starting from a uniform structure. This is simply due to the formulations and the resulting difference in values for the first iteration, because, from the second iteration on, the formulations are identical. If we take a non-uniform initial distribution for the second method, the solutions found will no longer be maximum likelihood estimators.

In the first example, it can be seen that, starting from a uniform initial age distribution (2.5 2.5), one arrives at the solution (0.84 1.16) after some 70 iterations, as with the IALK method. However, if the initial distribution is slightly different, the solution becomes widely different. For example, for an initial age group whose size varies from 2.35 to 3.25, the estimate of the final population for that group will fall from 5 to 0. Below 2.35, the final population of the first age group will remain at 5, and above 3.25, it will remain at zero. We see that the final estimate is highly sensitive to the initial distribution, which must be taken as precisely uniform.

It is instructive to use the same reference population for a larger observed population comprising 50 skeletons, for example. As long as the size of the observed population at the first stage is below 13, the population in the first age group is zero. But, as soon as it exceeds 13, the estimated population in the first age group increases regularly from 0 to 50, when the population at this stage reaches 34 individuals. It can be seen that the structure of the reference population can be used to estimate all possible combinations by age for the observed population. This contradicts the idea that some structures cannot be found with this method, as claimed by Bocquet-Appel (2005), as we shall see below. Naturally, this case needs to be extended to a larger number of age groups, but then the

Table 12.5 Reference population matrix by stage and age group

	Age group j									Stage totals
Stages (i)	n_{11}	.	.	.	n_{1j}	.	.	.	n_{1c}	$n_{1.}$
	.	.	.	.	.	.	.	.	.	.
	.	.	.	.	.	.	.	.	.	
	.	.	.	.	.	.	.	.	.	
	n_{i1}	.	.	.	n_{ij}	.	.	.	n_{ic}	$n_{i.}$
	.	.	.	.	.	.	.	.	.	.
	.	.	.	.	.	.	.	.	.	.
	.	.	.	.	.	.	.	.	.	.
	n_{l1}	.	.	.	n_{lj}	.	.	.	n_{lc}	$n_{l.}$
Age totals	$n_{.1}$	.	.	.	$n_{.j}$	.	.	.	$n_{.c}$	$n_{..}$

Table 12.6 Size of the observed population by stage

Stage totals
m_1
.
.
.
m_i
.
.
.
m_l
m

number of possible combinations becomes too high for the extension to be done properly. However, it can be clearly seen that in some cases we always find a population for which only one age group is represented. This coincides with Masset's result (1982), mentioned above, where certain age groups are estimated at zero.

12.2.4 Summary of the Two Methods

Now we examine the more general case of an observation of l stages in a population with c age groups. The reference population is given in Table 12.5, with numbers n_{ij} by age j and stage i.

We seek to estimate the age structure of a new population for which only the numbers by stage m_i are known, as given in Table 12.6.

Unlike the PFP estimate, where the probabilities to be estimated make no distinction between rows and columns, this method considers the initial reference table in an asymmetric fashion. The invariance hypothesis involves assigning a clear significance to the probability of belonging to stage i if we are dealing with a given age group j, denoted $p_{i\,|\,j}$, which is supposed to be applicable to any observed population. These probabilities then verify the relationship

$$\sum_j p_{i|j} = 1 \text{ for all values of } j.$$

To estimate these probabilities from Table 12.5, we shall see below the various assumptions that may be made.

For the observed population, we need to estimate the probability of belonging to one of the various stages p_i from the observed frequencies in Table 12.6. Using these probabilities and those estimated from Table 12.5, we can then estimate the age structure of this population $p_{.\,j}$, that we are seeking.

The initial matrix is used to calculate, for each age group j, a vector giving the distribution frequencies of biological indicator i:

$$f_{ij} = \frac{n_{ij}}{n_{.j}}.$$

If we assume that the observed numbers of individuals are high enough, the conditional age probabilities for a given stage will equal these frequencies $p_{i\,|\,j} = f_{ij}$. This gives us c vectors, for which we ignore any sampling errors in the reference population.

For the new observed population, we assume once again that the observed numbers are high enough. This gives us a vector estimating the probabilities that an individual from this population belongs either to the various stages π_i or, for stage i:

$$\pi_i = \frac{m_i}{m}.$$

From these estimates we can devise a method for estimating age structures.

12.2.4.1 Devising the Model

We wish to know if there is a set of weightings $(p_1, \ p_2, \ \cdots, p_c)$ representing the required age structure such that they verify the relationships:

$$\begin{aligned} \sum_j p_j f_{1j} &= \pi_1 \\ &\vdots \\ \sum_j p_j f_{ij} &= \pi_i \\ &\vdots \\ \sum_j p_j f_{lj} &= \pi_l \end{aligned} \qquad (12.5)$$

A final condition must be added:

$$\sum_j p_j = 1$$

which is a necessary one, because the weightings must sum to unity. So it is a mixture of distributions, of prior data, that must be estimated to find the distribution by stage of the observed population.

12.2.4.2 Solution for a Square Matrix

If the matrix is square ($l = c$), Cramer's rule applies, with generally one and only one solution. It is easy to see that the additivity condition for the p_j to be equal to one is necessarily satisfied.

This application of Cramer's rule can be expressed more simply in matrix notation.

If F is the square matrix

$$F = \begin{pmatrix} f_{11} & \cdot & f_{1i} & \cdot & f_{1c} \\ \cdot & & & & \cdot \\ f_{i1} & \cdot & f_{ii} & \cdot & f_{ic} \\ \cdot & & & & \cdot \\ f_{l1} & \cdot & f_{li} & \cdot & f_{lc} \end{pmatrix}$$

π the column vector

$$\pi = \begin{pmatrix} \pi_1 \\ \cdot \\ \cdot \\ \pi_l \end{pmatrix}$$

and p the column vector

$$p = \begin{pmatrix} p_1 \\ \cdot \\ p_j \\ \cdot \\ p_c \end{pmatrix}$$

The previous system of equations can be expressed concisely as

$$Fp = \pi$$

The estimator of p, $\hat{p}$, is then obtained by multiplying on the left the two members by the inverse of the F matrix, which is generally calculable:

$$\hat{p} = F^{-1}\pi$$

When the matrix is not square:

If $l < c$, the system is indeterminate and admits an infinity of solutions.
If $l > c$, the system usually has no solution, but $\hat{p}_j$ can be calculated using statistical methods.

12.2.4.3 Least Squares Method

This involves seeking those values of p_j that minimise the following sum of squares:

$$S = \left(\sum_j p_j f_{1j} - \pi_1\right)^2 + \ldots + \left(\sum_j p_j f_{ij} - \pi_i\right)^2 + \ldots + \left(\sum_j p_j f_{lj} - \pi_l\right)^2$$

with the constraint $\sum_j p_j = 1$. It can be seen, first, that where $l = c$, the solution $Fp = \pi$ given for a square matrix is always verified, because the solution is the same for both methods.

Where $l > c$, if we assume that we have values for p_j, then by introducing a variation ∂p_j, the differentials of the two previous equations give:

$$\frac{1}{2}\partial S = \left[f_{11}\left(\sum_j p_j f_{1j} - \pi_1\right) + \ldots f_{l1}\left(\sum_j p_j f_{lj} - \pi_l\right)\right]\partial p_1 + \ldots$$
$$+ \left[f_{1c}\left(\sum_j p_j f_{1j} - \pi_1\right) + \ldots f_{lc}\left(\sum_j p_j f_{lj} - \pi_l\right)\right]\partial p_c = 0$$
$$\text{and } \sum_j \partial p_j = 0$$

Now, multiplying the last equation by the arbitrary Lagrange multiplier λ and adding the two equations, we obtain:

$$\left(\frac{\partial S}{\partial p_1} + \lambda\right)\partial p_1 + \ldots + \left(\frac{\partial S}{\partial p_c} + \lambda\right)\partial p_c = 0$$

leading to a linear system for $(c + 1)$ equations with $(c + 1)$ unknowns $p_1, p_2, \ldots p_c, \lambda$:

$$\begin{cases} p_1\sum_i f_{i1}^2 + & \cdot\ \cdot & +p_j\sum_i f_{i1}f_{ij} + & \cdot\ \cdot & +p_c\sum_i f_{i1}f_{ic} & +\lambda & = \sum_i f_{i1}\pi_i \\ \vdots & \vdots & \vdots & \vdots & \vdots & \vdots & \vdots \\ p_1\sum_i f_{ic}f_{i1} + & \cdot\ \cdot & +p_j\sum_i f_{ic}f_{ij} + & \cdot\ \cdot & +p_c\sum_i f_{ic}^2 & +\lambda & = \sum_i f_{ic}\pi_i \\ p_1 + & \cdot\ \cdot & +p_j + & \cdot\ \cdot & +p_c & & = 1 \end{cases}$$

If the *invariance* condition and the assumption that the observed numbers are not subject to uncertainty are verified, this system of equations could be solved by Cramer's rule to obtain a system of weightings whose values all lie within [0, 1], corresponding to the structure of age at death. But, since the data are few, and therefore necessarily subject to uncertainty, and since this estimate is obtained by least squares, some estimates may fall outside [0, 1], even if the invariance hypothesis is verified. If so, the problem of estimating the structure of age at death will need to be solved allowing for these measurement errors, as we shall see below.

12.2.4.4 Maximum Likelihood Method

This method consists of considering that system (12.5) gives the probability π_i that an individual belongs to stage i. It is then possible to calculate the probability of observing a sample of m individuals of whom m_1 are at stage 1, m_2 at stage 2, etc., which will constitute its likelihood:

$$\frac{m}{\prod_{i=1}^{l} m_i!} \prod_{i=1}^{l} \left(\sum_{j=1}^{c} p_j f_{ij} \right)^{m_i}$$

with, as before, the constraint:

$$\sum_{i=1}^{l} m_i = m$$

Since the first fraction of likelihood is independent of the p_j, it suffices to maximise the logarithm of the second expression with the constraint. If this maximum satisfies the p_j positivity constraint, it will be found by setting the likelihood gradient to zero. Using the Lagrange multiplier, we then obtain the system of the following c equations to be solved:

$$\sum_{i=1}^{l} \frac{m_i f_{ij}}{\sum_{j=1}^{c} p_j f_{ij}} - \lambda = 0.$$

It can easily be seen that by multiplying each equation by p_j and summing the l equations, we obtain $\lambda = m$. The result is the non-linear system of c equations with c unknowns:

$$\sum_{i=1}^{l} \frac{\pi_i f_{ij}}{\sum_{j=1}^{c} p_j f_{ij}} = 1$$

It can be seen that where $l = c$, the solution $Fp = \pi$ is always verified and that this method gives the same result as the least squares method.

It can be demonstrated that this system can be solved by the following iterations (Hasselblad 1966):

$$p_j^n = \sum_{i=1}^{l} \frac{\pi_i p_j^{n-1} f_{ij}}{\sum_{j=1}^{c} p_j^{n-1} f_{ij}}$$

which are the same as those in formula (12.3). Starting from any positive values of p_j^0 that sum to unity, and setting a threshold beyond which the values of p_j^{n-1} and p_j^n are taken to be equivalent, we obtain the solution of the system by the maximum likelihood method.

It can be seen that these iterations are identical to those proposed by such palaeodemographers as Konigsberg and Frankenberg (2002) and Bocquet-Appel and Masset (1996). Note, however, that since it starts from positive values, this algorithm can only give positive or zero values. If the point of zero gradient on the likelihood curve does not fall within the domain of possible values, the IALK algorithm gives an estimate on the boundary of this domain (at least one of the p_j is zero), for which there is no proof that it is a maximum likelihood for $c > 2$ (although it is probably true in many cases). For that reason, in the following chapter, the maximum likelihood will be found by a procedure other than the IALK algorithm.

The least squares and maximum likelihood methods lead to different solutions where $l > c$, but these are generally close together. It is instructive to compare them in the application examples.

12.2.4.5 Application Examples

We shall now apply the two methods to four populations to see if they are always valid for palaeodemography.

Theoretical Example 1: Two Age Groups, Two Stages

We use once more the earlier simplified example, which provides the age structure (0.84 0.16), as we have shown. Since $l = c = 2$, it is easy to verify that the solutions given by the least squares and maximum likelihood methods are both identical to that given by direct calculation.

Note, however, that it is not possible to start from any marginal row value to find a column structure that fits with the two stage structures given for each age. It can easily be seen that the proportion for stage 1 of the observed population must fall between $\frac{1}{4}$ and $\frac{2}{3}$, in order to obtain solutions within the limits [0, 1] of the probabilities. In this particular case, this is so, because $\frac{3}{5}$ does lie in this interval $\left[\frac{1}{4}, \frac{2}{3}\right]$.

However, if we had an observed population with stage values of $\begin{pmatrix} 4 \\ 1 \end{pmatrix}$, which is perfectly possible if we observe, still from the same larger overall population, a sample of only five individuals, then the age structure estimated by the analytical method will be (1.32−0.32). It can be seen that the two solutions always sum to unity, but that they do not correspond to probabilities. Interestingly, the IALK method leads to a different solution (1 0), which does correspond to the maximum likelihood, but a maximum on the boundary of possible values, where the derivative of likelihood is not zero (see Box 12.1). In practice, we have here a case where none of the proposed solutions is appropriate.

Theoretical Example 2: Two Age Groups, Three Stages

Now we take the same number of age groups and one extra stage. Here we shall estimate the least squares and maximum likelihood solutions.

The reference population, established for this example, gives us the following matrix:

$$\begin{pmatrix} 40 & 10 \\ 20 & 30 \\ 4 & 40 \end{pmatrix}$$

and for the population whose age structure is to be estimated we have the following breakdown by stage:

$$\begin{pmatrix} 3 \\ 2 \\ 1 \end{pmatrix}$$

From this we deduce the two stage structures of the reference population for each age group:

$$S_1 = \begin{pmatrix} 0.625 \\ 0.3125 \\ 0.0625 \end{pmatrix} \text{ and } S_2 = \begin{pmatrix} 0.125 \\ 0.375 \\ 0.5 \end{pmatrix}$$

and the stage structure of the population for which we wish to estimate the age structure:

$$\pi = \begin{pmatrix} 0.5 \\ 0.3333 \\ 0.1667 \end{pmatrix}$$

Table 12.7 Transformation of Table 12.2 so column probabilities sum to unity

	23–34	35–46	47–58	59–70	71–82	83–94
I	0.088	0.013	0.000	0.000	0.000	0.000
II	0.242	0.126	0.040	0.000	0.000	0.000
III	0.516	0.443	0.347	0.131	0.114	0.000
IV	0.143	0.367	0.467	0.652	0.568	0.417
V	0.011	0.051	0.133	0.217	0.204	0.333
VI	0.000	0.000	0.013	0.000	0.114	0.250
Total	1.000	1.000	1.000	1.000	1.000	1.000

If we apply the least squares method, we obtain the following system of equations:

$$\begin{cases} 0.49219p_1 + 0.22656p_2 + \lambda = 0.42700 \\ 0.22656p_1 + 0.40625p_2 + \lambda = 0.27087 \\ p_1 + p_2 = 1 \end{cases}$$

This system is solvable and leads to the age structure of the observed population (0.7549 1.2451), with a parameter λ at the very low value of 0.0004.

If we now apply the maximum likelihood method, we obtain the following age structure (0.7576 0.2424), which is very close to the preceding one.

Example 3: Loisy-en-Brie Population

Here we take the example given by Bocquet-Appel (2005) of the Loisy-en-Brie population estimated from a reference population for which both age of individuals and stage in which their femurs are classified are known (Table 12.2).

Here we have a case where $l = c = 6$. The analytical solution is therefore applied, without turning to the least squares or maximum likelihood methods. The matrix F used is given in Table 12.7, where the column probabilities now sum to unity.

We calculate the inverse of the matrix:

$$\begin{pmatrix} 16.996 & -3.015 & 0.500 & -0.186 & 0.259 & -0.035 \\ -39.029 & 20.934 & -3.476 & 1.294 & -1.798 & 0.240 \\ 20.784 & -23.024 & 7.974 & -2.970 & 4.125 & -0.550 \\ -7.814 & 9.650 & -3.358 & 0.140 & 6.195 & -8.493 \\ 20.481 & -8.750 & -0.395 & 4.699 & -13.861 & 10.648 \\ -10.417 & 5.205 & -0.246 & -1.978 & 6.080 & -0.810 \end{pmatrix}$$

which, multiplied by the vector of the probabilities of belonging to each stage obtained from Table 12.8, finally gives the structure by age at death to be estimated:

$_{12}p_{23} = 0.219$, $_{12}p_{35} = 0.119$, $_{12}p_{47} = 0.380$, $_{12}p_{59} = 0.197$, $_{12}p_{71} = 0.040$, $_{12}p_{83} = 0.045$.

Table 12.8 Distribution by stage (combined suture closure coefficients) and age group observed in the Preference female reference population (established from three Portuguese collections, see Chap. 4)

Stages/ages	20–29	30–39	40–49	50–59	60–69	70–79	80+	Total
0–4	85	40	45	26	11	7	5	219
5–7	6	13	12	10	2	4	5	52
8–12	6	11	15	14	6	11	5	68
13–18	5	2	11	13	13	11	11	66
19–23	3	6	6	11	7	12	8	53
24–30	3	5	6	12	19	13	12	70
31–40	1	6	2	14	12	8	23	66
Total	109	83	97	100	70	66	69	594

Since $l = c = 6$, this solution is always identical to that obtained by the maximum likelihood or least squares methods.

However, these solutions differ sharply from that obtained by the PFP method:

$$_{12}p_{23} = 0.244,\ _{12}p_{35} = 0.225,\ _{12}p_{47} = 0.222,\ _{12}p_{59} = 0.142,\ _{12}p_{71} = 0.132,\ _{12}p_{83} = 0.035.$$

This last method provides an age distribution closer to the reference population than the previous one.

Example 4: The Maubuisson Nuns (Seventeenth–Eighteenth Centuries)

Here we use cranial suture closure as the age indicator for the female reference population established in Chap. 4, for which we have combined certain ages and stages (Table 12.8) so as not to take up too much space, but the result is the same when all values are considered.

The observed population is distributed by stage[10] (Table 12.9). Once again, since the matrix is square, the analytical method is applied. The corresponding matrix F is

$$\begin{pmatrix} 0.780 & 0.482 & 0.464 & 0.260 & 0.157 & 0.106 & 0.073 \\ 0.055 & 0.157 & 0.124 & 0.100 & 0.029 & 0.060 & 0.073 \\ 0.055 & 0.133 & 0.155 & 0.140 & 0.086 & 0.167 & 0.073 \\ 0.046 & 0.024 & 0.113 & 0.130 & 0.186 & 0.167 & 0.158 \\ 0.028 & 0.072 & 0.062 & 0.110 & 0.100 & 0.182 & 0.116 \\ 0.028 & 0.060 & 0.062 & 0.120 & 0.271 & 0.197 & 0.174 \\ 0.008 & 0.072 & 0.020 & 0.140 & 0.171 & 0.121 & 0.333 \end{pmatrix}$$

[10] The stage division adopted here is slightly different from that used in Sect. 4.1.2, since there needed to be at least as many stages and age groups (7).

Table 12.9 Distribution by stage observed in the archaeological population

Stages	Population
0–4	6
5–7	2
8–12	4
13–18	5
19–23	3
24–30	9
31–40	8
Total	37

Omitting intermediate calculations, we arrive at the structure of age at death to be estimated:

$$_{10}p_{20} = 0.091,\ _{10}p_{30} = -0.505,\ _{10}p_{40} = -5.018,\ _{10}p_{50} = 11.560,\ _{10}p_{60} = 1.709,\ _{10}p_{70} = -3.387,\ _{10}p_{80+} = -3.450$$

Although these figures sum to unity, they cannot now be used as a weighting system, since some are negative and others exceed unity. The numbers observed here are too small to give an age structure with all values lying within [0, 1]. The analytical method cannot provide an acceptable estimate here. Naturally, a positivity constraint could be introduced for the probabilities, but in that case some estimates would be zero, which would still not be satisfactory.

Whereas the least squares method always gives an identical solution, since $l = c = 7$, the maximum likelihood method, with its standard algorithm, leads to the following solution:

$$_{10}p_{20} = 0,\ _{10}p_{30} = 0.110,\ _{10}p_{40} = 0,\ _{10}p_{50} = 0,\ _{10}p_{60} = 0.614,\ _{10}p_{70} = 0,\ _{10}p_{80+} = 0.276.$$

As can be seen, this algorithm, which imposes positive probabilities, leads to a totally different estimate for the age structure, with four age groups with zero values. Once again, the solution is not acceptable.

However, in this case, where $l = c$, it is possible to estimate the maximum likelihood solution directly, because we obtain a square matrix that can be inversed. It can be verified that the age structure is the same as that obtained by the analytical solution or the least squares method.

12.2.5 Critique

First, some authors use this method where $l > c$. In some cases, the algorithm appears to converge towards a solution, but the variance of the estimates increases excessively and the method loses any potential interest . For example, Bocquet-

Appel and Bacro (1997) apply this method to estimate seven age classes though they have only six stages. Although the iteration results seem to converge properly, Konigsberg and Frankenberg (2002) estimate that the standard deviations of these estimators are respectively

$$(576 \quad 3.156 \quad 9.496 \quad 14.355 \quad 25.786 \quad 22.084 \quad 3.737),$$

demonstrating that the results are incoherent. An even more extreme example is Jackes' (2000) attempt to estimate 17 age classes with only six stages. Her results include a large number of zero-value age classes, clearly indicating that the model has not been identified. Furthermore, since this iteration procedure cannot provide negative probabilities, which should happen here, the positive results do not even verify the conditions in which the columns of the estimated matrix come closest to those of the reference matrix. Konigsberg and Frankenberg (2002) clearly state the conditions required to obtain a solution, as we have done above: the number of stages must be equal to or exceed the number of age groups considered. This condition is not verified in all the preceding examples, which explains the incoherencies.

Whereas Konigsberg and Frankenberg (1992) consider that they can correctly estimate the age structure of the observed population by this method, Bocquet-Appel and Masset (1996) believe that an estimate of this sort cannot yield any valid conclusion as to the shape of the sample age distribution. They think this is due to the random fluctuations of the ageing process, in both the reference population and the observed population. Consequently, they only use this method to provide estimates of the mean of the age distribution, whatever its actual shape. They consider this estimate of the mean as precise enough to be accepted with greater confidence. However, where the estimate of age structure is as implausible as in the example provided by Masset (1982), cited above, the average age calculated from this structure is unlikely to be more reliable.

As we have already said, the approximation method was devised empirically, after its supporters had noted that the method presented in the earlier section did not give satisfactory results, and had no clear mathematical justification. Not only did they doubt its validity for calculating an age structure, but they even thought that starting from a uniform distribution for age was likely to produce an unsatisfactory solution. One of the authors (Bocquet-Appel 2005) states:

> The uniform distribution was deliberately entrenched to eliminate the influence of the reference anthropological sample (Bocquet-Appel and Masset, 1996) but this prior distribution turns out to be a hindrance. It eliminates from the outset a large number of possible archaeological situations that are not represented by a uniform distribution of skeletons. (p. 279)

We have already shown, in the case of the simple example with two age groups and two stages, that this method covered all possible age distributions. Similarly, Konigsberg and Frankenberg (1992) state that the result of the iterations is independent of the age structure taken as a starting point: although a uniform structure is usually taken, this can be replaced by any other without changing the result of the iterations, since any distribution that sums to unity and has no zero cells is acceptable. The uniform distribution inserted in the proposed solution is no justification, in our opinion, for rejecting the IALK method, despite what Bocquet-Appel has suggested.

12.2.5.1 The IBFP Method

We turn to the iterative method he proposes (Bocquet-Appel 2005, 2008a; Bocquet-Appel et Bacro 2008) and calls IBPFP (Iterative Bayesian Proportional Fitting Procedure), an inappropriate name since the approach involves neither the methods usually called IPFP nor Bayesian methods. It aims to avoid the use of a uniform prior distribution.

Let us first describe the principle of the method. Since starting from a uniform age distribution appears to impose constraints in finding the age structure of a population for which only the stage structure is observed, it would seem to be more useful to start from more general distributions taken from a universe of possible demographic models of mortality. We use, therefore, the data from the reference population to calculate the probabilities for the stage indicator given the age. The reason is that from each of the prior age distributions it is possible to calculate the stage probabilities. We construct an indicator that indicates the variation between that probability and the one provided by the observed population. In the end we take the age distribution that minimises that indicator.

The problem is that from one publication to another, the method used for this purpose leads to different calculations. First, let us examine the presentation given by Bocquet-Appel in *La paléodémographie* (2005).

The algorithm he proposes is based on the probability of belonging to stage i, for individuals in age class j, f_{ij}, considered to be accurately known for the reference population. He then starts from each of the prior probabilities of belonging to age class j, p_j^0, provided by the various types of distribution considered here (Beta, non-central Beta, Weibull and Bi-Weibull). He attempts first to calculate the posterior probability of belonging to age class j, where the probability p_i of belonging to stage i is known from the observed population. He obtains (page 296)

$$p_j^1 = \sum_{i=1}^{l} \pi_i \frac{f_{ij} p_j^0}{\sum_{j=1}^{c} f_{ij} p_j^0}.$$

This estimate turns out to be identical with the first iteration of algorithm (12.3), corresponding to the IALK method, and not that of the approximation method (12.4). The result is that the subsequent iterations, given in greater detail (Bocquet-Appel 2005, p. 296), if taken far enough, should lead to the same age distribution whatever prior distribution one starts from. But the author ends the iterations when the variation ε between the observed and estimated frequencies by stage falls below the threshold of 10^{-5}, without giving any reason for this choice.

This manner of operating is confirmed by the example of the Loisy-en-Brie population (Bocquet-Appel 2005, pp. 297–298), presented above as Example 3 of the maximum likelihood method, with prior age probabilities

$$(0.10191\ \ 0.23279\ \ 0.26755\ \ 0.22668\ \ 0.1361\ \ 0.03497).$$

Taking the 10^{-5} threshold, we calculated that the IALK method leads to age structure

$$(0.212\ \ 0.155\ \ 0.332\ \ 0.201\ \ 0.061\ \ 0.039),$$

compared with Bocquet-Appel's estimate (op.cit., top of page 297) rounded to three decimal places

$$(0.212\ \ 0.156\ \ 0.332\ \ 0.201\ \ 0.061\ \ 0.038),$$

which is practically identical.

But in this case, whatever the values of the Beta distribution parameters, the calculated variations are all in principle very close to the 10^{-5} variation, and it is hard to see how the choice of parameters that give the smallest variation ε can lead to a final result. Nor is it clear what age distribution is to be chosen: the one we start from or the one we arrive at when the variation is 10^{-5}, which are quite different from each other.

Furthermore, if the iterations are continued, all these distributions will tend towards the IALK solution given above

$$(0.220\ \ 0.117\ \ 0.332\ \ 0.196\ \ 0.040\ \ 0.045).$$

This is exactly the solution obtained by continuing the iterations proposed in Bocquet-Appel's article, taking a variation of 10^{-10} after 5,587 iterations.

The result is that this first approach does not seem to lead to a fully satisfactory solution for estimating an age distribution, whatever family of demographic distributions is taken as a prior.

12.2.5.2 The Iterage Program

To understand more clearly the presentation by Bocquet-Appel (2008a) and Bocquet-Appel and Bacro (2008b) in more recent books, it is necessary to refer to the Fortran program Iterage.for, released in August 2007, which can be downloaded from Bocquet-Appel's site: http://www.evolhum.cnrs.fr/bocquet/index.html.

This time, the authors draw 1,000 equally probable samples with replacement by the bootstrap procedure from each of the age groups in the reference population. Unlike in the previous approach, this population is no longer fixed but can vary from one draw to another. Basically, this procedure is introduced so that the confidence intervals for the age distribution can be estimated. For each of these reference populations, the authors then use each of the prior probabilities by age, calculated from a mortality model including both usual and crisis mortality, to calculate a distance between the breakdown by stage of the observed population and that obtained by calculation, from each prior probability and the stage structure for each age in the reference population.

For that purpose, however, the authors appear to have returned to the method used in the earlier article by running iterations, in this case 1,000, and comparing in each iteration the stage structure of the observed population with that provided by the prior probabilities by age. Here we appear once more to come across the shortcoming mentioned above: for a reference population drawn by bootstrap, all the estimates taken from different prior age structures converge on the same solution. However, since this convergence is a slow one, the 1,000 iterations will produce solutions that appear to be different.

On closer inspection, it can be seen that the formulation of these iterations is almost identical to formula (12.3), except that the observed frequency π_i is replaced by its estimate, with the help of prior probabilities p_i:

$$p_i = \sum_{j=1}^{c} f_{ij} p_j^0$$

and its first iteration is therefore transformed into

$$p_j^1 = \sum_{i=1}^{l} p_i \frac{f_{ij} p_j^0}{\sum_{j=1}^{c} f_{ij} p_j^0}$$

In this case, it is easy to verify that p_j^1 equals p_j^0 and there is no point in continuing the iterations. This means that, contrary to what one might imagine from the complexity of this subroutine, one simple calculation leads directly to the result p_i. For each prior probability there corresponds one distance between p_i and π_i.

The authors state that this procedure does not permit any valid estimate of the terms in the matrix f_{ij}, which is now random, but can be used to choose for each sample the prior probability by age that provides the smallest distance from the stage structure of the observed population. They then calculate the mean of each of these probabilities and estimate a 95 % confidence interval with the various bootstrap estimates. However, although the bootstrap technique can be used when the model is properly specified, it is known that no theoretical result can be used to validate its results when, as here, an empirical model is used with no sufficient specification.

For this program, the authors slightly modified the Loisy-en-Brie data: but it is easy to use the stage data cited in Bocquet-Appel (2005). They also modified the age groups, using seven instead of six. We showed above that if a number of age groups greater than the number of stages is taken, it is impossible to solve the system of equations. The new program makes it possible and the results can still be compared with the previous ones by interpolation (failing which, the calculation of the 753 age probability vectors for six instead of seven age groups becomes unnecessarily cumbersome, and the authors do not supply the formulae used). For the Loisy-en-Brie example, the authors' program leads to the following solution:

$$_{10}p_{20} = 0.125,\ _{10}p_{30} = 0.133,\ _{10}p_{40} = 0.172,\ _{10}p_{50} = 0.199,\ _{10}p_{60} = 0.185,\ _{10}p_{70} = 0.125,\ _{10}p_{80+} = 0.061.$$

The program provides the mean chi-square distance between the observed and calculated stage distribution, 0.090. But this is not a chi-square distance, because it is calculated with the formula

$$\chi^2 = \sum_i m_i \left[\frac{(\pi_i - \hat{\pi}_i)^2}{\hat{\pi}_i} \right]$$

where m_i is the observed number of skeletons at stage i, and π_i, and $\hat{\pi}_i$ the frequencies by stage of the observed and estimated population, respectively. A standard chi-square distance should include the total number of observed skeletons m and not the number m_i by stage. The effect is that the value calculated by the authors' program underestimates the real chi-square distance. From these proportions of deaths, calculated for the first group over 7 years and the following over 10 years, it is possible to construct the curve that gives the survival probabilities of this population (Fig. 12.2).

From this curve, it is possible to calculate the 12-year survivors by interpolation (see Fig. 12.2) and thus the following structure of age at death:

$$_{12}p_{23} = 0.203,\ _{12}p_{35} = 0.182,\ _{12}p_{47} = 0.220,\ _{12}p_{59} = 0.213,\ _{12}p_{71} = 0.142,\ _{12}p_{83} = 0.040.$$

This structure will be used as a basis for comparison with the Bayesian estimates we provide below.

However, in the case of the Maubuisson nuns, it is possible to start from the matrix of prior probabilities for normal pre-industrial mortality, known as attritional (20–80 years and above, in seven 10-year groups), the female Lisbon reference population (see Chap. 4) and the distribution of observed deaths for seven stages, and use the Iterage.for program to estimate the distribution of death in the seven age groups. This gives

$$_{10}p_{20} = 0.025,\ _{10}p_{30} = 0.036,\ _{10}p_{40} = 0.073,\ _{10}p_{50} = 0.133,\ _{10}p_{60} = 0.210,\ _{10}p_{70} = 0.268,\ _{10}p_{80+} = 0.255.$$

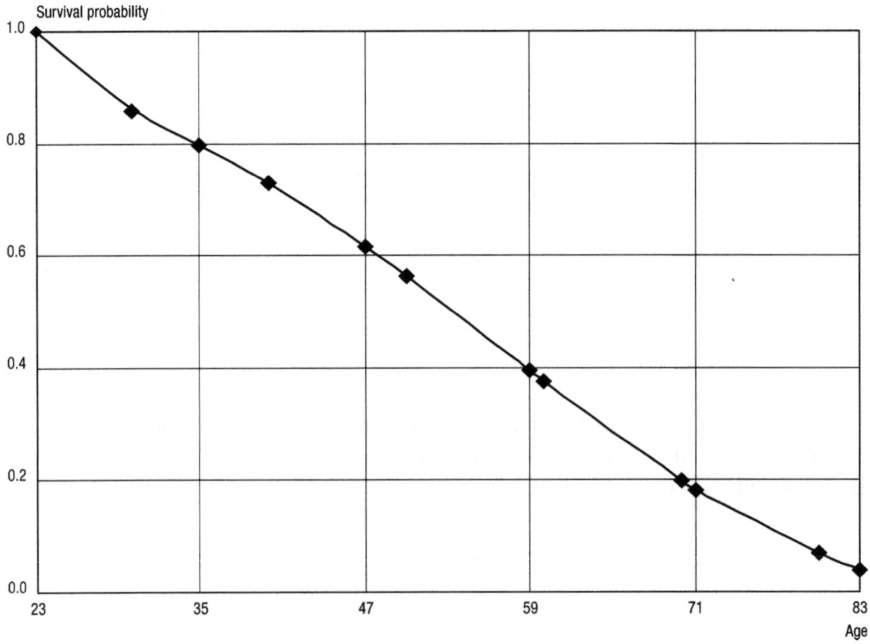

Fig. 12.2 Survival probabilities at Loisy-en-Brie, estimated by Bocquet-Appel and Bacro's Iterage.for program, interpolated to obtain survivors every 12 years

This breakdown will also be used for comparison with the Bayesian estimates we make below, particularly since we have an exhaustive estimate of these deaths during the period 1670–1789:

$$ {}_{10}p_{20} = 0.012,\ {}_{10}p_{30} = 0.025,\ {}_{10}p_{40} = 0.087,\ {}_{10}p_{50} = 0.170,\ {}_{10}p_{60} = 0.289,\ {}_{10}p_{70} = 0.210,\ {}_{10}p_{80+} = 0.207. $$

It can already be seen that, although the deaths in the first two ages are reasonably estimated, the quality of the estimates declines sharply for the following age groups.

Although this method does introduce a random element into the reference population, it is still not fully Bayesian, because the observed population is considered here to be non-random. The effect is that the variances calculated by the program must be strongly biased, and for this reason we have not addressed the point here. In fact, as we shall see below, it is more important to consider the observed frequencies as random than those of the reference population. And by choosing the age structure from a parametric model of mortality, this method, like the previous one, introduces a structure that is not necessarily verified by past populations. If the solution lies outside the proposed list, the authors have no way of verifying that fact.

12.3 Conclusions

After this overview of the methods used by palaeodemographers, we can see first that finding a table with observed stage totals where each term comes as close as possible to an initial table is not the best method. It assumes that the sum of the distances between each term in the reference and observed matrices must be minimised, without allowing for the asymmetric structure of the table, where it is the stage distribution for each given age which must be considered as invariant.

A much more satisfactory method is to find a table, still with observed stage totals, where the columns come as close as possible to those of an initial table. This complies with the invariance hypothesis that says that for a given stage the age distribution of individuals is independent of the population. However, a large number of conditions that validate this estimate by iteration are generally left unexplained by palaeodemographers, who also fail to mention that, even in acceptable conditions, these methods can lead to unacceptable solutions.

The general use of this method, which amounts to assuming that the observed population must follow one of the distributions in a universe of mortality models, may be of interest for providing a prior distribution for a genuinely Bayesian model. However, Bocquet-Appel and Bacro' approach is not strictly speaking a Bayesian generalisation of the second type of model but rather a highly empirical approach that does not consider the random nature of the reference data. Furthermore, it only holds if the observed population follows one of these distributions

In the following chapter, co-written with Henri Caussinus, we examine in greater detail the second type of model, generalising it as a fully Bayesian model that allows for uncertainty in all the starting data.

Chapter 13
A New Method for Estimating Age-at-Death Structure

Henri Caussinus and Daniel Courgeau

13.1 Introduction

The previous chapter presents the main methods hitherto recommended for estimating a population's age structure. This chapter proposes a new method based on a precise statistical model taking into consideration the essential specificity of the data upon which the estimation is based.

First the notation, which is basically that of the previous chapter. We denote p_{ij} the probability that an individual taken at random from the study population belongs to age class j ($j = 1, \ldots, c$) and stage i ($i = 1, \ldots, l$) of a given indicator; the sum of the p_{ij} with respect to i is denoted p_j (the probability that an individual is aged j), the sum of the p_{ij} with respect to j is denoted π_i (the probability that an individual is at stage i); the conditional probability of stage i being at age j is denoted $p_{i|j}$. These various probabilities are positive and satisfy the equations $\sum_i \pi_i = \sum_j p_j = 1$ and $\sum_i p_{i|j} = 1$ for all values of j. They are also connected by the following equation:

$$\sum_j p_i p_{i|j} = \pi_i \text{ for all values of } i = 1, \ldots, l \tag{13.1}$$

In practice, the estimation must be made with data n_{ij}, the number of observations of stage i and age j in a reference base ($i = 1, \ldots, l$ and $j = 1, \ldots, c$), and m_i, the number of observations of stage i at the site in question ($i = 1, \ldots, l$; $\sum_i m_i = m$): these data are shown in Tables 12.5 and 12.6 in the previous chapter. The invariance

H. Caussinus (✉)
Mathematic Institute, Toulouse, France
e-mail: henri.caussinus@math.univ-toulouse.fr

D. Courgeau
INED, Paris, France
e-mail: daniel.courgeau@wanadoo.fr

I. Séguy and L. Buchet, *Handbook of Palaeodemography*, INED Population Studies 2, 255
DOI 10.1007/978-3-319-01553-8_13, © Springer International Publishing Switzerland 2013

hypothesis assumes that the probability $p_{i|j}$ of stage i occurring at age j is the same for any population; it is possible therefore to calculate these conditional probabilities from the reference data even if they come from another population. Consequently, the model is parametered by $p_{i|j}$ and p_j, with the π_i being deduced if necessary from Eq. 13.1; the parameters of interest are clearly the p_j, whereas the $p_{i|j}$ are only of intermediate value.

The various proposals mentioned in the previous chapter do not take fully into consideration the variability of some of the observations: for example, the IALK method replaces each $p_{i|j}$ by $n_{ij}/n_{.j}$ as if this quantity were fixed and not random,[1] and Bocquet-Appel and Bacro's method (2008) does not fully consider the random nature of the m_i, since the results of the estimation depend solely on the m_i/m ratios (it is clear that the size of sample m affects the precision of the estimates, i.e., the confidence intervals). Furthermore, except for Bocquet-Appel and Bacro's method, the procedures proposed do not address the specific nature of the problem at hand. One does not estimate just any set of probabilities but rather a distribution of ages at death for a group of individuals about whom one may even have more precise specific information. If we also consider that the data available are usually scarce, this specific information is all the more valuable. This may also be concluded empirically from the high instability of most of the methods proposed up to now.

All these points have convinced us of the utility of introducing the Bayesian method we present in Sect. 13.2. A few simulations in Sect. 13.3 show that this method appears to effectively replace most of the former methods; they are also used in the discussion of certain questions of calibration. Section 13.4 addresses particular examples and provides comparisons with other approaches from a new angle.

One final point concerns vocabulary. Some methods in the literature may appear to be Bayesian in so far as they make use of the so-called Bayes formula or introduce a priori considerations into the resolution of estimation problems, but the paradigm on which they are based remains frequentist.[2] The method we present here, on the other hand, is Bayesian in the sense most often used in statistics: it is considered that the parameters themselves are random, with a probability distribution, called *prior*, chosen by the user to reflect his(her) knowledge (and ignorance) before the observation; this distribution is then corrected in response to the observations to achieve a *posterior* distribution, which is the observation-based probability distribution of the parameters, and, more specifically in our case, the posterior distribution of the parameters of interest p_j ($j = 1, \ldots, c$).

[1] The IALK method can be modified to take account of this randomness. One proposal is given in Box 13.2 (see p. 273); while this does improve the method in some ways, it may weaken it in others, which confirms our view that more radical changes in viewpoint are necessary.

[2] This is true of Bocquet-Appel and Bacro's method (2008), which takes account of the nature of the probabilities to be estimated by reducing the parametric space of the standard framework.

13.2 A Bayesian Estimation Method

13.2.1 Model and Principle

It is natural to suppose that the frequencies m_i ($i = 1, \ldots, l$) observed on the site for various stages are the observed values of a multinomial distribution whose parameters π_i are linked to the p_j and $p_{i|j}$ according to Eq. 13.1. We shall use these parameters to pursue the modelling.

We denote by G the prior density of parameters $p_{i|j}$, $i = 1, \ldots, l$ and $j = 1, \ldots, c$ (we shall see how G can be expressed in Sect. 13.2.2) and assume that the parameters p_j ($j = 1, \ldots, c$) have a prior density g (also discussed in Sect. 13.2.2) and are independent of the $p_{i|j}$.

If we denote by M the vector of m_i, P the vector of $p_{i|j}$ and p the vector of p_j, the joint density of (M, P, p) will be f given by

$$f(M,P,p) = g(p)G(P)\frac{m!}{\prod_i m_i!}\prod_i\left(\sum_j p_j p_{i|j}\right)^{m_i}$$

where the index i always goes from 1 to l and the index j from 1 to c.

The marginal density of the pair (M, p) is

$$\int f(M,P,p)\,dP$$

and the marginal density of M is

$$\iint f(M,P,p)\,dp\,dP$$

whereby the integrals are taken over the variation domains of P or p and P, which are a simplex (for p) or a product of simplexes (for P).

The conditional density of p, given M, is therefore

$$\frac{\int f(M,P,p)\,dP}{\iint f(M,P,p)\,dp\,dP}.$$

This is the *posterior* density of p_j ($j = 1, \ldots, c$) on which the Bayesian estimation will be based.

For example, one may have the posterior mean of p_j

$$\frac{\iint p_j f(M,P,p)\,dp\,dP}{\iint f(M,P,p)\,dp\,dP}.$$

More generally, the conditional expectation given M of a function φ of p will be given by

$$\frac{\iint \varphi(p) f(M,P,p)\, dp\, dP}{\iint f(M,P,p)\, dp\, dP}. \tag{13.2}$$

We thus obtain, for example, the kth-order moment of p_j with $\varphi(p) = p_j^k$. Taking for $\varphi(p)$ the function that equals 0 for $p_j > x$ and 1 for $p_j < x$ (indicator variable of the event $p_j < x$), we express the posterior distribution function for p_j at point x.

The various integrals in expression (13.2) may be evaluated by a Monte Carlo method as follows.

We denote $X = (X_1, \ldots, X_c)$ a random vector with density distribution g and Y a family of c vectors $Y_j = (Y_{1j}, \cdots, Y_{lj})$ $(j = 1, \ldots, c)$, whose joint distribution is independent of X and admits density G. We verify that expression (13.2) equals

$$\frac{E\left(\varphi(X) \prod_i \left[\sum_j X_j Y_{ij}\right]^{m_i}\right)}{E\left(\prod_i \left[\sum_j X_j Y_{ij}\right]^{m_i}\right)}.$$

Let us generate S independent sets of such random vectors (X,Y), with s $(s = 1, \ldots, S)$ representing the iterations. By virtue of the law of large numbers, if S is large enough, the expression above is approximated by

$$\frac{\sum_{s=1}^{S} \varphi(X_s) \prod_i \left(\sum_j X_{js} Y_{ijs}\right)^{m_i}}{\sum_{s=1}^{S} \prod_i \left(\sum_j X_{js} Y_{ijs}\right)^{m_i}}.$$

This supplies the posterior expectation of each p_h $(h = 1, \ldots, c)$ – which can be taken as a point estimate – or the posterior variance useful for characterising the accuracy of the estimate. The same principle can be applied to evaluate cross-moments, such as the covariance matrix of the posterior distribution of the p_h parameters. The posterior distribution function of a p_h parameter can be used, for example, to calculate intervals containing that p_h with a given probability, known as credibility intervals, which are the Bayesian equivalent of the confidence intervals of the standard system.

13.2.2 Use in Practice

13.2.2.1 Choice of Prior Distributions

Density G

The only source of information on the conditional probabilities $p_{i|j}$ is the reference data. If they are raw data merely obtained by recording the stage frequencies on a sample of skeletons of known ages, we can logically conclude that, for each age class j ($j = 1,\ldots,c$), the frequencies n_{ij} are the observed values of a multinomial distribution with a total n_j and probabilities $p_{i|j}$ ($i = 1,\ldots,l$). Adopting a prior distribution for the $p_{i|j}$ probabilities, we deduce the conditional distribution given the reference data. We take it, in turn, as the prior distribution of the $p_{i|j}$ probabilities in the final model. Given the absence of supplementary information on these $p_{i|j}$ probabilities beyond what is contained in the reference data, it makes sense to adopt a uniform distribution as the prior distribution of the $p_{i|j}$ probabilities for each j. For a given j, the posterior distribution of the $p_{i|j}$ probabilities is a Dirichlet distribution (see Box 13.1) of parameters $\alpha_{ij} = n_{ij} + 1$ ($i = 1,\ldots,l$). Density G is then the product of c Dirichlet densities, namely

$$G(p) = \frac{\prod_j \Gamma(\alpha_{.j})}{\prod_i \prod_j \Gamma(\alpha_{ij})} \prod_i \prod_j p_{i|j}^{\alpha_{ij}-1}.$$

In practice, the raw data may be "processed" in various ways (for example, in order to achieve the right weighting between male and female samples), so that their distribution is no longer strictly multinomial. However, the prior G as defined above appears still to hold, since the multinomial nature of the reference data is more an indication than a necessity for arriving at that distribution.

The choice of G may be refined in various ways. For example, in order to avoid excessive confidence in the reference data, the α_{ij} may be multiplied by a "reducing" coefficient r ($0 < r < 1$) with a choice of $\alpha_{ij} = r(n_{ij} + 1)$, which does not affect the prior mean values of $p_{i|j}$ but increases the prior variances, thereby expressing the degree of doubt. These variances are roughly multiplied by $\frac{1}{r}$; note that it is very broadly equivalent to assume that the n_{ij} are multiplied by r, another way of reducing the information contained in the reference data since it amounts to assuming that the relative frequencies observed in the data reference are retained but taken from a smaller sample.

Box 13.1 The Dirichlet Distribution

Let D be the subset of $\Re^k$ defined by:

$$x = (x_1, \ldots, x_k) \in D \quad \Leftrightarrow \quad x_i > 0 \ \ for \ \ all \ \ i = 1, \ldots, k \ \ and \ \ \sum_{i=1}^{k} x_i = 1$$

and $a = (a_1, \ldots, a_k)$ a vector of strictly positive real numbers.

The random vector $X = (X_1, \ldots, X_k)$ follows a Dirichlet distribution with parameter a if its probability density d is such that:

$$d(x) = \begin{cases} \dfrac{\Gamma(a_.)}{\prod_{i=1}^{k} \Gamma(a_i)} \prod_{i=1}^{k} x_i^{a_i - 1} & for \ \ x \in D \\ 0 & for \ \ x \notin D \end{cases}$$

where $a_. = \sum_{i=1}^{k} a_i$ and Γ is the Euler's Gamma function defined by $\Gamma(p) = \int_0^{\infty} e^{-x} x^{p-1} dx.$

Note that d is constant over D (uniform distribution) when $a_i = 1$ for all i. The marginal distribution of X_i is Beta with parameters $(a_i, \ a_. - a_i)$. The moments of X_i are:

$$E(X_i) = \frac{a_i}{a_.}$$

$$E\left(X_i^2\right) = \frac{a_i(a_i + 1).}{a_.(a_. + 1)}, \qquad Var(X_i) = \frac{a_i(a_. - a_i)}{a_.^2(a_. + 1)}.$$

More generally, the moment of order h ($h \geq 1$) is:

$$E\left(X_i^h\right) = \prod_{j=0}^{h-1} \frac{a_i + j}{a_. + j}$$

We have also, for $i \neq j$: $E\left(X_i X_j\right) = \frac{a_i a_j}{a_.(a_.+1)}$ and $Cov\left(X_i X_j\right) = -\frac{a_i \ a_j}{a_.^2(a_.+1)}$.

(continued)

Box 13.1 (continued)

Remarks

1. The X_i means, $e_i = \frac{a_i}{a_.}$, are proportional to the a_i; they remain unchanged if the a_i are all multiplied by the same positive number s. We can write: $Var(X_i) = \frac{e_i(1-e_i)}{a_.+1}$; hence, for equal means, the variances are larger when the a_i are smaller: if a_i is multiplied by s, then $Var(X_i)$ is multiplied by $\frac{1+a_.}{1+sa_.}$, a decreasing function of s (if $a_.$ is large enough, this is almost equivalent to multiplying $Var(X_i)$ by $\frac{1}{s}$).
2. In the Bayesian statistical framework, the Dirichlet distribution is said to be conjugated to the multinomial one; when the prior distribution of the parameters of a multinomial law is Dirichlet, the posterior distribution is another Dirichlet distribution whose parameters are obtained by adding the vector of observed numbers to the parameter vector of the prior.

Density g

The choice of the prior distribution for the p_j parameters is a trickier matter. We give our preferred method first, which will be systematically used in this chapter. But we shall also briefly mention other possibilities, some of which merit further examination.

As there is no clearly designated "class" of distributions from which to select the prior distribution, the most sensible course is to opt for a Dirichlet distribution, which is well suited to probability vectors. This leaves the problem of choosing the distribution parameters, say ($\beta_1, \ldots, \beta_c$). In the absence of specific information, we can, as above, choose a uniform distribution and take $\beta_j = 1$ for all j. This is a "neutral" choice and may sometimes be justified. It also yields reasonable results with simple examples. However, in palaeodemography, other choices would appear to be preferable as certain information is naturally available. We can, for example, take a "standard" mortality distribution and calculate the probabilities for each of its age classes. The class probabilities become the means of the prior distribution. This gives the parameters β_j up to a proportionality coefficient (see Box 13.1), i.e. the $\beta_j/\beta_.$ values, where $\beta_.$ is the sum of the β_j parameters over $j = 1, \ldots, c$. The remaining step is to choose $\beta_.$, i.e. in practice, the prior variances. Note that the variances need to be relatively large in order to express the fact that the prior means are not very reliable and that the prior distribution should not play a dominant role – in other words, that the family of possibilities envisaged covers a broad field. Hence the β_j should be fairly small, say, below unity or barely above. We shall see that this is indeed the case in the simulations examined below.

Note that the prior means may be seen as "test" values: if the data are scarce and the estimates consequently imprecise, it is helpful to use the posterior distribution qualitatively by observing in which direction these means move, i.e. how the data "correct" the prior values.

This principle for the choice of the prior distribution may be extended in a number of ways. For example, instead of choosing a standard mortality distribution as the basis for constructing the prior distribution, one may choose a mix of two "standard" distributions, leading to a mix of two Dirichlet distributions. These might be the mix (in carefully chosen proportions) of a routine mortality distribution (attrition) and a catastrophic distribution.

Clearly, quite different approaches are also possible, such as, along the lines of Bocquet-Appel and Bacro's proposals (2008), defining the prior distribution as a uniform distribution on a finite set of distributions corresponding to standard mortality distributions. We have examined this in Caussinus and Courgeau (2010), together with the comparison of our method with that of Bocquet-Appel and Bacro. As standard mortality distributions, rather than the "artificial" distributions proposed by these authors, one may consider the distributions used to construct the pre-industrial standard (Table 8.8 at Springer Extra). This deserves further research; however, this type of prior distribution is likely to place too much weight on routine mortality and be less effective in identifying specific situations of interest. Even if a Dirichlet prior as described is not necessarily optimal, it does provide a flexible general approach that is easy to implement: it is therefore the only one we consider below.

13.2.2.2 Posterior Distribution and Credibility Intervals

Earlier, we saw how to calculate the posterior distribution function for each p_j point by point. The posterior density for each p_j can be numerically derived and then appropriately smoothed. A graphical display of the densities may help in interpreting the numerical results. In some cases, the posterior density of each p_j can be approximated by a Beta density with the same mean and variance, which, for example, simplifies the evaluation of densities. Approximation quality can be controlled to a certain extent via higher-order moments: one can check the proximity of the beta distribution's third- and fourth-order moments with respect to the corresponding moments of the "true" posterior distribution, easily calculable by simulation, as seen above. Note, however, that this type of approximation is not always valid and must be used with care, and avoided in those cases where exact calculations can readily be performed.

After calculating the posterior distribution function for each p_j, we can determine *α-credible intervals* (Robert 2006, p. 278) in which a p_j parameter has a probability $1 - \alpha$ conditional upon the observations. It is preferable to use the exact posterior distribution function, but in some cases the approximation by beta distribution mentioned above[3] is acceptable.

[3] One example is the set of data processed in Sect. 13.4. But there are cases where this approximation is highly unsatisfactory: an example of a bimodal posterior distribution is even given in Séguy et al. (2012).

Finally, note that it is extremely inadvisable to use an interval of the "mean plus or minus one (or two) standard deviations" type because the posterior distribution is, in most cases, highly asymmetric.

13.2.2.3 Size of the Reference Data Table

System (13.1) described in Sect. 13.1 is undetermined if the number of rows (stages) l is smaller than the number of columns c (ages). In other words, the parameters of interest are not identifiable, given that several values lead to the same distribution of observable samples. The Bayesian method avoids the difficulty by starting with a prior distribution, and the aim is simply to make it change by means of the data. The posterior distribution steers us towards a distribution of the unknown parameters, which is wholly compatible with the fact that they are not completely determined. This method can therefore be used with $l < c$. Clearly, the posterior distribution can be somewhat dispersed, which merely reflects the indeterminacy inherent in the situation.

13.3 Brief Simulation Study

The following examples, taken from a wider study, are intended to illustrate the properties of the recommended Bayesian method and to specify certain points in the choice of parameters for the prior distributions. The first two examples are elementary and do not refer specifically to the nature of the underlying application, although they are described in the "language of palaeodemography" (ages, stages) for the sake of consistency and to simplify explanation. The third and fourth examples are more directly connected to applications in palaeodemography. In order to compare the Bayesian method with frequentist methods, we only consider the point estimates it provides with the posterior mean.

In all events, we start from a population distributed by two discrete features with l (in lines) and c (in columns) classes respectively. We assume as known the probabilities p_{ij} that an individual will be located in line i and column j. We then simulate a large number R of situations, taking each time a multinomial sample with l categories and probabilities π_i in order to simulate the site data and also c multinomial samples with, for the jth, the probability $p_{i|j}$ for the ith class, in order to simulate the reference data. Each repetition leads to estimates of the probabilities p_j (the desired structure by age) by various methods.

In Examples 1, 2 and 3, we evaluate the least-squares regression method, the IALK method (which we prefer to call "Maximum Likelihood 1"), the Maximum Likelihood 2 method described in Box 13.2 (see p. 273), and the proposed Bayesian method, by comparing the estimates found with the true values of the parameters, known in this case. These results are given both graphically in the form of frequency histograms for the estimated probability of being in one of the age groups, and in the form of standard summaries: mean, standard deviation and mean squared error. We know that the mean squared error of an estimator X of the real parameter θ equals the expected value of the square of the $X - \theta$ difference, or $E[(X - \theta)^2] = Var(X) + E[(X - E(X))^2]$; it accounts therefore both for the variance of the estimator, the first term in the sum above, and for its bias, the second term in the sum.

In Examples 3 and 4, we add the comparison with the Bocquet-Appel and Bacro method (2008), since its restriction of parametric space only becomes fully meaningful with palaeodemographic data; in both examples we have chosen a breakdown into age classes compatible with the authors' "prior" datasets. In these larger examples, the quality of results is examined with an overall criterion of distance between the vector of true probabilities and the vector of estimated probabilities. In fact, two criteria are used: the sum of mean squared errors obtained for the various age classes ("total MSE") and an analogous sum weighted by the true probabilities [as in a chi-square test], ("relative MSE").

Example 1. We first take the two-row two-column example from the previous chapter, drawing multinomial samples of 20 by the probabilities considered in that chapter: as conditional probability (reference) for Line 1 we have 0.667 for Column 1 and 0.25 for Column 2; the marginal probability for Line 1 is 0.6. There is only a single parameter to be estimated in this case, for example, the probability for Column 1, which we know to be 0.84. We run 1,000 iterations, obtaining different samples. Since $l = c$, equivalent results are obtained for the "corrected" regression (i.e., least squares subject to positivity constraint) and Maximum Likelihood 1 (IALK).

"Ordinary" regressions with no positivity constraint run on each of these samples gave 347 "estimates" greater than unity, which is understandable since the true probability is fairly close to unity, and also 11 negative values, which is more surprising. In fact, there is a wide dispersion of results (standard deviation 0.63) around a mean close to the true value, but which hardly make sense. Correcting the higher estimates to unity and the negative ones to zero, the mean obtained is 0.78, standard deviation 0.25, and mean squared error 0.06. The histogram of estimated values is given in Fig. 13.1 (left).

For the Bayesian model, we first take "neutral" prior parameters $\beta_1 = \beta_2 = 1$. We obtain estimates with mean 0.64, standard deviation 0.12, and mean squared error 0.05. The corresponding histogram is given in Fig. 13.1 (centre).

It can be seen that the Bayesian method is more satisfactory, even when the situation is highly unfavourable for it, with a probability to be estimated relatively close to unity and a prior distribution that allocates a mean of 0.5.

If it is known in advance that the probability to be estimated is "fairly high", this may be allowed for in the value of β_1; to check the impact we repeat the estimation with $\beta_1 = 1.5$ and $\beta_2 = 0.5$. We obtain a mean estimate of 0.78, standard deviation 0.11 and mean squared error 0.01. The estimates are consequently much better, and this is also illustrated in Fig. 13.1 (right). In general terms, therefore, if one has some idea of the age distribution of the observed population, it should be introduced

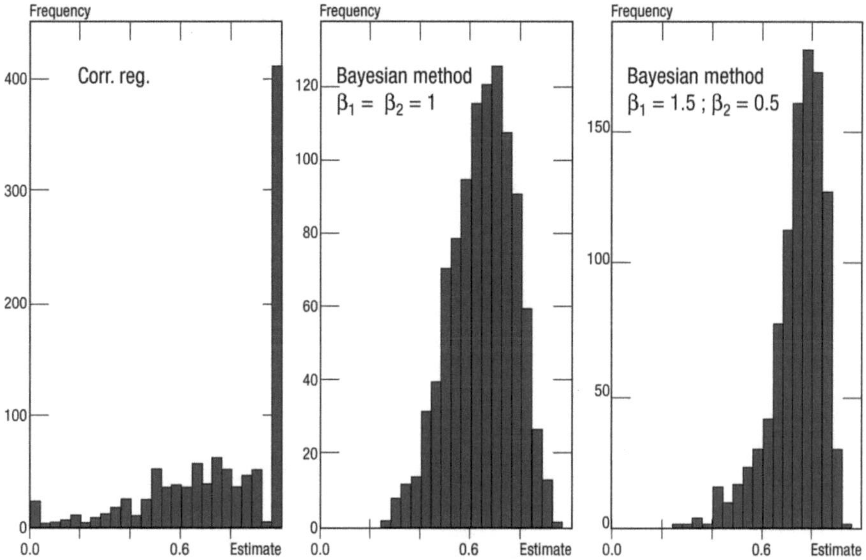

Fig. 13.1 Simulations – Example 1: histograms of estimates of the probability of belonging to the first age group obtained in 1,000 iterations. *Left*: "corrected" regression; centre: Bayesian method with $\beta_1 = \beta_2 = 1$; *right*: Bayesian method with $\beta_1 = 1.5$ and $\beta_2 = 0.5$

into the model without hesitation. However, in practice, it must be borne in mind that the choice of the prior distribution must be justified.

Example 2. This example corresponds very closely to the second theoretical example (three stages and two ages) in the previous chapter. The conditional probabilities of the three stages are (0.6250; 0.3125; 0.0625) for age class 1 and (0.125; 0.375; 0.500) for age class 2; the marginal probabilities for the three stages are (0.500; 0.328; 0.172) and the marginal probabilities for the ages are 0.755 and 0.245. These last two probabilities are to be estimated from multinomial samples of stages of size t (chosen as 20). The estimated results are given for the first probability (here p), whose true value is 0.755.

Since in this case l and c are different, the IALK method and the corrected regression do not necessarily give the same results, so it is instructive to compare them. In this comparison, we also introduce the Maximum Likelihood 2 method (Box 13.2, see p. 273) and our Bayesian method with $\beta_1 = \beta_2 = 1$. The histograms of the results are shown in Fig. 13.2 and the key features in Table 13.1.

As we said above, we use the term "Maximum Likelihood 1" for what is more generally known as the IALK method. This is because the name IALK confuses the concept with a numerical solution technique which is incomplete in any case. The I in IALK basically refers to an "iterative" process that only gives a clear result if the maximum likelihood lies within the set of possible solutions and, in that case, corresponds to the zero point on the likelihood gradient; we prefer to take the

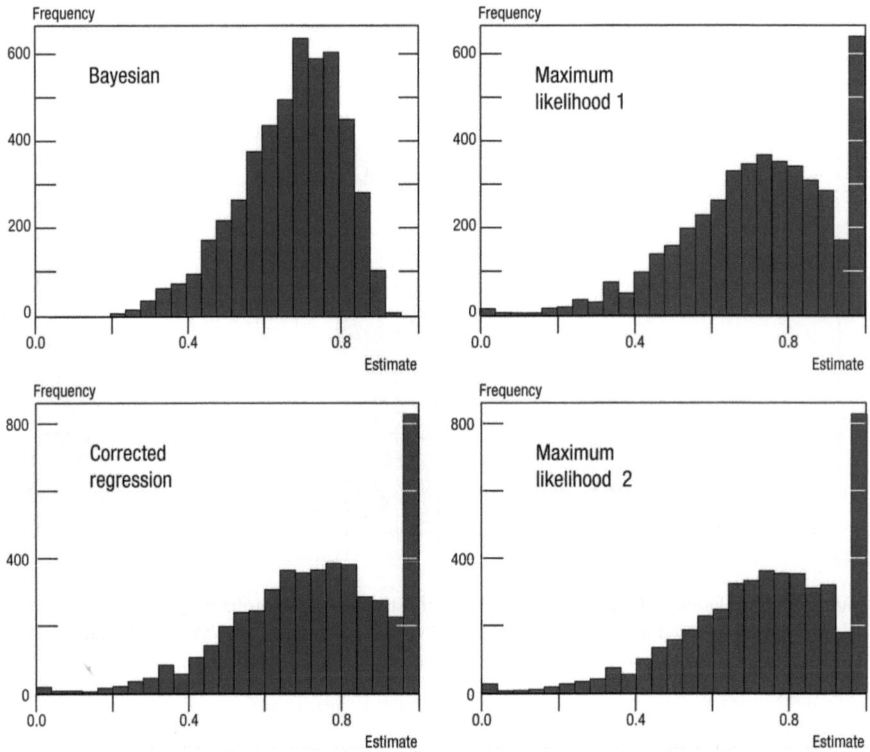

Fig. 13.2 Simulations – Example 2: histograms of estimates obtained by four methods (with $\beta_1 = \beta_2 = 1$ for the Bayesian method)

Table 13.1 Simulations – Example 2 – 5,000 iterations. Characteristics of the estimates of p obtained by four different methods

	Mean	Standard deviation	MSE
Bayesian method	0.672	0.134	0.025
Corrected regression	0.724	0.201	0.041
Maximum Likelihood 1	0.727	0.197	0.039
Maximum Likelihood 2	0.737	0.205	0.042

principle of the maximum likelihood method to its logical conclusion and also consider a maximum at the boundary. Although the IALK iterative process probably provides the maximum, this has not, to our knowledge, been rigorously proven, so we look for the maximum likelihood *in all cases* by using the constrOptim procedure from the R package (R Development Core Team 2008). The same procedure was used to find the maximum likelihood in the more general model underlying Maximum Likelihood 2.

Table 13.1 shows that the mean squared errors are similar for the corrected regression and the two maximum likelihood methods, while that of the Bayesian method is significantly lower, although the prior distribution uses no indication as

Table 13.2 Simulations – Example 2. Characteristics of p estimates by the Bayesian method for prior values of β_1 and β_2

	Mean	Standard deviation	MSE
$\beta_1 = \beta_2 = 1.2$	0.660	0.123	0.024
$\beta_1 = \beta_2 = 1$	0.672	0.129	0.024
$\beta_1 = \beta_2 = 0.75$	0.688	0.149	0.027
$\beta_1 = \beta_2 = 0.50$	0.707	0.163	0.029

to the true value of p. This in fact explains why this method gives the largest bias, fortunately corrected by a much lower variance.

Maximum Likelihood 1 gives slightly better results than corrected regression, which is to be expected, since it accounts better for the nature of the sampling errors. However, it may at first glance seem surprising that Maximum Likelihood 2 is not better than ML1 since, here again, it accounts better for the nature of the errors; the reason is probably that the number of parameters to be estimated is less parsimonious, which does reduce the bias but increases the variance more, and thus increases the mean squared error. Note finally that the individual differences between the estimates of the last three methods are quite small, rarely more than 0.1; in this particular example, regression and Maximum Likelihood 1 relatively often provide estimates of p "at the boundary", with unity in more than 10 % of cases and even a certain number of zero values (approximately 0.2 %). Maximum Likelihood 2 has the advantage of leading less often to these results (in our simulations we obtained no zero estimates and only 1.6 % at unity).

Up to now we have considered the Bayesian method with $\beta_1 = \beta_2 = 1$ and have seen that it performs better than methods based on other principles. However, we still need to examine the influence of the parameters of the prior distribution. To avoid giving undue advantage to the method we are comparing, we have taken various values for these prior parameters, staying within "neutral" prior means (0.50 0.50) but varying the confidence levels, with successively 1, 0.75 and 0.50 as common values for β_1 and β_2. A higher value for β_j (1.2) was also envisaged.

The characteristics of the distribution of estimates obtained for p are given in Table 13.2.

Figure 13.3 also compares the histograms obtained with $\beta_1 = \beta_2 = 1$ (left) and $\beta_1 = \beta_2 = 0.5$ (right).

In both the numerical values in Table 13.2 and the histograms in Fig. 13.3, it can be seen that, just as with the previous example, reducing the β_j slightly reduces the bias (which is due to a mean choice of prior probability lower than the true value), but at the cost of a noticeable increase in variance; in all, this ultimately gives a slight deterioration in mean squared error (MSE). Increasing the β_j above unity increases the bias and gives an equivalent mean squared error. There appears to be no strong arguments for any particular choice of β_j, but the simple option $\beta_1 = \beta_2 = 1$ can no doubt be recommended with no great risk.

Example 3. Now we apply simulation to an example that comes closer to the problems encountered in palaeodemographic practice. We began with the 7×7 example of the Maubuisson nuns with conditional reference probabilities deduced

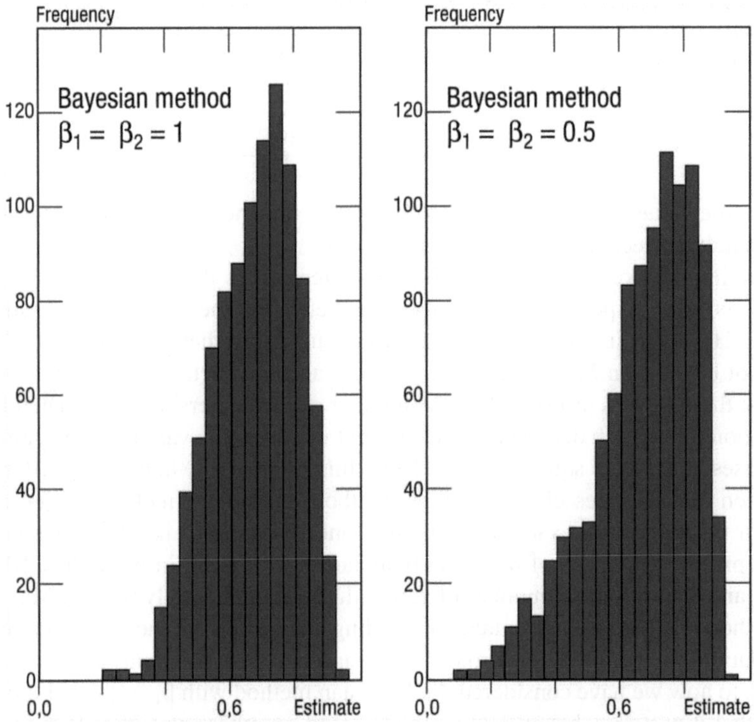

Fig. 13.3 Simulations – Example 2. Histograms of estimates obtained for p by the Bayesian method with $\beta_1 = \beta_2 = 1$ (*left*) and $\beta_1 = \beta_2 = 0.5$ (*right*)

from the frequencies given in Table 12.8 of the previous chapter, and row (stage) probabilities of

$$(0.180 \;\; 0.068 \;\; 0.115 \;\; 0.159 \;\; 0.119 \;\; 0.188 \;\; 0.171).$$

These probabilities comply with probabilities of dying in each age class of

$$(0.012 \;\; 0.025 \;\; 0.087 \;\; 0.170 \;\; 0.289 \;\; 0.210 \;\; 0.207)$$

as calculated from an exhaustive evaluation of deaths in the period 1670–1789 recorded in the registers available.

The simulation results are therefore to be compared with the second set. The simulation involves $R = 1{,}000$ iterations, the number of trials of the multinomial distributions is 37 for the site, and equal to the number of trials of the reference data for the latter.

We first compare the estimates obtained by the Bayesian method (posterior means) with $\beta = (1, 1, 1, 1, 1, 1, 1)$, by regression (with a positivity constraint), and by Maximum Likelihood 1 and 2. Table 13.3 shows the mean of estimates for

Table 13.3 Simulations – Example 3. Means of probability estimates for each age class obtained by four methods, and total and relative MSEs

	Age class							Total	Relative
Method	20–29	30–39	40–49	50–59	60–69	70–79	80+	MSE	MSE
Bayes, uniform prior	0.078	0.098	0.102	0.143	0.194	0.182	0.203	0.033	0.751
Regression	0.001	0.069	0.061	0.232	0.032	0.402	0.203	0.233	1.526
Max. L. 1	0.035	0.056	0.063	0.137	0.253	0.232	0.222	0.253	2.204
Max. L. 2	0.174	0.071	0. 155	0. 257	0.168	0.125	0. 051	0.170	3.409
True values	0.012	0.025	0.087	0.170	0.289	0.210	0.207		

each age class and each method, and, for each method, the total over seven age classes of mean squared errors (total MSE) and this same total weighted by the true values of the probabilities (relative MSE).

To round out the raw numerical data above, Fig. 13.4 gives the histograms of the frequencies of estimates obtained by the four methods for the probability of a single age-class, class 4 ("true" value 0.17).

It can be seen that the Bayesian method clearly outranks the other three in terms of mean squared error (whether total or relative). For the other three methods, theory predicts that where $l = c$, regression with positivity constraint and Maximum Likelihood 1 should provide equivalent results, at least when the estimates are not at the boundary of admissible values. And this is indeed observed with other examples not described here. In this case, however, the two methods provide fairly dissimilar results; most likely because the results are all at the boundary (one estimated probability is zero). It is also clear that the optimisation of the functions concerned is highly unstable, which is a further argument against these methods. With respect to zero estimates, Fig. 13.4 shows that there are a fair number of them with regression and Maximum Likelihood 1, even for Class 4, where the value to be estimated (0.17) is not particularly low.

Although the Bayesian method is clearly superior to the others, it presents notable biases, particularly in estimating the low probabilities of classes 1 and 2. The method could be improved in two ways. One is to adjust the "weighting" of the prior distribution by modifying the variance while keeping the same means (equal for each class). It is clear, however, that the observed biases are produced by a prior distribution highly unfavourable in its means since it allocates the same probability of death to each age class. It would be more realistic to allocate to each class a prior mean equivalent to standard mortality rates, and the data collected at a given site would serve to modify the standard for that site.

To test this first attempted improvement, we adjust the variances of the prior distribution without changing the means; the β_j remains the same for all j but takes successive values 1.25, 1, 0.75 and 0.50, so that the variances gradually increase. The means of the estimates obtained for these four prior distributions with 1,000 iterations are given in Table 13.4, with total and relative mean squared errors in the last two columns as before.

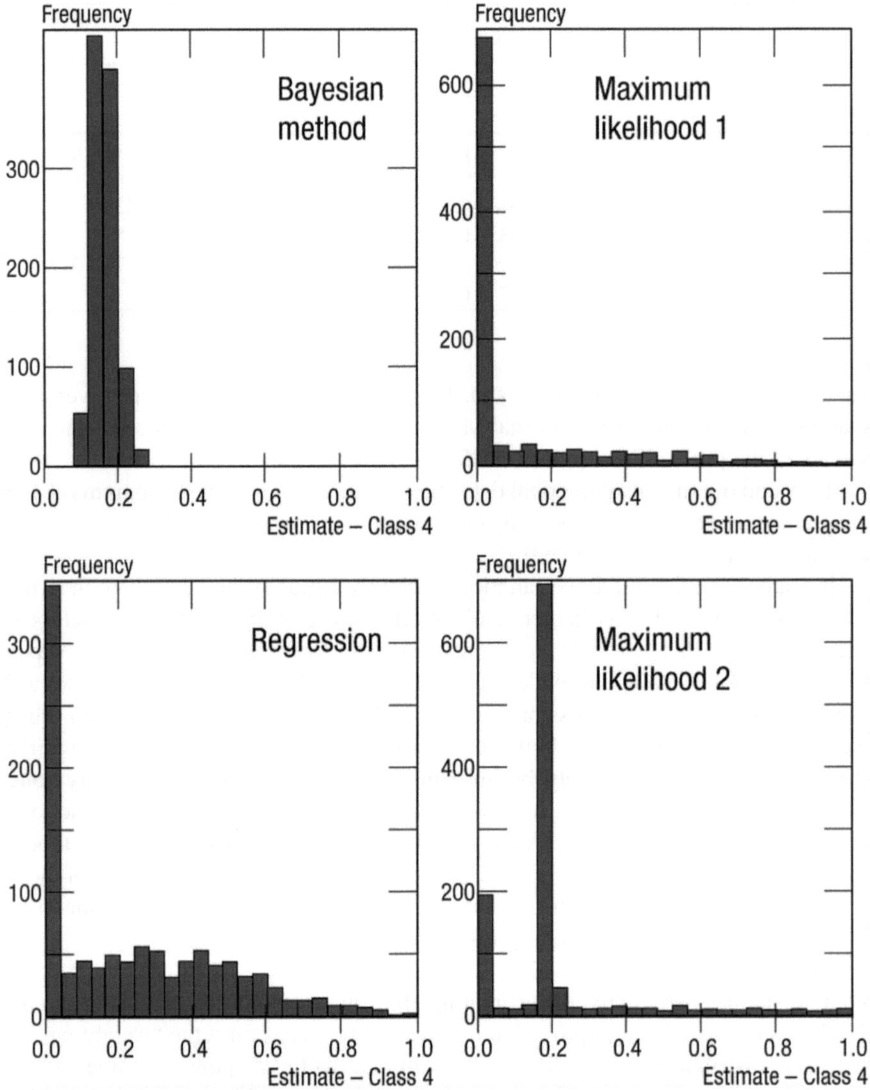

Fig. 13.4 Simulations – Example 3. Histograms of estimates obtained from 1,000 iterations by four methods for probability p_4 (with uniform prior for the Bayesian method)

It can be seen that $\beta_j = 1$ is a reasonable compromise. Low probabilities tend to be overestimated and high probabilities underestimated; the overestimation of the low probabilities is less marked when the β_j are smaller, giving a lower relative mean squared error, but what is gained in one place is lost in another, so the total mean squared error is higher.

We now attempt to use standard mortality rates instead of equal prior means. In view of the nature of this example, we take the pre-industrial standard mortality for

Table 13.4 Simulations – Example 3. Mean of the estimates by the posterior mean from identical β_j varying from 0.50 to 1.25, and total and relative mean squared errors

	Age class							Total	Relative
Value of β_j	20–29	30–39	40–49	50–59	60–69	70–79	80+	MSE	MSE
1.25 for all j	0.082	0.102	0.105	0.143	0.190	0.177	0.200	0.030	0.811
1 for all j	0.078	0.098	0.102	0.143	0.194	0.182	0.203	0.033	0.751
0.75 for all j	0.0678	0.087	0.093	0.141	0.205	0.190	0.216	0.036	0.672
0.5 for all j	0.059	0.079	0.085	0.141	0.215	0.200	0.221	0.042	0.598
True probabilities	0.012	0.025	0.087	0.170	0.289	0.210	0.207		

Table 13.5 Simulations – Example 3. Comparison of mean estimates obtained from a uniform prior distribution (row 1) and three prior distributions deduced from the female pre-industrial standard, with total and relative mean squared errors

	Age class							Total	Relative
Value of β_j	20–29	30–39	40–49	50–59	60–69	70–79	80+	MSE	MSE
1 for all j	0.078	0.098	0. 102	0.143	0.194	0.182	0.203	0.033	0.751
Standard $\beta_. = 5$	0.054	0.074	0.084	0.149	0.260	0.239	0.140	0.028	0.428
Standard $\beta_. = 7$	0.059	0.079	0.089	0.148	0.254	0.235	0.135	0.024	0.451
Standard $\beta_. = 10$	0.065	0.084	0.094	0.148	0.246	0.233	0.130	0.022	0.498
True probabilities	0.012	0.025	0.087	0.170	0.289	0.210	0.207		

women, which, with considerable rounding, gives the following proportions for the seven classes:

$$0.10 \quad 0.11 \quad 0.12 \quad 0.15 \quad 0.21 \quad 0.21 \quad 0.10.$$

The β_j are calculated in proportion to these values, with a sum $\beta_.$ varying around 7 as suggested by the previous study.

The results obtained for $\beta_.$ successively equal to 5, 7 and 10, are given in Table 13.5, with comparative figures for a uniform prior distribution.

The first point to note is that the new prior results in a substantial improvement, in both absolute and relative terms, particularly in relative errors because of the major bias in estimating low probabilities with a uniform prior distribution. Comparison of the three prior distributions deduced from the standard shows relatively similar behaviour, with an advantage for a smaller $\beta_.$ if focusing on relative errors, for a larger $\beta_.$ if focusing on absolute ones. The choice of $\beta_. = 7$ (number of columns) appears to be a good compromise and our conclusion is to recommend it.

Actually, for the Maubuisson nuns whose ages at death are simulated in this example, the fact that the site is a convent cemetery provides important supplementary information because the young nuns were probably in better health on average than the general population and not exposed to certain major mortality risks, particularly death in childbirth. The method can incorporate this prior information by modifying the β_j parameters. For example, we may consider that mortality in the 20–29 age class is probably more than halved and that mortality in the

Table 13.6 Simulations – Example 3. Comparison of the Bayesian method using two prior distributions (*PI* female pre-industrial standard, *MPI* modified pre-industrial standard, see text) with the Bocquet-Appel and Bacro method

Method	Total MSE	Relative MSE
Bayes (PI)	0.024	0.451
Bayes (MPI)	0.019	0.183
Bocquet-Appel	0.021	0.304

following age class is also halved. This leads us to consider a new β_j vector $(\frac{7}{6.21})$ (0.30 0.40 0.84 1.05 1.47 1.47 0.70) (the coefficient $\frac{7}{6.21}$ is used to bring total $\beta_.$ to 7 as recommended above). This produces the following mean estimates:

$$0.033 \quad 0.051 \quad 0.107 \quad 0.164 \quad 0.262 \quad 0.245 \quad 0.138$$

with a total mean squared error of 0.019 and a relative mean squared error of 0.183.

The improvement is significant, most clearly for the low probabilities in Classes 1 and 2, which are most affected by the change in prior distribution, and consequently for the relative mean squared error.

Finally, it is reasonable in this case to use the Bocquet-Appel and Bacro method (2008) with the ProbAtri20-90 set of 756 base vectors. Table 13.6 gives the total and relative mean squared errors for the results obtained with that method and the Bayesian method using the β_j above (MPI means "modified" pre-industrial); for a comprehensive comparison we also give the results as in Table 13.5 for the β_j corresponding to the pre-industrial (PI) standard.

Comparing our method and that of Bocquet-Appel and Bacro on this example, it can be seen that their performances in this case are similar, with a slight advantage to the Bayesian method if an informed choice of prior distribution is possible.

Example 4. Here we take an example where the bone stages are subdivided into five categories, with the same seven age classes as before. Traditional frequentist methods cannot be used because there are more columns than rows, but it is possible to use the Bocquet-Appel and Bacro method with the ProbAtri20-90 set of 756 vectors. This example serves to continue the comparison between that method and ours.

The reference data table is the 5 × 7 table for both sexes as follows:

38	68.8	58.2	33.2	13.4	7.0	5.0
42	58.6	54.6	48.8	24.2	26.4	14.6
18	25.4	35.4	38.2	31.0	33.8	24.0
12	17.0	20.4	35.6	49.2	42.0	36.6
4	16.4	12.4	24.6	42.2	32.4	42.8

We took in turn three different probability vectors for the age classes

Table 13.7 Simulations – Example 4. Comparison of the Bayesian method (with pre-industrial standard prior) and the Bocquet-Appel and Bacro method

	Case 1		Case 2		Case 3	
	Total MSE	Relative MSE	Total MSE	Relative MSE	Total MSE	Relative MSE
Bayes (PI)	0.001	0.006	0.004	0.028	0.035	0.231
Bocquet-Appel	0.020	0.155	0.022	0.182	0.034	0.248

$$(0.166 \ \ 0.115 \ \ 0.150 \ \ 0.178 \ \ 0.173 \ \ 0.134 \ \ 0.084)$$

$$(0.10 \ \ 0.10 \ \ 0.15 \ \ 0.15 \ \ 0.20 \ \ 0.20 \ \ 0.10)$$

$$(0.35 \ \ 0.09 \ \ 0.09 \ \ 0.10 \ \ 0.18 \ \ 0.11 \ 0.08)$$

They were chosen to represent realistic situations, assumed to be favourable to one method or the other. The first is the vector mean of the ProbAtri20-90 set of 756 vectors; the second a vector close to the pre-industrial standard; and the third the estimate made for the Frénouville site. We consider that no particular prior information is available: the Bayesian method is therefore used with a prior distribution complying with the pre-industrial standard and $\beta_. = 7$.

Table 13.7 gives the mean squared errors observed from a simulation of 1,000 repetitions. It can be seen that the two methods perform more or less equally for the last case, while ours is clearly better for the other two (including the first case where intuition might have suggested otherwise).

Conclusion to the Simulated Examples. The examples provide initial data for a discussion of the practical aspects of choosing the prior distribution. They also show that our method is clearly preferable to any method that does not address the specific features of the problem posed. Compared with the Bocquet-Appel and Bacro method, which makes wide use of these features, our method appears on the whole to be perfectly competitive, and much simpler to apply: the choice of a prior distribution is clearly easier and more flexible than constructing a set of base vectors. However, this comparison merits further examination (on this point, see Caussinus and Courgeau 2010).

Note that the comparisons above concern the effectiveness of the method in producing point estimates. In the section below, we demonstrate a few more of its advantages.

Box 13.2 Another Maximum Likelihood Approach: ML2

Let us consider the following statistical model which takes into account the random character of all the data. The set of parameters is the set of probabilities p_{ij} ($i = 1, \ldots, l$ and $j = 1, \ldots, c$); their sums over i are denoted p_j while the sums over j are denoted π_i. For each j ($j = 1, \ldots, c$), the n_{ij} ($i = 1, \ldots, l$) are the observed frequencies of a multinomial distribution with l categories, the total

(continued)

Box 13.2 (continued)

number of trials being n_j and the cell probabilities $p_{i/j} = p_{ij}/p_j$ $(i = 1, \ldots, l)$; the frequencies m_i $(i = 1, \ldots l)$ follow a multinomial distribution with l categories, total number of trials m and cell probabilities π_i $(i = 1, \ldots, l)$; these $c + 1$ multinomial distributions are independent. Up to an additive constant, the log-likelihood is:

$$\sum_i \sum_j n_{ij}(\ln(p_{ij}) - \ln(p_j)) + \sum_i m_i \ln(\pi_i).$$

Under the usual constraints on the set of probabilities p_{ij} it can be shown that this function has a unique maximum, either inside the parameter space or on its boundary. In both cases, this maximum likelihood solution can be obtained by a suitable algorithm, for instance by the constrOptim procedure in the R package (R Development Core Team 2008). We shall call this estimating method "Maximum Likelihood 2" (ML2).

In practice, the parameters of interest are the p_j $(j = 1, \ldots, c)$. It is worth noting that, even if the maximum of the likelihood is reached on the boundary (i.e. at least one of the p_{ij} is equal to zero), this does not mean that the corresponding p_j vanishes since this is only the case if $p_{ij} = 0$ *for all* $i = 1, \ldots, l$.

13.4 Examples of Archaeological Application

We now apply our Bayesian method to the two archaeological examples addressed differently in the previous chapter. The choice of the prior distribution will use the principles described in Sect. 13.3. And a further question will be addressed: how to weight the reference data. This was not relevant in the above simulations because the invariance model was assumed to be valid by definition.

Example 1: Loisy-en-Brie Population. The data are those considered in the previous chapter, for which regression was used (or, equivalent in this case, IALK). There are six age classes of equal duration and six stages. If we have no precise prior information, we can first apply the Bayesian method with $\beta = (1, 1, 1, 1, 1, 1)$. Table 13.8 gives the estimated proportions for each class obtained for two values of coefficient r (weighting of reference data): 1 and 0.75. The standard deviations of the posterior distributions are also given.

The differences in expected posterior probabilities according to r are negligible if one takes account of the standard deviations (note that the standard deviations themselves differ little). There seems no reason, therefore, to weight the reference data and we continue the study with $r = 1$.

From the posterior means and standard deviations we can establish the Beta distributions approximating the posterior distribution of each of the 6 probabilities of belonging to each age class (see Sect. 13.2.2.2). We first examine how

Table 13.8 Loisy-en-Brie example. Estimated parameters ${}_{12}p_{23}, \ldots, {}_{12}p_{83}$ with $\beta_1 = \beta_2 = \ldots = \beta_6 = 1$

	Age class	23–34	35–46	47–58	59–70	71–82	83–94
$r = 1$	Expected posterior probability	0.148	0.252	0.245	0.174	0.114	0.067
	Standard deviation	0.099	0.162	0.171	0.122	0.094	0.059
$r = 0.75$	Expected posterior probability	0.153	0.247	0.238	0.175	0.116	0.071
	Standard deviation	0.101	0.159	0.165	0.122	0.097	0.062

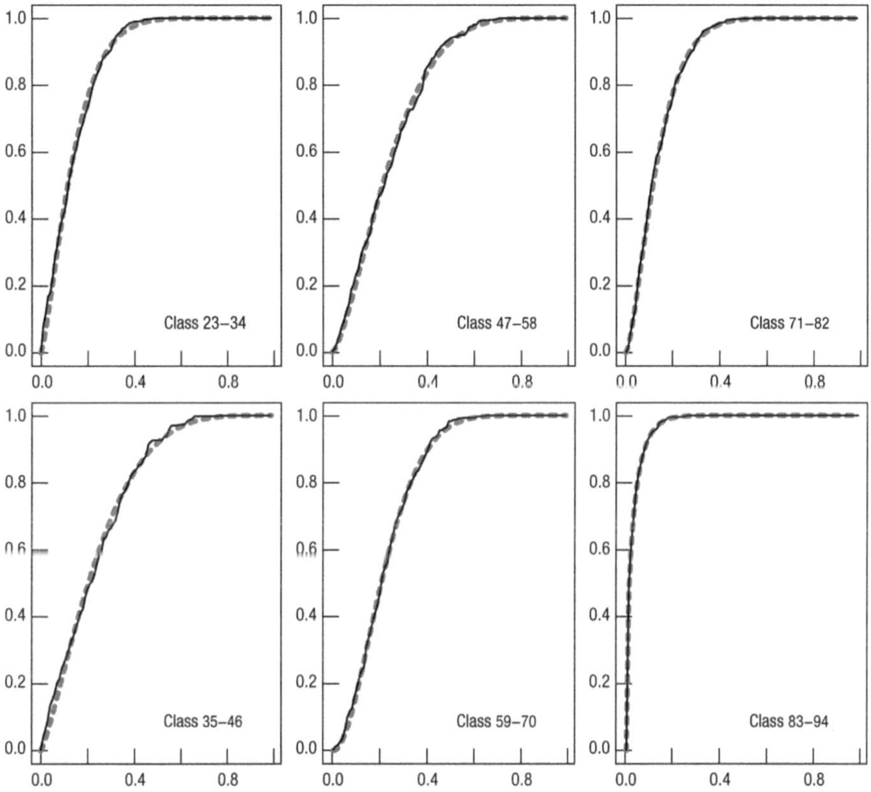

Fig. 13.5 Loisy-en-Brie example (six age classes). Exact (*solid line*) and Beta approximate (*dotted line*) posterior distribution functions

satisfactory this approximation was by also calculating the exact distribution functions by the method given in Sect. 13.2.1. Figure 13.5 compares the exact and approximate distribution functions. It can be seen that the approximation is extremely close, as is also suggested by the relative difference between the 3rd-order (to the power 1/3) and 4th-order (to the power ¼) moments: they are all less than 1.8 % for the 3rd order and less than 3.8 % for the 4th order. In practice the Beta distribution approximation appears mainly to provide an opportune sort of smoothing.

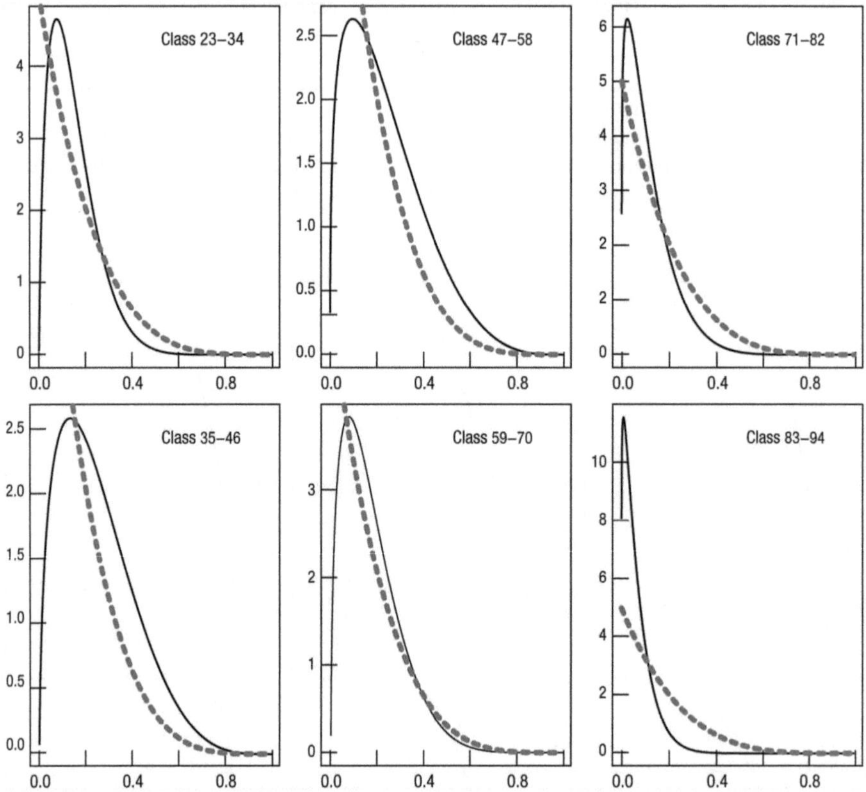

Fig. 13.6 Loisy-en-Brie example. Posterior probability density (*solid line*) for each age class, compared with prior density (*dotted line*), in the case of uniform prior distribution

In the light of these results, we consider the Beta approximations of the posterior distributions. The corresponding densities are shown in Fig. 13.6, compared with the prior densities (all Beta (1; 5) densities, mean 0.167 and standard deviation 0.141).

We shall not enter here into a detailed discussion of these initial results because the prior distribution considered here does not integrate the fact that we are attempting to estimate a mortality distribution. Since we have no other specific a priori argument, it is reasonable to fit the prior distribution on the pre-industrial standard. We did this by assuming that parameters β_j are proportional to that standard and sum to 6 (see Sect. 13.3) as follows:

$$\beta = (0.77\ 0.90\ 1.16\ 1.53\ 1.25\ 0.39)$$

The posterior means and standard deviations are now the ones shown in Table 13.7.

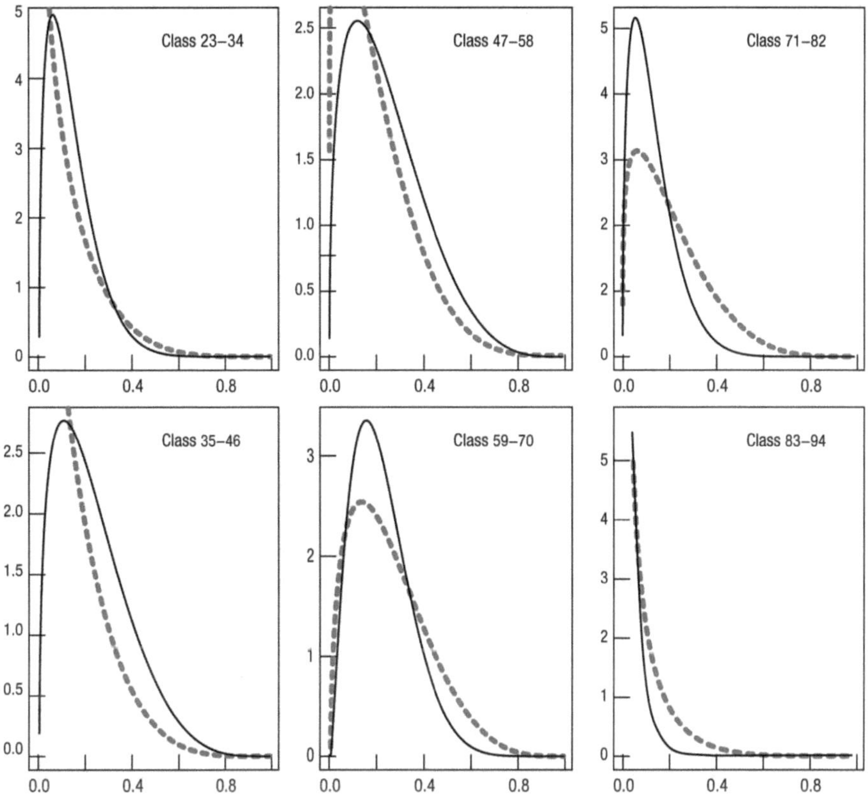

Fig. 13.7 Loisy-en-Brie example. Posterior probability density (*solid line*) for each age class, compared with prior density (*dotted line*), in the case of a prior complying with the pre-industrial standard

Table 13.9 Loisy-en-Brie example. Estimated parameters ${}_{12}p_{23}, \ldots, {}_{12}p_{83}$ with prior deduced from pre-industrial standard

Age class	23–34	35–46	47–58	59–70	71–82	83–94
Expected posterior probability	0.139	0.233	0.250	0.220	0.132	0.026
Standard deviation	0.100	0.157	0.165	0.123	0.096	0.038

As before, the Beta approximation of the posterior distributions is excellent. We used it therefore for the densities shown in Fig. 13.7.

Comparing the numerical values in Tables 13.8 and 13.9 as well as the posterior densities in Figs. 13.6 and 13.7, most of the estimated probabilities are quite stable, demonstrating the limited influence of this prior on the result of the estimation. The greatest difference is observed for the 83–94 age class, which is to be expected since this class will clearly have a low probability, a feature allowed for by the second prior distribution but not the uniform one. With the uniform prior, the posterior deviates

Table 13.10 Loisy-en-Brie example. Estimates by posterior mean and quantiles for 90 % and 50 % credibility intervals

Estimated probabilities	${}_{12}p_{23}$	${}_{12}p_{35}$	${}_{12}p_{47}$	${}_{12}p_{59}$	${}_{12}p_{71}$	${}_{12}p_{83}$
Posterior mean	0.139	0.233	0.250	0.220	0.132	0.026
Quantile 0.05	0.019	0.031	0.035	0.053	0.018	0.000
Quantile 0.95	0.334	0.536	0.566	0.452	0.319	0.104
Quantile 0.25	0.062	0.108	0.119	0.126	0.059	0.002
Quantile 0.75	0.193	0.329	0.353	0.296	0.187	0.035

quite significantly from it, showing that the data impose a serious downward revision; the same is true, though to a lesser extent, for the second prior distribution (taking only the mean, it falls from 0.39/6 = 0.065 to 0.026), confirming that the corresponding probability is not only low but in all likelihood lower than the pre-industrial standard. The next greatest difference between the two estimates can be seen in the 59–70 age class, where the posterior mean rises with the pre-industrial standard, as does the prior mean; the posterior standard deviation also increases (in fact the posterior density is quite clearly more "open"). But these differences are limited, if we consider the wide dispersion of the posterior distributions, due to the small sample size and the structural instability of the problem considered.

For a clearer idea of the accuracy of the estimates, credibility intervals can be calculated. As we have already pointed out, because of the considerable asymmetry of the distributions involved, it is highly inadvisable to calculate symmetric confidence intervals of the "mean plus or minus so many standard deviations" type. It is better to stay with the Bayesian paradigm and give quantiles of the posterior distribution. Table 13.10 gives quantiles 0.05 and 0.95, which provide a 90 % credibility interval and quantiles 0.25 and 0.75 (quartiles), which provide a 50 % credibility interval.

Now we shall compare these results with those obtained by the various methods presented in the previous chapter. Rather than work with deaths observed for various age groups, it is useful to observe a quantity more generally used in demography: the probabilities of death per age group, which can be estimated from mortality data on the assumption of a stationary population. We examine the various estimates made for the Loisy-en-Brie population. Figure 13.8 gives these various probabilities.

The figure clearly shows that the curve of regression method estimates, which, it will be recalled, gives the same results as the IALK method, is highly erratic. This confirms what is seen in the simulations, where the regression method leads to widely dispersed results (see Sect. 13.3). On the other hand, the Bayesian methods, with either a uniform or a pre-industrial standard prior, exhibit a regular increase in the probabilities of death with age. The difference in prior distribution only affects the last two age groups, and then only slightly. We enter the probabilities of death for this standard, clearly showing its effect on the Bayesian estimates for the oldest age group. Despite this slight effect, the Bayesian estimated distributions differ significantly from the standard and are similar to each other, showing the robustness of this method, whatever the prior distribution chosen. We explained above why we prefer the second estimate (PI). We also entered the probabilities of death obtained

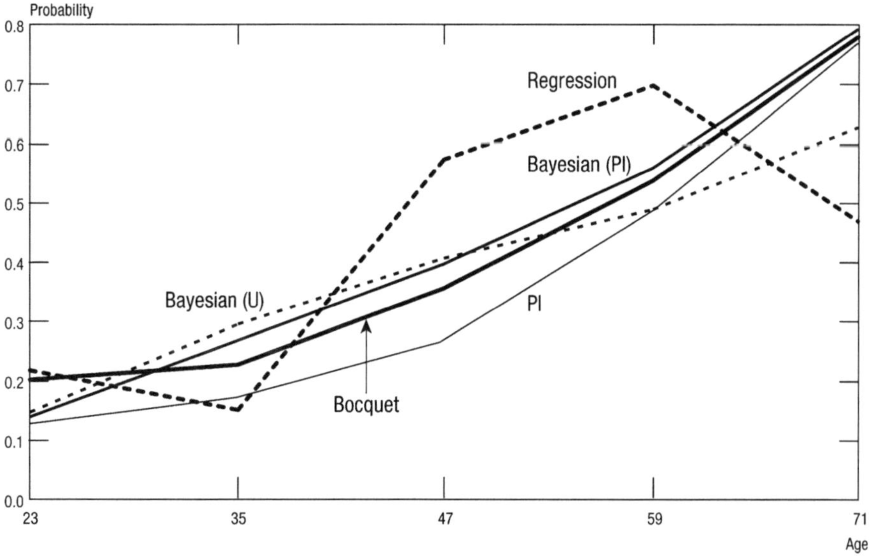

Fig. 13.8 Probabilities of death for Loisy-en-Brie estimated by the Bayesian method with pre-industrial standard prior (*Bayesian PI*), uniform prior (*Bayesian U*), regression, and the method proposed by Bocquet-Appel and Bacro in 2008 (*Bocquet*), compared with the pre-industrial standard (PI)

by the method proposed in Bocquet-Appel and Bacro (2008: see presentation and results in the previous chapter): their distribution is fairly close to the Bayesian solution with the pre-industrial standard. But it is higher for the youngest age group and lower for the others. On this precise point, the fact that our estimates are stable with respect to the prior distribution and that the dispersion of the posterior distribution is relatively low suggests that they can be ascribed to a high confidence level. As for the overall results, it can be seen that the Bocquet-Appel and Bacro method with the choice of a parametric space comprising a distribution of age distributions close to the Bayesian solution, leads to results similar to the Bayesian method, but at the cost of a significantly more cumbersome technique.

Example 2: The Maubuisson Nuns (Seventeenth–Eighteenth Centuries). We now turn again to the Maubuisson example, which gave unacceptable results with the ordinary least squares method, such as negative values and values above unity. Even if the least squares can be forced to meet the necessary constraints, the estimates obtained are on the boundary of the parametric space (zero values) and clearly unrealistic. We adopt now the same division into seven stages and seven age classes as before.[4] The numbers observed for the various stages in a sample of 37 skulls are (6 2 4 5 3 9 8).

[4] A study of this site with five suture stages instead of seven and 5-year age classes (a total of 13) has been carried out and is given in Séguy et al. (2012). The results tally completely, showing in particular that the method described here works effectively with significantly more age classes than stages.

We have a large amount of prior information about this site, particularly useful since the sample is fairly small (37). These are nuns, and consequently all women, all theoretically older than 20. We can therefore opt for specific reference data, namely those already used in the previous chapter and Sect. 13.3 of this chapter. If we stick to this information, using the experience from Sect. 13.3, Example 3, we shall take for the seven parameters to be estimated a Dirichlet prior probability distribution with β_j parameters proportional to the values of the pre-industrial standard (women) and summing to 7, namely (0.70 0.77 0.84 1.05 1.47 1.47 0.70). We present here these initial estimates, more for comparative purposes than for a conclusion. The fact that these women were nuns gives us further information: on admission they were for many reasons in better health than the mean of the general population; they were then protected from various major mortality risks, particularly death in childbirth. These factors can be considered to reduce the mortality of the 20–29 age class by just over 50 % and of the 30–39 class by just under 50 %, thus replacing the parameters of the prior distribution by (0.30 0.40 0.84 1.05 1.47 1.47 0.70) or rather by the proportional values (0.337 0.449 0.944 1.180 1.652 1.652 0.786) summing to 7, as recommended in Sect. 13.3. From this prior distribution we propose a second estimation: it appears prima facie to be the one that should be adopted in practice, and we shall see how far the results obtained confirm this.

Finally, as mentioned in the previous chapter, there is a further major source of information in this case: the convent records give direct evidence of the actual ages at death; as a result the age class probabilities may be evaluated as follows: (0.012 0.025 0.087 0.170 0.289 0.210 0.207). We have therefore an objective way of judging the effectiveness of the method, although some caution is necessary, because the evaluation is probably only approximate and the 37 skulls are only a small and possibly biased sample.[5]

We begin with an analysis using a prior distribution that complies with the pre-industrial standard for women. We applied various "reduction coefficients" to the reference data; since the results were observed to be stable, we shall stick with coefficient 1. The posterior expected values and standard deviations are given in Table 13.11.

It is instructive to compare the posterior means with the prior means, in this case (0.10 0.11 0.12 0.15 0.21 0.21 0.10). It can be seen that the data significantly revise downwards the probabilities for the "youngest" classes, and upwards only the two oldest ones; which is consistent with the discussion above. We leave this analysis as it stands and move on to the Bayesian analysis with a prior distribution with parameters (0.337 0.449 0.944 1.180 1.652 1.652 0.786) corresponding to a modified pre-industrial standard as discussed above. The posterior means and standard

[5] Note, however, that with the reference probabilities we are using, the sample is fully compatible with the documented values. If we calculate theoretical frequencies for stages from these data and compare them with the observed values by chi-squared test, we obtain 1.93 with 6 degrees of freedom.

Table 13.11 Maubuisson example. Estimated parameters ${}_{10}p_{20}$, to p_{80+}: posterior means and standard deviations for a "standard" prior distribution

	${}_{10}p_{20}$	${}_{10}p_{30}$	${}_{10}p_{40}$	${}_{10}p_{50}$	${}_{10}p_{60}$	${}_{10}p_{70}$	p_{80+}
Posterior expected value	0.048	0.067	0.071	0.135	0.301	0.219	0.159
Posterior standard deviation	0.050	0.068	0.069	0.114	0.166	0.142	0.135

Table 13.12 Maubuisson example. Estimated parameters ${}_{10}p_{20}$, to p_{80+}: posterior means and standard deviations for a "modified pre-industrial standard" prior distribution

	${}_{10}p_{20}$	${}_{10}p_{30}$	${}_{10}p_{40}$	${}_{10}p_{50}$	${}_{10}p_{60}$	${}_{10}p_{70}$	p_{80+}
Posterior expected value	0.025	0.041	0.083	0.151	0.311	0.230	0.159
Posterior standard deviation	0.037	0.054	0.074	0.119	0.163	0.142	0.132

Table 13.13 Maubuisson example. Posterior medians and quantiles for 90 % and 50 % credibility intervals

Estimated probability	${}_{10}p_{20}$	${}_{10}p_{30}$	${}_{10}p_{40}$	${}_{10}p_{50}$	${}_{10}p_{60}$	${}_{10}p_{70}$	p_{80+}
Median	0.010	0.020	0.065	0.125	0.300	0.208	0.130
Quantile 0.05	0.002	0.002	0.006	0.015	0.069	0.042	0.008
Quantile 0.95	0.106	0.156	0.232	0.387	0.604	0.501	0.417
Quantile 0.25	0.003	0.005	0.028	0.061	0.186	0.121	0.053
Quantile 0.75	0.034	0.058	0.121	0.216	0.427	0.319	0.239

deviations obtained are given in Table 13.12. We now go on to examine various other parameters of the posterior distribution relating to each age class. For example, we may calculate quantiles: a selection is given in Table 13.13. Figure 13.9 is a graphical representation of the 50 % credibility intervals, comparing posterior means and medians with the values in the records.

The posterior means in this case are

$$(0.048 \quad 0.064 \quad 0.135 \quad 0.169 \quad 0.236 \quad 0.236 \quad 0.112)$$

and the target values given in the records are

$$(0.012 \quad 0.025 \quad 0.087 \quad 0.170 \quad 0.289 \quad 0.210 \quad 0.207).$$

First, it is clear that with so small a sample it is not possible to obtain very precise estimates, as can be seen from the rather wide credibility intervals. But some relevant information may be inferred from the analysis of the data. It can be seen that this analysis leads to a further downward revision of probabilities for the first three classes and a further upward revision for the fifth and seventh. The most noticeable differences (particularly in relative terms) between posterior means and target values are to be seen for the first two probability figures, where the values taken from the records are much lower than would have been expected. In fact, with the highly asymmetric distributions corresponding to extremely low probabilities, the mean can be

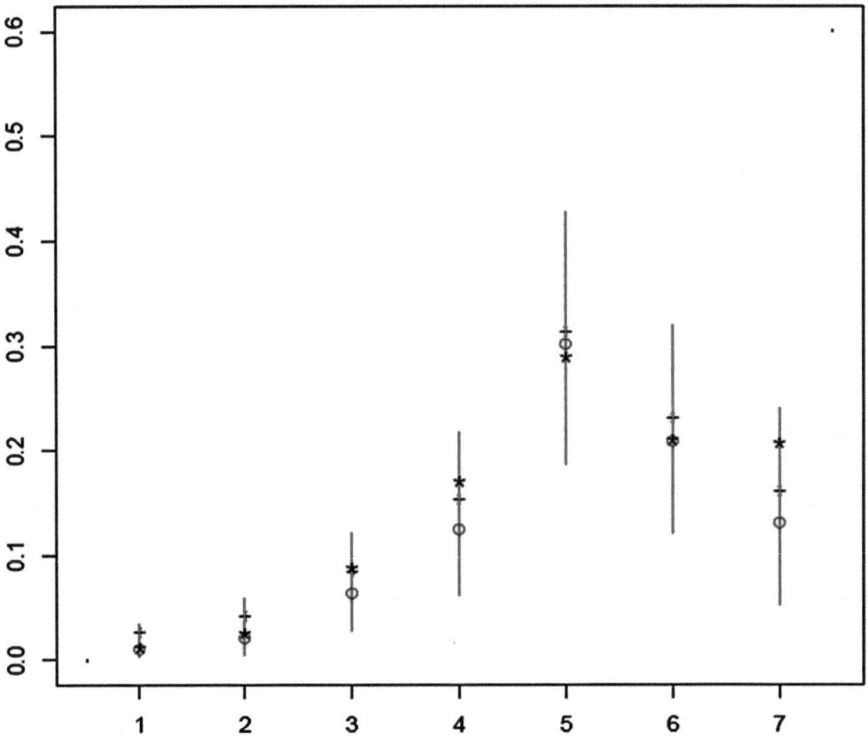

Fig. 13.9 Maubuisson example. Fifty-percent credibility intervals for seven age classes (*black lines*), posterior means (*dashes*) and medians (*circles*) compared with values from the records (*asterisks*)

deceptive; if we examine class 1, for example, we can see that 50 % of the posterior probability lies in the interval [0.003 0.034], whose mean 0.0185 is close to the target value, as is, indeed even more so, the posterior median 0.010; the same holds for class 2. It is therefore justifiable to suppose that for these two classes the posterior means overestimate the true values, an indication that turns out to be realistic. In more general terms, it can be seen that the 50 % credibility intervals do indeed cover the target values, which are always close to the posterior mean and median, and the greatest discrepancy, albeit quite understandable given the sample size, is to be found in class 7. To sum up, we may say that the posterior means provide base information likely to be usefully supplemented by a range of considerations concerning various features of posterior distributions (medians, credibility intervals, etc.).

Here our results can be compared with those obtained by Bocquet-Appel and Bacro's Iterage algorithm, because the division into classes adopted corresponds to the ProbAtri20-90 file of "prior" vectors they provide. Their algorithm gives the following estimates:

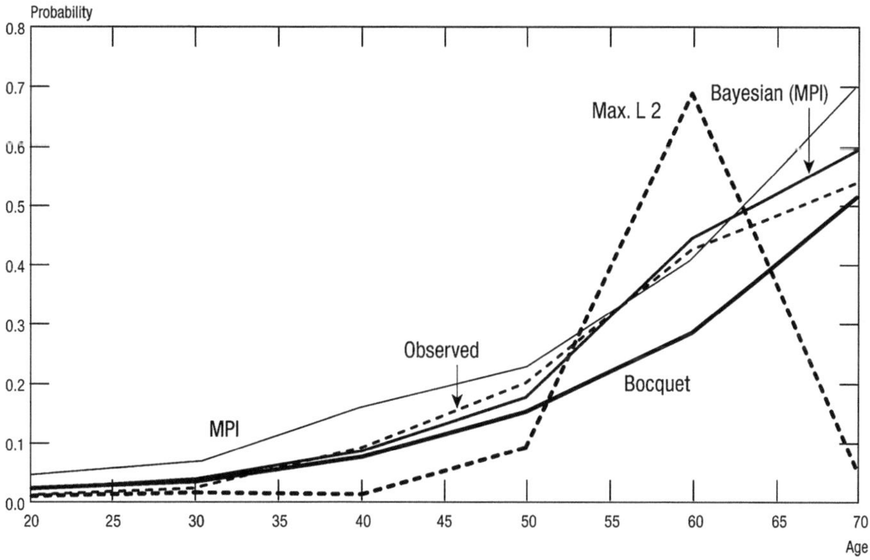

Fig. 13.10 Probabilities of death for the Maubuisson nuns as estimated by the Bayesian method with prior from the modified pre-industrial standard (*MPI*), the method proposed by Bocquet-Appel and Bacro (*Bocquet*) and the Maximum Likelihood 2 method (*Max. L. 2*), compared with the modified pre-industrial standard (*MPI*) and the values actually observed (*Observed*) for these nuns

0.025 0.036 0.073 0.133 0.209 0.268 0.255

It can immediately be seen that the estimates for the first two classes are closer to the target values (probably because we did not wish to give too low a prior mean for these classes), but that the estimates for the other classes are on the whole better with our method. To have a closer idea, we calculated the distances between estimates and target values in two ways: sum of squared deviations and sum of squared deviations weighted by the target value. The figures are

- For the Bayesian method: 0.003 and 0.040
- For the Bocquet-Appel and Bacro method: 0.014 and 0.078

demonstrating a significant advantage for our method.

It must be said, however, that if we had merely used the Bayesian analysis without taking account of the information provided by the particular situation of the nuns (our first analysis), we would have obtained distances 0.007 and 0.199. The method would have retained its advantage in crude deviation but lost it in relative deviation because of too large an "error" in the low probabilities.

We look now at the calculation of probabilities of death for the Maubuisson nun population. Figure 13.10 shows these probabilities calculated under various hypotheses.

This figure compares the 10-year probabilities of death estimated by the Bayesian method described here with as prior the modified pre-industrial standard and the method proposed by Bocquet-Appel and Bacro (2008) with the 756 vectors proposed by these authors (combination of Gompertz-Makeham distributions and extreme values – see previous chapter). For the purposes of comparison, the graph also shows the observed probabilities of death distribution for the nuns as estimated for the period 1640–1889 (blue dashed line). The IALK method (see previous chapter) gave results outside the limits [0, 1] for the proportion of deaths by age group. This method can, however, be used with the addition of a positivity constraint. Here, we used the Maximum Likelihood 2 method (see Box 13.2).

It can be seen straightaway that this last method (Max. L. 2) provides rather unlikely probabilities of death close to zero except for the 50–60 and 60–70 age groups. Even if they are all positive, they are hardly acceptable. The Bocquet-Appel and Bacro method only gives probabilities of death close to the observed values for the first two age groups. They differ widely for the later groups, systematically underestimating the probabilities of dying. Conversely, the Bayesian method we propose gives quite accurate estimated probabilities for all ages, slightly overestimating for the first two age groups and underestimating for the last age group.

13.5 Conclusion

The previous chapter presented a detailed critical examination of the main approaches used by palaeodemographers to estimate the age structure of a population for which they only have biological indicators measured from skeletons. Palaeodemographers often call these methods Bayesian because they use Bayes' theorem and introduce a priori considerations into their method of estimation, but the paradigm upon which they are based is frequentist in nature. This chapter has rather proposed a strictly Bayesian approach in the sense habitually used in statistics, as specified in the introduction, in order to solve this major and recurrent problem for paleodemographers. This conclusion provides an overview of the main advantages of this approach compared with the previous ones.

First, the previous approaches considered that the data taken from the observed groups (frequencies of data from reference population, frequencies of data from observed population) were entirely or partially fixed quantities. The probability vectors method and the IALK method took all these parameters as fixed in order to estimate the age structure of the observed population; the method proposed by Bocquet-Appel and Bacro (2008) considered the frequencies of data by stage taken from the observed population as fixed when establishing recommended confidence intervals. Since the numbers of skeletons, especially for the observed population, are often small, these hypotheses do not hold. The IALK method thus yields estimates that are incorrect or totally unrealistic (age-groups with zero probability) for the age structure of the observed population. The confidence intervals provided by Bocquet-Appel's Iterage software, which we ran with the Lisbon reference data for various site data frequencies, appear to be erroneous and much too small, and therefore over-optimistic: in some cases the interval is zero, some do not contain the estimated parameter value or that value is at one extremity of the interval. We shall see below a further possible reason for these inadequacies in the algorithm used.

Our approach considers all observed frequencies (both site and reference data) to be random; the same is true for the model in Box 13.2, but we have seen that this in itself is not necessarily an advance. We continue by considering the unknown parameters (reference conditional probabilities and age probabilities) to be random under the Bayesian paradigm. Estimates are thus obtained in the form of posterior distributions of probabilities of various age classes, from which can be deduced point estimates (such as posterior means) and "Bayesian confidence intervals", more commonly known as "*credibility intervals*", whereas methods that do not allow for the randomness of the data cannot provide confidence intervals since these are based, by definition, on the uncertainty caused by the randomness.

Second, earlier approaches, except for that of Bocquet-Appel and Bacro, do not take into account the specific nature of the palaeodemographic question, although the demographic knowledge accumulated over many years and information about the living conditions of the populations concerned can provide information about their mortality that is potentially usable. Establishing networks of model life tables has made it possible to hypothesise a standard pre-industrial mortality that can be used to select a prior age distribution for the observed population that is more satisfactory than the uniform distribution. Similarly, in work on the Maubuisson nuns, say, one can use the fact that these women, because of their monastic lives, were not exposed to the same mortality risk as the general French population: in particular they avoided the risk of death in childbirth. Bocquet-Appel and Bacro's approach (2008) also uses prior information taken from palaeodemographic research, but presented in a different manner. As we have said, the Iterage program introduces a restriction of the parametric space to make up for the small number of data. In some respects, their method is a regression, but instead of considering that the vector of the parameters to be estimated is in a space with as many dimensions as the number of age groups considered, with the sole restriction that its components sum to unity, it is confined within the convex envelope of vectors defined ex-ante by a mix of Gompertz-Makeham distributions and extreme values, for which the four parameters vary within restricted intervals. Although this mode of calculation is justifiable as it reduces the variability of the estimators, it may in return introduce a considerable bias. If the age distribution sought falls outside this restricted space, the estimate obtained may be at a considerable distance. This may result in some confidence intervals that never contain the true values, as pointed out above: this occurs if the intervals are confined within the same limits as the point estimates, as is the case with those provided by the Iterage algorithm.

The Bayesian approach differs in a third way from the IALK and frequentist approaches in general (except for the method proposed by Bocquet-Appel and Bacro 2008). Whereas many authors have stressed that when the number of age groups is greater than the number of stages considered, no valid estimate of the age structure of the observed population is possible with a frequentist method, use of a

Bayesian method removes this obstacle. We have seen that such an estimate is always possible in Bayesian terms; using examples we have even shown that the quality of the estimate may be improved with a finer division into age classes (see Séguy et al. 2013).

A fourth feature sets the Bayesian approach apart. In the frequentist approach, it is usually considered sufficient to evaluate the mean and variance of the estimator of a parameter because, when the number of observations is large, its probability distribution often tends towards a normal distribution characterised by these two values. For the estimates in which we are interested (low probabilities and small samples), however, the distribution is generally highly asymmetric, casting doubt on this approach. The same asymmetry can be found in the posterior distributions of the Bayesian approach, but these distributions are easy to determine, making it possible to calculated reliable credibility intervals.

A fifth difference, this time between the Bayesian approach and the Iterage algorithm, is its simplicity of operation. Selecting a prior distribution is much easier and more flexible that the construction of a mortality model whose parameters are supposed to represent the most varied conditions of mortality, by attrition as well as disaster. Furthermore, once the palaeodemographic sample differs from the model conditions, both in number of age classes and in selected intervals, the model has to be reset and all the prior probability vectors recalculated, whereas, to our knowledge, Bocquet-Appel and Bacro do not provide anywhere their full parametric formulation. It is clear that some situations will always fall outside such a mortality model, whereas the Bayesian method can be used to address all possible cases, albeit with varied efficacy.

In parallel to the "theoretical" considerations we have presented here, it is important to recall the results of our empirical studies. The simulations we have run can be used to measure the quality of results obtained under the various methods proposed. Calculation of their mean squared errors provides a comparison between the estimates produced and the true values of the parameters, which in this case are known. The method we propose outperforms the other methods in almost all cases, often with substantial gains in accuracy. Only some of the results obtained with the Iterage algorithm manage to equal its performance, but in other cases Iterage introduces noticeable biases because it requires a restricted parametric space, whereas the age structure of the population under study falls outside that space.

We trust that this chapter has clearly demonstrated all the advantages of using a fully Bayesian estimation of the age structure of historic populations for which there is no record of age at death and where the records are replaced by the measurement of biological indicators. We hope that many palaeodemographers will use it, providing further insight into its application and encouraging any improvements that may be necessary, so that their experience will complement our own.

Authors and Contributors

Luc Buchet

A historian and anthropologist by training, Luc Buchet developed a multidisciplinary approach at CNRS to bring biological anthropology, via the excavation of burial grounds, into the mainstream of historical research. His research topics cover methods and applications in palaeodemography, interactions between health conditions and demographic parameters, a dynamic approach to settlement via the combined study of osteological material and archaeological and historical data. He is head of the biological anthropology laboratory at CEPAM (Cultures et Environnements. Préhistoire, Antiquité, Moyen Âge), associate researcher at INED, and lectures in anthropology and palaeodemography at the University of Nice Sophia Antipolis.

luc.buchet@cepam.cnrs.fr

Henri Caussinus

A statistician, Henri Caussinus is emeritus professor at the Toulouse Mathematics Institute (Paul Sabatier University). In addition to his contribution to the work of palaeodemographers, his most recent research has covered multidimensional analysis (projection pursuit) as well as change-points and outliers detection with applications for climatology.

henri.caussinus@math.univ-toulouse.fr

I. Séguy and L. Buchet, *Handbook of Palaeodemography*, INED Population Studies 2,
DOI 10.1007/978-3-319-01553-8, © Springer International Publishing Switzerland 2013

Daniel Courgeau

Emeritus director of research at INED, Daniel Courgeau is also collection editor, with Robert Franck, of the Methodos Series at Springer. His earlier work covered migration modelling, population genetics and the development of event-history and multilevel analysis. His current research focuses on the methodology and epistemology of the social sciences: examination of the methods used in palaeodemography, advanced study of the links between probabilities and social sciences since the seventeenth century, analysis of the notion of cumulativity in demography and critical examination of the methods used in population genetics.

daniel.courgeau@wanadoo.fr

Isabelle Séguy

An archaeologist and historical demographer by training, Isabelle Seguy is a researcher at INED and CEPAM, where she works with Luc Buchet. She studies the demographic behaviours of populations from written and material evidence. Her research focuses on the analysis of factors determining the dynamics of historical populations, governed to a large extent by their natural environment and by prevailing health and social conditions. Together with specialists from a range of disciplines, she is currently working on plague epidemics of the Middle Ages and the early modern Period, the demographic consequences of climatic and environmental variations, and on methods for estimating age in the absence of civil records.

seguy@ined.fr

Bibliography

Acsádi, G. (1965). La paléodémographie en Hongrie, définitions, sources, méthodes. *Annuaire de statistique historique 1963–1964*. Budapest, Office de la statistique.

Acsádi, G., & Nemeskéri, J. (1957). Palaeodemographische probleme. *Homo, 8*, 133–148.

Acsádi, G., & Nemeskéri, J. (1970). *History of human life span and mortality*. Budapest: Akadémiai Kiadó.

Adalian, P., Piercecchi-Marti, M.-D., Bourlière-Najean, B., Panuel, M., Fredouille, C., Dutour, O., & Leonetti, G. (2001). Postmortem assessment of fetal diaphyseal femoral length: Validation of a radiographic methodology. *Journal of Forensic Sciences, 46*(2), 215–219.

Adalian, P., Piercecchi-Marti, M.-D., Bourlière-Najean, B., Panuel, M., Leonetti, G., & Dutour, O. (2002). New formula for the determination of fetal age. *Comptes rendus biologies, 325*(3), 261–269.

Adalian, P., Piercecchi-Marti, M.-D., Lalys, L., Leonetti, G., Dutour, O. (2006). Détermination de l'âge fœtal: Evaluation des formules "classiques". In L. Buchet, C. Dauphin, I. Séguy (Dir.). *La paléodémographie. Mémoire d'os, mémoire d'hommes* (APDCA éd., pp. 271–279).

Akaike, H. (1973). Information theory and the maximum likelihood principle. In B. N. Petrov & F. Csaki (Eds.), *Second international symposium on information theory* (pp. 268–281). Budapest: Hungarian Academy of Sciences.

Akaike, H. (1992). Information theory and an extension of the maximum likelihood principle. In S. Kotz & N. Johnson (Eds.), *Breakthroughs in statistics* (pp. 610–624). New York: Springer Verlag.

Akane, A., Seki, S., Shiono, H., Nakamura, H., Hasegawa, M., Kagawa, M., Matsubara, K., Nakahori, Y., Nagafuchi, S., & Nakagome, Y. (1992). Sex determination of forensic samples by dual PCR amplification of an X-Y homologous gene. *Forensic Science International, 52*, 143–148.

Alduc-Le Bagousse, A. (1988). Estimation de l'âge dentaire des non-adultes: maturation dentaire et croissance osseuse. Données comparatives de deux nécropoles bas-normandes. In L. Buchet (Ed.), *Anthropologie et histoire ou anthropologie historique?* (Coll. Notes et monographies techniques, Vol. 24, pp. 81–95). Paris: CNRS Éditions.

Alduc-Le Bagousse, A. (1994). Maturation osseuse-majorité légale: la place des adolescentes en paléoanthropologie. In L. Buchet (Ed.), *La femme pendant le Moyen Âge et l'Époque moderne (Actes des 6es Journées anthropologiques de Valbonne, juin 1992)* (Coll. Dossiers de documentation archéologique, Vol. 17, pp. 31–39). Paris: CNRS Éditions.

Alduc-Le Bagousse, A., Metz, F., & Azencoth, S. (1980). Quelques données sur les conditions de vie d'une population mérovingienne de Basse-Normandie: Verson (Calvados). *Archives suisses d'anthropologie générale, 44*(2), 197–206.

I. Séguy and L. Buchet, *Handbook of Palaeodemography*, INED Population Studies 2, DOI 10.1007/978-3-319-01553-8, © Springer International Publishing Switzerland 2013

Alekseeva, T. I., & Buzhilova, A. P. (1997). Medieval urban Russians according to anthropological data: Origins, palaeodemography, palaeoecology. *Anthropology and Archeology of Eurasia, 35*, 42–62.

Alesan, A., Malgosa, A., & Simo, C. (1999). Looking into the demography of an iron age population in the western Mediterranean. I. Mortality. *American Journal of Physical Anthropology, 110*, 285–301.

Alexandre-Bidon, D., & Lett, D. (1997). *Les enfants au Moyen Âge (ve- xve siècles)* (Coll. La Vie quotidienne). Paris: Hachette Littératures.

Ammermann, A., & Cavalli-Sforza, L. (1984). *The neolithic transition and the genetics of populations in Europe*. Princeton: Princeton University Press.

Angel, J. L. (1972). Ecology and population in the Eastern Mediterranean. *World Archaeology, 4* (1), 88–105.

Ardagna, Y. (2004). *La conservation des archives biologiques et des documents associés en anthropologie biologique. Applications à des collections anthropologiques françaises et hongroises*. Ph.D. thesis from Université de la Méditerranée-Aix-Marseille II, Faculté de médecine, and the University of Szeged.

Armelagos, G. J. (1990). Health and disease in prehistoric populations in transition. In A. Swedlung & G. J. Armelagos (Eds.), *Disease in populations in transition: Anthropological and epidemiological perspectives* (pp. 127–144). New York: Bergin and Gravey.

Armelagos, G. J., Goodman, A. H., & Jacobs, K. H. (1991). The origins of agriculture: Population growth during a period of declining health. *Population and Environment, 1*, 9–22.

Aykroyd, R. G., Lucy, D., Pollard, A. M., & Solheim, T. (1997). Technical note: Regression analysis in adult age estimation. *American Journal of Physical Anthropology, 104*, 259–265.

Aykroyd, R. G., Lucy, D., Pollard, A. M., & Roberts, C. A. (1999). Nasty, brutish, but not necessarily short: A reconsideration of the statistical methods used to calculate age at death from adult human skeletal and dental age indicators. *American Antiquity, 64*, 55–70.

Aymard, I. (2004). La détermination du sexe à partir du squelette: évaluation de l'efficacité de certaines méthodes appliquées à la collection Simon. DEA dissertation, université de la Méditerranée-Aix-Marseille II, Faculté de médecine.

Aymard, I., Signoli, M., Buchet, L., & Perréard-Lopreno, G. (2005). La détermination du sexe à partir du squelette: évaluation de l'efficacité de certaines méthodes appliquées à la collection Simon, poster presented at the *Société d'Anthropologie de Paris* (January 2005).

Bagnall, R. S., & Frier, B. W. (1994). *The demography of Roman Egypt* (Cambridge studies in population, economy and society in past time, Vol. 23). Cambridge: Cambridge University Press.

Balard, M. (1992). *Bibliographie de l'histoire médiévale en France (1965–1990)*. Paris: Publications de la Sorbonne.

Bass, W. M. (1971). *Human osteology*. Columbia: Missouri Archaeological Society.

Bats, M., Dedet, B., & Garmy, P. (Eds.). (2003). Peuples et territoires en Gaule méditerranéenne: hommage à Guy Barruol. *Numéro spécial of Revue archéologique de Narbonnaise*, Suppl. 35.

Baud, C. A., & Gossi, M. (1980). Degree of mineralization of bone tissue as determined by quantitative micro-radiography: Effect of age, sex and pathological conditions. In *Proceedings of the 4th international conference on bone measurement* (Ontario, 1978) (Pub. No. 80-1938, pp. 345–351). Washington, DC: US National Institutes of Health.

Bedford, M., Russell, F., Lovejoy, C., Meindel, R., Simpson, S., & Stuart-Macadam, P. (1993). Test of the multifactorial aging method using skeletons with known age-at-death from grant collection. *American Journal of Physical Anthropology, 91*, 287–297.

Behar, C. L. (1976). Des tables de mortalité aux xviie- xviiie siècles. Histoire, signification. *Annales de démographie historique*, 173–200. Paris: La Haye, Mouton & Cie.

Bello, S., Dutour, O., Signoli, M., Thomann, A., & Andrews, P. (2006). Age and sex bias in the reconstruction of past population structures. *American Journal of Human Biology, 129*(1), 24–38.

Benedictow, O. J. (1996). (1st ed. 1993). *The medieval demographic system of the Nordic Countries*. Oslo, Middelalderforlaget.

Bennett, K. A. (1973). On the estimation of some demographic characteristics on a prehistoric population from the American Southwest. *American Journal of Physical Anthropology, 39*, 223–232.

Berger, J.-F., Bertoncello, F., Braemer, F., Davtian, G., & Gazenbeek, M. (Eds.). (2005). *Temps et espaces de l'homme en société: analyses et modèles spatiaux en archéologie (Actes des Rencontres des 21–23 octobre 2004)*. Antibes: APDCA.

Bergot, C., & Bocquet, J.-P. (1976). Étude systématique en fonction de l'âge de l'os spongieux et de l'os cortical de l'humérus et du fémur. *Bulletins et mémoires de la Société d'anthropologie de Paris, 3*, 215–242.

Bertoncello, F. (1999). *Le Peuplement de la basse vallée de l'Argens et de ses marges (Var) de la fin de l'âge du Fer à la fin de l'Antiquité*. Ph.D. thesis from Université de Provence Aix-Marseille 1, Archéologie, Histoire et civilisation de l'Antiquité et du Moyen Âge.

Bideau, A. (1983). Les mécanismes régulateurs des populations traditionnelles. *Annales, économies, sociétés, civilisations, 38*(5), 1040–1057.

Billard, M. (1994). Les lignes de Harris en ostéoarchéologie: évolution d'un stigmate événementiel au cours de la croissance et du vieillissement squelettique: implications en paléo-épidémiologie. In L. Buchet (Ed.), *La femme pendant le Moyen Âge et l'Époque moderne (Actes des 6es Journées anthropologiques de Valbonne, juin 1992)* (pp. 135–155). Paris: CNRS Éditions.

Bintliff, K., & Sbonias, J. (Eds.). (1999). *Reconstruction of past population trends in Mediterranean Europe (3000 BC-AD 1800)* (The archaeology of the Mediterranean landscape, Vol. 1). Oxford: Oxbow Books.

Biraben, J. N. (1969). Durée de la vie dans la population de Columnata. *Population, 24*(3), 487–500.

Biraben, J.-N. (1971). Les méthodes de la démographie préhistorique. In *Actes du congrès international de la population (Londres, 1969)* (Vol. 4, pp. 2315–2320). Liège: UIESP.

Biraben, J.-N. (1979). Essai sur l'évolution du nombre des hommes. *Population, 34*(1), 13–25.

Biraben, J.-N. (1982). L'évolution récente du temps biologique dans les sociétés industrielles. 1re partie: aspects physiologiques. In *Les âges de la vie (Actes du VIIe colloque national de dé mographie, Strasbourg, mai 1982)* (Coll. "Travaux et documents", Vol. 96, pp. 3–10). Paris: INED.

Biraben, J.-N. (1988). Préhistoire I: le peuplement et son évolution. In J. Dupâquier (Ed.), *Histoire de la population française, tome I* (Des origines à la Renaissance, pp. 19–64). Paris: Puf.

Biraben, J.-N. (2003). The rising numbers of humankind. *Population and Societies, 394*(9), 38–42.

Bishop, Y. M., Fienberg, S. E., & Holland, P. W. (1975). *Discrete multivariate analysis: Theory and practice*. Cambridge: MIT Press.

Bizot, B., Castex, D., Reynaud, P., & Signoli, M. (2005). *La saison d'une peste (avril-septembre 1590): le cimetière des Fédons à Lambesc (Bouches-du-Rhône)*. Paris: CNRS Éditions.

Blaizot, F., & Castex, D. (2005). Du bon usage des outils anthropologiques à l'étude des sociétés historiques. In O. Dutour, J.-J. Hublin, & B. Vandermeersch (Eds.), *Origine et évolution des populations humaines* (Orientation et méthodes, Vol. 8, pp. 259–279). Paris: Comité des Travaux historiques et scientifiques.

Blaizot, F., Tranoy, L., Le Bot-Helly, A., Sellier, P., & Valentin, F. (1997). La paléoanthropologie funéraire aujourd'hui. *Les nouvelles de l'archéologie, 66*, 43–53.

Blaizot, F., Alix, G., & Ferber, E. (2003). Le traitement funéraire des enfants décédés avant un an dans l'antiquité: Études de cas. *Bulletins et Mémoires de la Société d'anthropologie de Paris, 15*(1–2), 49–77.

Blayo, Y. (1975). La mortalité en France de 1740 à 1829. *Population, 30*, 123–142. Démographie historique.

Blondiaux, J. (1989). *Essai d'anthropologie physique et de paléopathologie des populations du Nord de la Gaule au haut Moyen Âge*. Ph.D. thesis, Université de Lille III.

Blondiaux, J., Gabart, N., Alduc-Le Bagousse, A., Niel, C., & Tyler, E. (2006). Relevance of cement annulations to palaeopathology. *Palaeopathology Newsletter, 135*, 4–13.

Bock, R. D., & Gibbons, R. D. (1996). High-dimensional multivariate probit analysis. *Biometrics, 52*, 1183–1194.

Bocquet, J.-P. (1977a). Palaéodémographie: ce que nous apprend la Nubie soudanaise. *Annales, économies, sociétés, civilisations*, 32e année, n°1, 54–69.

Bocquet, J.-P. (1977b). *Perspectives paléodémographiques, thèse d'Anthropologie historique*. Paris: Éditions de l'EHESS.

Bocquet, J.-P. (1978). Méthodes d'estimation de l'âge au décès des squelettes adultes et structure démographique des populations du passé. *Simposio de Antropología Biológica de España, 1*, 37–47.

Bocquet, J.-P. (1979). Une approche de la fécondité des populations inhumées. *Bulletins et mé moires de la Société d'anthropologie de Paris, XIII*, 261–268.

Bocquet-Appel, J.-P. (1985). Small populations: Demography and palaeoanthropological inferences. *Journal of Human Evolution, 14*, 683–691.

Bocquet-Appel, J.-P. (1986). Once upon a time: Palaeodemography. In B. Herrmann (Ed.), *Innovative Trends in der prähistorischen Anthropologie* (Mitteilungen der Berliner Gesellschaft für Anthropologie, Ethnologie und Urgeschichte, Vol. 7, pp. 127–133). Berlin: Berliner Gesellschaft für Anthropologie, Ethnologie und Urgeschichte.

Bocquet-Appel, J.-P. (1994). L'hypogée néolithique de Loisy-en-Brie (Marne), lieu-dit les Gouttes d'Or. L'interprétation démographique. *Préhistoire et protohistoire en Champagne-Ardenne, 18*, 55–60.

Bocquet-Appel, J.-P. (2002). Palaeoanthropological traces of a neolithic demographic transition. *Current Anthropology, 43*(4), 637–650.

Bocquet-Appel, J.-P. (2005). La paléodémographie. In O. Dutour, J.-J. Hublin, & B. Vandermeersch (Eds.), *Objets et méthodes en paléoanthropologie* (pp. 271–313). Paris: Éditions du Comité des Travaux historiques et scientifiques. Version corrected since publication available at. http://www.evolhum.cnrs.fr/bocquet/cths05.pdf

Bocquet-Appel, J.-P. (2008a). *La paléodémographie*. Paris: Éditions Errance.

Bocquet-Appel, J.-P. (Ed.). (2008b). *Recent advances in palaeodemography. Data, techniques, patterns*. Dordrecht/London: Springer Verlag.

Bocquet-Appel, J.-P., & Arsuaga, J. L. (1999). Age distributions of hominid samples at Atapuerca (H) and Krapina could indicate accumulation by catastrophe. *Journal of Archaeological Science, 26*, 327–338.

Bocquet-Appel, J.-P., & Bacro, J.-N. (1997). Brief communication: estimates of some demographic parameters in a neolithic rock-cut chamber (approximately 2000 BC) using iterative techniques for aging and demographic estimators. *American Journal of Physical Anthropology, 102*, 569–575.

Bocquet-Appel, J.-P., & Bacro, J.-N. (2008). Estimation of an age distribution with its confidence intervals using an iterative Bayesian procedure and a bootstrap sampling approach. In J.-P. Bocquet-Appel (Ed.), *Recent advances in palaeodemography. Data, techniques, patterns* (pp. 63–82). Dordrecht: Springer Verlag.

Bocquet-Appel, J.-P., & Bar-Yosef, O. (Eds.). (2008). *The neolithic demographic transition and its consequences*. Dordrecht/London: Springer.

Bocquet-Appel, J.-P., & Demars, P.-Y. (2000). Neanderthal contraction and modern human colonization of Europe. *Antiquity, 74*, 544–552.

Bocquet-Appel, J.-P., & Dubouloz, J. (2003). Traces paléoanthropologiques et archéologiques d'une transition néolithique en Europe. *Bulletin de la Société préhistorique française, 100*(4), 699–714.

Bocquet-Appel, J.-P., & Dubouloz, J. (2004). Expected palaeoanthropological and archaeological signal from a Neolithic demographic transition on a worldwide scale. *Documenta Praehistorica, XXXI*, 25–33.

Bocquet, J.-P., & Masset, C. (1977). Estimateurs en paléodémographie. *L'Homme, 17*(4), 65–90.
Bocquet-Appel, J.-P., & Masset, C. (1982). Farewell to palaeodemography. *Journal of Human Evolution, 11*, 321–333.
Bocquet-Appel, J.-P., & Masset, C. (1985). Palaeodemography: Resurrection or ghost? *Journal of Human Evolution, 14*, 107–111.
Bocquet-Appel, J.-P., & Masset, C. (1995). L'âge au décès dans les populations inhumées: comparaison de méthodes et de résultats. *Antropologia Portuguesa, 13*, 39–48.
Bocquet-Appel, J.-P., & Masset, C. (1996). Palaeodemography: Expectancy and false hope. *American Journal of Physical Anthropology, 99*, 571–583.
Bocquet-Appel, J.-P., & Naji, S. (2006). Testing the hypothesis of a worldwide Neolithic demographic transition. Corroboration from American cemeteries (with comments). *Current Anthropology, 47*(2), 341–365.
Bocquet-Appel, J.-P. & Xavier de Morais (1987). *Anthropologie et histoire. Un essai de reconstitution de la variation biologique de la population portugaise au XIXe siècle* (Fontes documentais portuguesas, Vol. XXII). Paris: Fondation Calouste, Gulbenkian, Centre culturel portugais.
Bocquet, J.-P., Maia Neto, A., Tavares Da Rocha, A., & Xavier de Morais, H. (1978). Estimation de l'âge au décès des squelettes d'adultes par régressions multiples. *Contribuições para o Estudo da Antropologia Portuguesa, 10*(3), 107–167.
Bocquet, J.-P., Tavares de Rocha, M.-A. D. A., et al. (1980). Peut-on estimer l'âge au décès à l'aide du remaniement osseux? *Biométrie humaine, 15*, 51–56.
Bocquet-Appel, J.-P., Demars, P.-Y., Noiret, L., & Dobrowsky, D. (2005). Estimates of upper palaeolithic meta-population size in Europe from archaeological data. *Journal of Archaeological Science, 32*, 1656–1668.
Bodzsár, R. E., & Susanne, C. (Eds.). (1998). *Secular growth changes in Europe*. Budapest: Eötvös University Press.
Bogin, B. (2001). *The growth of humanity*. New York: Wiley-Liss.
Bogin, B., & Smith, B. H. (1996). Evolution of the human life cycle. *American Journal of Human Biology, 8*, 703–716.
Boldsen, J. (1988). Two methods for reconstructing the empirical mortality profiles. *Human Evolution, 3*(5), 335–342.
Boldsen, J. L., & Paine, R. R. (1995). Defining extreme longevity from Mesolithic to the Middle Ages: Estimates based on skeletal data. In B. Jeune & J. W. Vaupel (Eds.), *Exceptional longevity from prehistory to the present* (pp. 25–36). Odense: Odense University Press.
Boldsen, J. L., Milner, G. R., Konigsberg, L. W., & Wood, J. W. (2002). Transition analysis: A new method for estimating age from skeletons. In R. D. Hoppa & J. W. Vaupel (Eds.), *Palaeodemography* (pp. 73–106). Cambridge: Cambridge University Press.
Bonneuil, N. (2005). Fitting to a distribution of deaths by age with application to palaeodemography. *Current Anthropology, 46*(suppl), S29–S45.
Boserup, E. (1965). *The conditions of agricultural growth. The economics of agriculture under population Pressure*. London: Earthscan.
Boucher, B.-J. (1957). Sex differences in the fetal pelvis. *American Journal of Physical Anthropology, 15*, 581–600.
Bourgeois-Pichat, J. (1966). *Le concept de population stable. Application à l'étude des populations ne disposant pas de bonnes statistiques démographiques* (Études démographiques, Vol. 39). New York: UN.
Bourgeois-Pichat, J. (1990). Théorie des populations semi-stables et application. *Population, 45* (4–5), 821–847.
Bourgeois-Pichat, J. (1994). *La dynamique des populations. Population stables, semi-stables, quasi stables* (Travaux et documents, Vol. 133). Paris: INED.
Bouvier, M., & Ubelaker, D. H. (1977). A comparison of two methods for the microscopic determination of age at death. *American Journal of Physical Anthropology, 46*, 391–394.

Boyaval, B. (1977). Épigraphie antique et démographie: problèmes de méthode. *Revue du Nord*, 163–191.

Brass, W. (1971). On the scale of mortality. In W. Brass (Ed.), *Biological aspects of demography* (pp. 69–110). London: Taylor and Francis.

Brass, W. (1975). *Methods for estimating fertility and mortality from limited and defective data*. Chapel Hill: Carolina Population Center.

Brass, W. (1977). Notes on empirical mortality models. *United Nations Population Bulletin, 9*, 38–42.

Brass, W., & Coale, A. J. (1968). Methods of analysis and estimation. In W. Brass et al. (Eds.), *The demography of tropical Africa* (pp. 88–139). Princeton: Princeton University Press.

Broca, P. (1875). Instructions craniologiques et craniométriques. *Bulletins et mémoires de la Socié té d'anthropologie de Paris, 2*, 204–207.

Brooks, S. T. (1955). Skeletal age at death: The reliability of cranial and pubic age indicators. *American Journal of Physical Anthropology, 13*, 567–597.

Brooks, S. T., & Suchey, J. M. (1990). Skeletal age determination based on the os pubis: A comparison of the Acsádi-Nemeskéri and Suchey-Brooks method. *Human Evolution, 5*, 227–238.

Brothwell, D. R. (1971). Palaeodemography. In W. Brass (Ed.), *Biological aspects of demography* (pp. 111–130). London: Taylor and Francis Ltd.

Brothwell, D. R. (1972). Palaeodemography and earlier British populations. *World Archaeology, 4* (1), 75–87.

Brothwell, D. R. (1981). (3rd ed., illustrated and revised; 1st ed. 1965, London, British Museum). *Digging up the bones: The excavation, treatment, and study of human skeletal remains*. New York: Cornell University Press.

Brunner, C., Kissling, R., & Jacob, H. A. (1991). The effects of morphology and histopathologic findings on the mobility of the sacro-iliac joint. *Spine, 16*, 1111–1117.

Bruzek, J. (1991). *Fiabilité des procédés de détermination du sexe à partir de l'os coxal. Application à l'étude du dimorphisme sexuel de l'homme fossile*. Ph.D. thesis, Muséum national d'Histoire naturelle, Institut de Palaéontologie humaine, Paris.

Bruzek, J. (1992). Fiabilité des fonctions discriminantes dans la détermination sexuelle de l'os coxal. Critiques et propositions. *Bulletins et mémoires de la Société d'anthropologie de Paris, 1–2*, 67–104.

Bruzek, J. (2002). A method for visual determination of sex, using the human hip bone. *American Journal of Physical Anthropology, 117*, 157–168.

Bruzek, J., & Ferembach, D. (1992). Fiabilité de la méthode visuelle de détermination du sexe à partir du bassin, proposée par le Groupe de travail d'Anthropologues européens. Application à l'os coxal. *Archivio per l'Antropologia e la Etnologia, CXXII*, 145–161.

Bruzek, J., & Murail, P. (2006). Methodology and reliability of sex determination from the skeleton. In A. Schmitt, E. Cunha, & J. Pinheiro (Eds.), *Forensic anthropology and medicine: Complementary sciences from recovery to cause of death* (pp. 225–242). New Jersey: Humana Press.

Bruzek, J., Castex, D., & Majo, T. (1996). Évaluation des caractères morphologiques de la face sacro-pelvienne de l'os coxal. Proposition d'une nouvelle méthode de diagnose sexuelle. *Bulletins et mémoires de la Société d'anthropologie de Paris, 8*(3–4), 491–502.

Bruzek, J., Houet, F., & Cunha, E. (2003). DSP: un outil de diagnose sexuelle probabiliste à partir des données métriques de l'os coxal. Application à l'os coxal. *Bulletins et mémoires de la Socié té d'anthropologie de Paris, 17*(3–4), 167–176.

Bruzek, J., Schmitt, A., & Murail, P. (2005). Identification biologique individuelle en paléoanthropologie. Détermination du sexe et estimation de l'âge au décès à partir du squelette. In O. Dutour, J.-J. Hublin, & B. Vandermeersch (Eds.), *Objets et méthodes en paléoanthropologie* (Orientations et méthodes, Vol. 7, pp. 217–246). Paris: Comité des travaux historiques et scientifiques.

Buchet, L. (1978). La nécropole gallo-romaine et mérovingienne de Frénouville (Calvados). Étude anthropologique. *Archéologie médiévale, VIII*, 5–53.

Buchet, L. (1998). Les habitants de la Gaule du Ier au VIIIe siècle. Apports de l'anthropologie physique, habilitation dissertation. Université de Paris I Panthéon-Sorbonne, 4 vol.

Buchet, L., & Séguy, I. (1999). *Ancien cimetière d'Antibes. Réflexions méthodologiques en palé odémographie*. Rapport d'analyse, SRA de PACA.

Buchet, L., & Séguy, I. (2002). La paléodémographie: bilan et perspectives. *Annales de dé mographie historique, 2*, 161–212.

Buchet, L., & Séguy, I. (2003). Les fouilles de l'ancien cimetière d'Antibes (06): aspects paléodémographiques. *Mémoires de l'Institut de préhistoire et d'archéologie Alpes Mé diterranée, XLV*, Nice, 309–318.

Buchet, L., & Séguy, I. (2008). L'âge au décès des enfants: âge civil, âge biologique, âge social?. In F. Gusi, S. Muriel, & C. Olària (coord.) (Eds.), *Nasciturus, Infans, Puerulus Vobis Mater Terra. La muerte en la infancia/La mort dans l'enfance/La mort a l'infancia/The death in the childhood* (pp. 25–39), Castelló de la Plana, Servicio de Investigaciones Arqueológicas y Prehistóricas, Colección de Prehistoria y Arqueología.

Buchet, L., Séguy, I., Boulle, E.-L., Gallien, V., & Wabon, M. (2003). Mort probable, mort certaine. Réflexions en paléodémographie à partir de squelettes modernes et contemporains. In *Sens dessus dessous – La recherche du sens en Préhistoire, recueil d'études offert à Jean Leclerc et Claude Masset, Revue archéologique de Picardie,* 21 (ns), (pp. 101–111).

Buchet, L., Séguy, I., & Gallien, V. (2004). À quel âge sont-ils morts? Âge "individuel" et âge "collectif. *Archéopages, 14*, 16–23.

Buchet, L., & Séguy, I. et al. (2005). L'âge au décès des enfants: nouvelles approches paléodémographiques. L'Homme et ses images. In *Actes du XXVe colloque du Groupe des anthropologistes de langue française* (Marseille, July 2001), CD-ROM, UMR 6578, CNRS-Faculté de médecine de Marseille.

Buchet, L., Séguy, I., & Darton, Y. (2006a). Incidences du travail des enfants au Moyen Âge. In *Enfants d'aujourd'hui, diversité des contextes, pluralité des parcours. Actes du Colloque international de Dakar, 2002,* Paris, Puf/Aidelf, tome 2, pp. 927–936.

Buchet, L., Dauphin, C., & Séguy, I. (Eds.). (2006b). *La paléodémographie. Mémoire d'os, mé moire d'hommes (Actes des 8es journées anthropologiques de Valbonne, juin 2003)*. Antibes: Éditions APDCA.

Buchet, L., Séguy, I., Suchecki, M., & Belaigues-Rossard, M. (2006c). Estimation de l'âge au décès des immatures par le degré de minéralisation. In L. Buchet, C. Dauphin, & I. Séguy (Eds.), *La paléodémographie. Mémoire d'os, mémoire d'hommes (Actes des 8es Journées anthropologiques de Valbonne, juin 2003)* (pp. 289–296). Antibes: Éditions APDCA.

Buck, C. E., Cavanagh, W. G., & Litton, C. D. (1996). *Bayesian approach to interpreting archaeological data*. Chichester: Wiley.

Budnik, A., Liczbińska, G., & Gumna, I. (2004). Demographic trends and biological status of historic populations from central Poland: The Ostrów Lednicki microregion. *American Journal of Physical Anthropology, 125*, 369–381.

Buikstra, J. E. (1997). Palaeodemography: Context and promise. In R. R. Paine (Ed.), *Integrating archaeological demography: Multidisciplinary approaches to prehistoric population* (Occasional papers, Vol. 24, pp. 367–380). Carbondale: Center for Archaeological Investigations.

Buikstra, J. E., & Konigsberg, L. W. (1985). Palaeodemography: Critiques and controversies. *American Journal of Physical Anthropology, 87*, 316–333.

Buikstra, J. E., & Ubelaker, D. H. (Eds.). (1994). *Standards for data collection from human skeletal remains: Proceedings of a seminar at the field museum of natural history* (Arkansas archaeological survey, research series, Vol. 44). Fayetteville: Arkansas Archaeological Society.

Buikstra, J. E., Konigsberg, L. W., & Bullington, J. (1986). Fertility and the development of agriculture in the prehistoric Midwest. *American Antiquity, 51*, 528–546.

Burnham, K. P., & Anderson, D. R. (1998). *Model selection and inference: A practical information-theoretic approach*. New York: Springer Verlag.

Burnouf, J., Bravard, J. P., & Chouquer, J. P. (Eds.). (1997). *La dynamique des paysages protohistoriques, antiques, médiévaux et modernes*. Antibes: APDCA.

Cadwell, J. C., & Cadwell, B. K. (2003). Pretransitional population control and equilibrium. *Population Studies, 57*, 199–215.

Campbell, K. L., & Wood, J. W. (1988). Fertility in traditional societies. In P. Diggory, M. Potts, & S. Teper (Eds.), *Natural human fertility: Social and biological determinants* (pp. 36–69). London: Macmillan.

Cardoso, H. F. V. (2006). Brief communication: The collection of identified human skeletons housed at the Bocage museum (National Museum of Natural History). Lisbon, Portugal. *American Journal of Physical Anthropology, 129*, 173–176.

Carrier, N. H. (1958). A note on the estimation of mortality and other population characteristics given deaths by age. *Population Studies, 12*, 149–163.

Carru, D., Boccacino, C., Borgard, P., Bouillot, J., Buchet, L., Buisson-Catil, J., & Vattéoni, S. (1991). Une nécropole de l'Antiquité tardive à Vaison-la-Romaine. Les fouilles des quartiers du Colombier et de Pommerol. *Documents d'archéologie vauclusienne, 2*, 96.

Caselli, G. (1995). Parcours et étapes des études sur la mortalité dans la dynamique de la recherche démographique. *Population, 6*, 1519–1544.

Castex, D. (1994). *Mortalité, morbidité et gestion de l'espace funéraire au cours du haut Moyen Âge. Contribution spécifique de l'anthropologie biologique*. Ph.D. thesis, Université de Bordeaux I, 2 vol.

Castex, D. (2005). Le cimetière de peste des Fédons: des données anthropologiques aux interprétations. Des particularités démographiques propres au site des Fédons. In B. Bizot, D. Castex, P. Reynaud, & M. Signoli (Eds.), *La saison d'une peste (avril-septembre 1590). Le cimetière des Fédons à Lambesc* (pp. 46–53). Paris: CNRS Éditions.

Castex, D. (2007). Les anomalies démographiques: clefs d'interprétation des cimetières d'épidémies en archéologie. In D. Castex & I. Cartron (Eds.), *Épidémies et crises de mortalité du passé. Actes des séminaires (année 2005) de la Maison des Sciences de l'Homme d'Aquitaine* (Études, Vol. 15, pp. 109–138). Bordeaux: Ausonius Éditions.

Castex, D., & Cartron, I. (Eds.). (2007). *Épidémies et crises de mortalité du passé. Actes des sé minaires (année 2005) de la Maison des Sciences de l'Homme d'Aquitaine* (Études, Vol. 15). Bordeaux: Ausonius Éditions.

Castex, D., & Drancourt, M. (2005). D'un gisement funéraire à la détection d'une crise épidémique, identité biologique et patrimoine génétique. In N. Gualde, J.-P. Goubert et S. Barry (Eds.), *Actes du colloque "Épidémies et sociétés dans le monde occidental, xive-xxie siècles", Revue sociologie et santé, les études hospitalières* (pp. 191–205).

Castex, D., & Reveillas, H. (2007). Investigation sur une crise de mortalité à Boulogne-sur-Mer (Pas-de-Calais, XVIIIe siècle). Hypothèses d'interprétation. *Bulletins et mémoires de la Société d'anthropologie de Paris, 19*(1–2, ns), 21–37.

Castex, D., Courtaud, P., & Hambucken-Bontempi, A. (1993). La détermination sexuelle des séries archéologiques: la validité de certains caractères osseux 'extra-coxaux'. *Bulletins et mé moires de la Société d'anthropologie de Paris, 5*(1–2), 225–235.

Castex, D., Sellier, P., Bruzek, J., Velemínský, P., Bessou, M., & Dobisíková, M. (2005). Saint-Benoît de Prague (République tchèque, XVIIe s.): nouvelles données démographiques pour l'interprétation d'un cimetière de peste. *Bulletins et mémoires de la Société d'anthropologie de Paris, 16* (3–4, ns), 227.

Caussinus, H. (1965). Contribution à l'analyse statistique des tableaux de corrélation. *Annales de la Faculté des Sciences de l'université de Toulouse, 29*, 77–182.

Caussinus, H., & Courgeau, D. (2010). Estimating age without measuring it: A new method in palaeodemography. *Population English edition, 65*(1), 117–144.

Chaillet, N. (2003). *Applications anthropologiques de l'approche bayésienne dans la détermination de l'âge des immatures*. Ph.D. thesis, Université de la Méditerranée-Aix-Marseille II, Faculté de médecine.

Chamberlain, A. C. (2000). Problems and prospects in palaeodemography. In M. Cox & S. Mays (Eds.), *Human osteology in archaeology and forensic science* (pp. 101–115). London: Greenwich Medical Media.

Chamberlain, A. C. (2001). Palaeodemography. In R. D. Brothwell & A. M. Pollard (Eds.), *Handbook of archaeological sciences* (pp. 259–268). Chichester: Wiley.

Chamberlain, A. (2006). *Demography in archaeology* (Manuals in archaeology). Cambridge: Cambridge University Press.

Charlet, G. (1984). Âge dentaire. Âge osseux". *Revue d'odonto-stomatologie, XIII*(1), 19–31.

Chesnais, J.-C. (1986). *La transition démographique: étapes, formes, implications économiques: é tude de séries temporelles (1720–1984) relatives à 67 pays* (Travaux et documents, Vol. 113). Paris: INED.

Chiang, C. L. (1978). *The life table and mortality analysis*. Geneva: WHO.

Chiang, C. L. (1984). *The life table and its applications*. Malabar: Krieger Publishing Company.

Chib, S., & Greenberg, E. (1998). Analysis of multivariate probit models. *Biometrika, 85*, 347–361.

Chikuni, S. (1975). Biological study on the population of the Pacific Ocean perch in the North Pacific. *Bulletin of the Far Seas Fisheries Research Laboratory, 12*, 1–19.

Cipriano-Bechtle, A., Grupe, G., & Schröter, P. (1996). Ageing and life expectancy in the early middle ages. *Homo, 46*(3), 267–279.

Coale, A. J. (1957). How the age distribution of a human population is determined. *Cold Spring Harbor Symposia on Quantitative Biology, 22*, 83–88.

Coale, A. J. (1972). *The growth and structure of human populations*. Princeton: Princeton University Press.

Coale, A. J., & Brass, W. (1968). Methods of analysis and estimation. In W. Brass, A. J. Coale, et al. (Eds.), *The demography of tropical Africa* (pp. 88–139). Princeton: Princeton University Press.

Coale, A. J., & Demeny, P. (1966). *Regional model life-tables and stable populations*. Princeton: Princeton University Press.

Coale, A. J., & Trussell, J. (1996). The development and use of demographic models. *Population Studies, 50*(3), 469–484.

Coale, A. J., & Watkins, S. C. (Eds.). (1986). *The decline of fertility in Europe*. Princeton: Princeton University Press.

Coale, A. J., Demeny, P., & Vaughan, P. (1983). *Regional model life-tables and stable populations*. New York: Academic.

Cocquerelle, S. (1993). *Étude paléodémographique des sépultures de Saint-Maclou, Val-d'Oise*. M.Sc. dissertation, Université Paris 1-Panthéon Sorbonne.

Cohen, M. N. (1977). *The food crisis in prehistory*. New Haven: Yale University Press.

Cohen, M. N. (1994). The osteological paradox reconsidered. *Current Anthropology, 35*(5), 629–631.

Cohen, M. N. (1997). Does palaeodemography measure community health? A rebuttal of 'The osteological paradox' and its implication for world history. In R. Paine (Ed.), *Integrating archaeological demography: Multidisciplinary approaches to prehistoric population* (Occasional papers, Vol. 24, pp. 242–260). Carbondale: Center for Archaeological Investigations.

Cohen, M. N., & Armelagos, G. J. (Eds.). (1984). *Palaeopathology at the origins of agriculture*. New York: Academic.

Condé, J., Fleury-Brousse, M., & Waltisperger, D. (1980). *La mortalité dans les pays en dé veloppement* (4th ed.). Paris: OECD.

Coppa, A., Cucina, A., Chiarelli, B., Luna Calderon, F., & Mancinelli, D. (1995). Dental anthropology and palaeodemography of the precolumbian populations of hispaniola from the third millennium BC to the Spanish conquest. *Human Evolution, 10*, 153–167.

Corruccini, R. S., Brandon, E. M., & Handler, J. S. (1989). Inferring fertility from relative mortality in historically controlled cemetery remains from Barbados. *American Antiquity, 54*, 609–614.

Corvisier, J.-N. (2001). L'état présent de la démographie historique antique: tentative de bilan. *Annales de démographie historique, 2*, 101–140.

Corvisier, J.-N., & Bettancourt-Valdher, M. (Eds.). (1999). *La démographie historique antique (Actes du premier colloque international de démographie historique, Arras, 22–23 nov. 1996)*, Arras, Artois presses université, coll. "Cahiers scientifiques de l'université d'Artois".

Corvisier, J.-N., & Suder, W. (1996). *Polyanthropia-Oliganthropia: bibliographie de la dé mographie du monde grec*. Wrocław/Paris: Arboretum/De Boccard.

Courgeau, D. (1988). *Méthodes de mesure de la mobilité spatiale* (Les Manuels). Paris: INED.

Courgeau, D., & Lelièvre, E. (1989). *Analyse démographique des biographies* (Les Manuels). Paris: INED.

Courgeau, D., & Lelièvre, E. (1992). *Event history analysis in demography*. Oxford: Oxford University Press.

Cox, R. D. (1972). Regression models and life tables. *Journal of the Royal Statistical Society, Serie B, 34*, 187–202.

Cribari-Neto, F., & Zarkos, S. G. (1999). Bootstrap methods for heteroskedastic regression models: Evidence on estimation and testing. *Econometric Reviews, 18*(2), 211–228.

Crubézy, E., Duchesne, S., & Arlaud, C. (Eds.). (2006). *La mort, les morts et la ville: Saints-Côme-et-Damien, Montpellier, x^e- xvie siècles*. Paris: Errance.

Danion, B. (1991). *Abbaye cistercienne de Maubuisson (Saint-Ouen-l'Aumône, Val-d'Oise). Étude anthropologique des sépultures de la galerie ouest du cloître*. M.Sc. dissertation, Université de Paris 1 Panthéon-Sorbonne.

Danion, B., Dumont, C., & Langlois, J.-Y. (1994). Les moniales cisterciennes de l'abbaye de Maubuisson (Val-d'Oise), à travers les différents lieux d'inhumations. In L. Buchet (Ed.), *La femme pendant le Moyen Âge et l'Époque moderne (Actes des 6es Journées anthropologiques de Valbonne, juin 1992)* (Dossiers de documentation archéologique, Vol. 17, pp. 13–29). Paris: CNRS Éditions.

Davis, K. (1945). The world demographic transition. *Annals of the American Academy of Political and Social Science, 237*, 1–11.

de Jonge Cohen, T. E. (1920). *Mühlreiter's Anatomie des menschlichen Gebisses*. Leipzig: Arthur Felix Verlag.

de La Rochebrochard, É. (2000). Age at puberty of girls and boys in France. Measurements from a survey on adolescent sexuality. *Population An English Selection, 12*, 51–60.

de Vito, C., & Saunders, S. R. (1990). A discriminant function analysis of deciduous teeth to determine sex. *Journal of Forensic Sciences, 35*, 845–858.

Deming, W. E., & Stephan, F. F. (1940). On a least squares adjustment of a sampled frequency table when the expected marginal totals are known. *The Annals of Mathematical Statistics, 11*, 427–444.

Demirjian, A., Goldstein, H., & Tanner, J. M. (1973). A new system of dental age assessment. *Human Biology, 45*, 211–227.

Dempster, A. P., Laird, N. M., & Rubin, D. B. (1977). Maximum likelihood from incomplete data via the EM algorithm. *Journal of the Royal Statistical Society. Series B (Methodological), 39*(1), 1–38.

Devroey, J.-P. (1981). Méthodes d'analyse démographique des polyptyques du Haut Moyen Âge. *Acta Historica Bruxellensia, IV*, 71–88.

Devroey, J.-P. (1984). *Le polyptyque et la liste des cens de l'abbaye de Saint-Remi de Reims (IXe-XIe siècles)*. Reims: Académie nationale de Reims.

Donat, P., & Ullrich, H. (1971). Einwohnerzahlen und Siedlungsgröße der Merowingerzeit. *Zeitschrift für Archäologie, 5*, 234–265.

Drancourt, M., Aboudharam, G., Signoli, M., Dutour, O., & Raoult, D. (1998). Detection of 400-year-old Yersinia pestis DNA in human dental pulp: An approach to the diagnosis of ancient septicemia. *Proceedings of the National Academy of Sciences, 95*, 12637–12640.

Drancourt, M., Signoli, M., Dang, L. V., Bizot, B., Roux, V., Tzortzis, S., & Raoult, D. (2007). Yersinia pestis orientalis in remains of ancient plague patients. *Emerging Infectious Diseases, 13*, 332–333.

Drusini, A. G., Carrara, N., Orefici, G., & Rippa Bonat, M. (2001). Palaeodemography of the Nasca valley: Reconstruction of the human ecology in the southern Peruvian coast. *Homo, 52* (2), 157–172.

Duchêne, J. (1999). Les tables types de mortalité. In *Démographie: analyse et synthèse* (Actes du Séminaire international de Roma-Napoli-Pozzuoli, 26–29 mai 1999), vol. 2, pp. 155–169.

Duchêne, J. (2006). Les tables-types de mortalité. In C. Graziella, J. Vallin, & G. Wunsch (Eds.), *Démographie: analyse et synthèse, Vol. VIII: observation, méthodes auxiliaires, enseignement et recherche* (Les Manuels, pp. 267–286). Paris: INED.

Duday, H. (1978). Archéologie funéraire et anthropologie. Application des relevés et de l'étude ostéologique à l'interprétation de quelques sépultures pré- et protohistoriques du Midi de la France. *Cahiers d'anthropologie, 1*, 55–101.

Duday, H. (1990). Observations ostéologiques et décomposition du cadavre: sépultures colmatées ou en espace vide. *Revue archéologique du Centre de la France, 29*(2), 193–196.

Duday, H. (1997). Mais, au fait, quelle anthropologie? *Les nouvelles de l'archéologie, 66*, 54–56.

Duday, H. (2005). L'archéothanatologie ou l'archéologie de la mort. In O. Dutour, J.-J. Hublin, & B. Vandermeersch (Eds.), *Objets et méthodes en paléoanthropologie* (Orientations et méthodes, Vol. 7, pp. 153–217). Paris: Éditions du Comité des travaux historiques et scientifiques.

Duday, H., & Masset, C. (1987). *Anthropologie physique et archéologie: méthodes d'étude des sépultures (actes du Colloque de Toulouse, 4, 5 et 6 novembre 1982)*. Paris: CNRS.

Duday, H., & Monzeglio, E. (2006). *Lezioni di archeotanatologia. Archeologia funeraria e antropologia di campo*. Roma: Soprintendenza archeologica di Roma, École française de Rome.

Dumont, C. (1989). *Étude anthropologique de la salle capitulaire de l'abbaye de Maubuisson (Saint-Ouen-l'Aumône, Val-d'Oise)*. M.Sc. dissertation, Université de Paris 1 Panthéon-Sorbonne.

Dupâquier, J. (1972). "De l'animal à l'homme: le mécanisme autorégulateur des populations traditionnelles". *Revue de l'Institut de sociologie, 2*, 177–211.

Dupâquier, J. (1977). Les tables de mortalité de Wargentin. *Annales de démographie historique*, 385–415. Paris: La Haye, Mouton & Cie.

Dupâquier, J. (1996). *L'invention de la table de mortalité: de Graunt à Wargentin 1662–1766*. Paris: Puf.

Durand-Daste, F., Favory, F., Fiches, J.-L., Mathian, H., Pumain, D., Raynaud, C., Sanders, L., & van der Leeuw, S. (1998). *Des oppida aux métropoles. Archéologues et géographes en vallée du Rhône*. Paris: Anthropos.

Dutour, O., Hublin, J.-J., & Vandermeersch, B. (Eds.). (2005a). *Objets et méthodes en palé oanthropologie* (Objets et méthodes, Vol. 7). Paris: Éditions du Comité des Travaux historiques et scientifiques.

Dutour, O., Hublin, J.-J., & Vandermeersch, B. (Eds.). (2005b). *Origine et évolution des populations humaines* (Objets et méthodes, Vol. 8). Paris: Éditions du Comité des Travaux historiques et scientifiques.

Edwards, A. W. F. (1972). *Likelihood*. London: Cambridge University Press.

Eliason, S. R. (1993). *Maximum likelihood estimation: Logic and practice*. Newbury Park: Sage.

Eshed, V., Gopher, A., Gage, T. B., & Hershkovitz, I. (2004). Has the transition to agriculture reshaped the demographic structure of prehistoric populations? New evidence from the Levant. *American Journal of Physical Anthropology, 124*(4), 315–329.

Espenshade, T. J., Bouvier, L. F., & Arthur, W. B. (1982). Immigration and the stable population model. *Demography, 19*(1), 125–133.

Ewbank, D., Gómez de León, J., & Stoto, M. (1983). A reducible four-parameter system of model life table. *Population Studies, 37*, 105–127.

Faerman, M., Filon, D., Kahila, G., Greenblatt, C. L., Smith, P., Oppenheim, A. (1995). Sex identification of archaeological remains based on amplification of the X and Y amelogenin alleles. *Gene*, 327–332.

Fairgrieve, S. I., & Oost, T. S. (1995). On a test of the multifactorial aging method by Bedford et al., 1993. *American Journal of Physical Anthropology, 97*, 83–85.

Falys, C. G., Schutkowski, H., & Weston, D. A. (2006). Auricular surface aging: Worse than expected? A test of the revised method on a documented historic skeletal assemblage. *American Journal of Physical Anthropology, 130*(4), 508–513.

Farley, J., & Basson, M. (2005). *Developing age-length keys for the Australian STB surface fishery based on direct age estimation using otoliths.* CSIRO Division of Marine and Atmospheric research, Australian Fisheries Management Authority.

Fattorini, P., Caccio, S., Gustinch, S., Florian, F., Altamura, B. M., & Graziosi, G. (1993). Sex identification by polymerase chain reaction of α-satellite in aged tissue samples. *Electrophoresis, 14*, 23–26.

Fazekas, I. G., & Kósa, F. (1978). *Forensic fetal osteology*. Budapest: Akadémiai Kiadó.

Ferembach, D., Schwidetsky, I., & Stloukal, M. (1979). Recommandations pour déterminer l'âge et le sexe sur le squelette. *Bulletins et mémoires de la Société d'anthropologie de Paris*, XIII° Série, *6*(1), 7–45.

Fienberg, S. E. (1968). The geometry of an *r* x *c* contingency table. *The Annals of Mathematical Statistics, 39*, 1186–1190.

Fienberg, S. E. (1970). An iterative procedure for estimation in contingency tables. *The Annals of Mathematical Statistics, 41*, 907–917.

Fitzgerald, C. M., & Rose, J. C. (2008). Reading between the lines: Dental developmental and subadult age assessment using the microstructural growth markers of teeth. In M. A. Katzenberg & S. R. Saunders (Eds.), *Biological anthropology of the human skeleton* (pp. 237–263). New York: Wiley.

Forfar, D. (2006). Five parameter approach to mortality in ancient times. In L. Buchet, C. Dauphin, & I. Séguy (Eds.), *La paléodémographie. Mémoire d'os, mémoire d'hommes (Actes des 8es Journées anthropologiques de Valbonne, juin 2003)* (pp. 297–302). Antibes: Éditions APDCA.

Fossier, R. (Ed.). (1995). *Population et démographie au Moyen Âge (Actes du 118e congrès national des sociétés savantes)*. Paris: CTHS.

Foti, B., Adalian, P., Signoli, M., Ardagna, Y., Dutour, O., & Leonetti, G. (2001). Limits of the lamendin method in age determination. *Forensic Science International, 122*(2–3), 101–106.

Foti, B., Adalian, P., Lalys, L., Chaillet, N., Leonetti, G., & Dutour, O. (2003a). Approche probabiliste de l'estimation de l'âge chez l'enfant à partir de la maturation dentaire. *Comptes rendus biologies, 326*, 441–448.

Foti, B., Lalys, L., Adalian, P., Giustiniani, J., Maczel, M., Signoli, M., Dutour, O., & Leonetti, G. (2003b). New forensic approach to age determination in children based on tooth eruption. *Forensic Science International, 132*, 49–56.

Frederic, J. (1906). Untersuchungen über die normale Obliteration der Schädelnähte. *Zeitschrift für Morphologie und Anthropologie, 9*, 373–456.

Freter, A. C. (1997). The question of time: the impact of chronology on Copán prehistoric settlement demography. In R. Paine (Ed.), *Integrating archaeological demography: Multidisciplinary approaches to prehistoric population* (Occasional papers, Vol. 24, pp. 21–42). Carbondale: Center for Archaeological Investigations.

Friðriksson, A. (1934). On the calculation of age distribution within a stock of cods by means of relatively few age-determinations as a key to measurements on a large scale. *Rapports et procès-verbaux des réunions du Conseil permanent international pour l'exploration des mers, 86*, 1–14.

Friedlander, D. (1961). A technique for estimating a contingency table, given the marginal totals and some supplementary data. *Journal of the Royal Statistical Society Series A, 124*, 412–420.

Frucht, S., Schnegelsberg, C., Schulte-Mönting, J., Rose, E., & Jonas, I. (2000). Dental age in Southwest Germany. *Journal of Orofacial Orthopedics, 61*(5), 318–329.

Gabard, N., Colard, T., Bartoli, C., Blondiaux, J., Leonetti, G., & Foti, B. (2007). Intérêt médico-légal et paléopathologique du cément/forensic and palaeopathological interest of cementum. *Journal de médecine légale, droit médical, 50*(4), 181–190.

Gabriel, K. R., & Ronan, I. (1958). Estimates of mortality from infant mortality rates. *Population Studies, 12*, 64–169.

Gage, T. B. (1985). Demographic estimation from anthropological data: New methods. *Current Anthropology, 26*, 644–647.

Gage, T. B. (1988). Mathematical hazard models of mortality: An alternative to model life tables. *American Journal of Physical Anthropology, 76*, 429–441.

Gage, T. B. (1989). Biomathematical approaches to the study of human variation in mortality. In *Yearbook of Physical Anthropology,* suppl. 10 to the American Journal of Physical Anthropology, pp. 185–214.

Gage, T. B. (1990). Variation and classification of human age patterns of mortality: Analysis using competing hazards models. *Human Biology, 62*(5), 589–617.

Gage, T. B. (1998). The comparative demography of primates: With some comments on the evolution of life histories. *Annual Reviews in Anthropology, 27*, 197–221.

Gage, T. B., & Dyke, B. (1986). Parameterizing abridged mortality tables: The siler three-component hazard model. *Human Biology, 58*, 275–291.

Gage, T. B., & Leslie, P. (1989). Demography and human population biology: Problems and progress. In M. Little & J. D. Haas (Eds.), *Human biology: A transdisciplinary science* (pp. 15–44). Oxford: Oxford University Press.

Gage, T. B., & Mode, C. J. (1993). Some laws of mortality: How well do they fit? *Human Biology, 65*, 445–461.

Gallien, V. (1992). *Deux populations du haut Moyen Âge à Saint-Denis. Histoire, archéologie et anthropologie.* Ph.D. thesis, Université de Paris-Sorbonne Paris IV.

Gallien, V. (2004). Évolution et développement d'un cimetière urbain normand: le cimetière Saint- Jean à Rouen du ixe au xvie siècle. In C. Balmelle, P. Chevalier, & G. Ripoll (Eds.), *Mélanges d'Antiquité Tardive, Studiola in honorem Noël Duval* (Bibliothèque de l'Antiquité tardive, Vol. 5, pp. 199–208). Turnhout: Association pour l'Antiquité Tardive, Brepols Publishers.

Garland, A. N. (1987). A histological study of archaeological bone decomposition. In G. Grupe & A. N. Garland (Eds.), *Histology of ancient human bone: Methods and diagnosis* (pp. 109–126). Berlin: Springer.

Garn, S. M., & Rohmann, C. G. (1966). Interaction of nutrition and genetics in the timing of growth and development. *Pediatric Clinics of North America, 13*, 353–379.

Gemmerich, I. (1999). *Création d'une collection anthropologique de référence et application des caractères discrets dans le cas de généalogies connues*. Ph.D. thesis, Université de Genève.

Gere, C. (1999). Bones that matter: Sex determination in palaeodemography 1948–1995. *Studies in History and Philosophy of Science Part C: Biological and Biomedical Sciences, 30*(4), 455–471.

Gilbert, B. M., & McKern, T. W. (1973). A method for aging the female os pubis. *American Journal of Physical Anthropology, 38*, 31–38.

Giles, E. (1964). Sex determination by discriminant function analysis of the mandible. *American Journal of Physical Anthropology, 22*, 129–136.

Giles, E., & Elliot, O. (1963). Determination by discriminant function analysis of crania. *American Journal of Physical Anthropology, 21*, 53–68.

Gill, P., Ivanov, P. L., Kimpton, C., et al. (1994). Identification of the remains of the Romanov family by DNA analysis. *Nature Genetics, 6*(2), 130–135.

Giovannini, F. (2002). Archeologia e demografia. *Popolazione e Storia, 2*, 63–81.

Gloor, P.-A., & Chamla, M.-C. (1986). Variations diachroniques depuis trois siècles. Données et facteurs responsables. In D. Ferembach, C. Suzanne, & M.-C. Chamla (Eds.), *L'Homme, son é volution, sa diversité. Manuel d'anthropologie physique* (pp. 463–490). Paris: CNRS Éditions.

Gompertz, B. (1825). On the nature of the function expressive of the law of human mortality, and on a new mode of determining the value of life contingencies. *Philosophical Transactions of the Royal Society of London, 115*, 513–583.

Goodman, A. H. (1993). On the interpretation of health from skeletal remains. *Current Anthropology, 34*(3), 281–288.

Goodman, A. H., & Martin, D. L. (2002). Reconstructing health profiles from skeletal remains. In R. H. Steckel & J. C. Rose (Eds.), *The backbone of history. Health and nutrition in the western world* (pp. 11–60). Cambridge: Cambridge University Press.

Goudineau, C. (1980). Y a-t-il une ville protohistorique? La Gaule méridionale. In G. Duby (Ed.), *Histoire de la France urbaine* (pp. 143–193). Paris: Seuil.

Gowland, R. L., & Chamberlain, A. T. (2002). A Bayesian approach to ageing perinatal skeletal material from archaeological sites: Implications for the evidence for infanticide in Roman-Britain. *Journal of Archaeological Science, 29*, 677–685.

Gowland, R. L., & Chamberlain, A. T. (2005). Detecting plague: Palaeodemographic characterization of a catastrophic death assemblage. *Antiquity, 79*, 146–157.

Grande, R. M., & Garralda, M. D. (1983). Algunos problemas de la palaeodemografía en España. *Homenaje al prof. Martín Almagro Basch, 1*, 417–429.

Graunt, J. (1662). *Natural and political observations made upon the bills of mortality*. London.

Greene, D. L., Van Gerven, D. P., & Armelagos, G. J. (1986). Life and death in ancient populations: Bones of contention in palaeodemography. *Human Evolution, 1*(3), 193–207.

Grøn, A. M. (1962). Prediction of tooth emergence. *Journal of Dental Research, 41*, 573–585.

Guillon, M. (1998). *Anthropologie de terrain et paléodémographie: études méthodologiques sur les grands ensembles funéraires. Applications au cimetière médiéval de Tournedos-Portejoie (Eure)*. Anthropology thesis, Université de Bordeaux I, 3 vol.

Gustafson, G. (1947). Microscopic examination of teeth as a means of identification in forensic medicine. *Journal of American Dental Association, 35*(10), 720–724.

Gustafson, G. (1950). Age determination of teeth. *Journal of American Dental Association, 41*, 45–54.

Gustafson, G. (trans: Brabant H., & Smet R. M. de) (1969). *Odonto-stomatologie médico-légale*. Brussels, SC Ed.

Guy, H. (1995). Principes méthodologiques appliqués à la paléodémographie d'un cimetière du haut Moyen Âge. *Les Nouvelles de l'Archéologie, 59*, 39–45.

Guy, H. (2003). Contrôle d'un estimateur en paléodémographie à partir des sources historiques (registres paroissiaux de Saint-Maclou de Pontoise, Val-d'Oise), in edited volume, Sens dessus dessous- La recherche du sens en Préhistoire. Recueil d'études offert à Jean Leclerc et Claude Masset. *Revue archéologique de Picardie, 21* (ns), 113–116.

Guy, H., & Masset, C. (1997). Particularités taphonomiques des os d'enfants. In L. Buchet (Ed.), *L'enfant, son corps, son histoire (Actes des 7es Journées anthropologiques de Valbonne, juin 1994)* (pp. 35–43). Antibes: APDCA.

Guy, H., Masset, C., & Baud, C.-A. (1997). Infant taphonomy. *International Journal of Osteoarchaeology, 7*, 221–229.

Halcrow, S. E., Tayles, N., & Buckley, H. R. (2007). Age estimation of children from prehistoric Southeast Asia: Are the dental formation methods used appropriate? *Journal of Archaeological Science, 34*(7), 1158–1168.

Halley, E. (1693). An estimate of the degrees of the mortality of mankind. *Philosophical Transactions, 17*(196), 579 sq.

Hänni, C. (1994). Utilisation de l'ADN ancien en anthropologie. *Bulletins et Mémoires de la Socié té d'Anthropologie de Paris, 1–2*, 5–28.

Haraven, T. K. (2000). Le jeu de la norme et de la subjectivité dans les étapes de la vie. La construction sociale et culturelle des parcours de vie et son évolution au cours du vingtièm^e^ siècle. In A. Bideau, P. Bourdelais, & J. Légaré (Eds.), *De l'usage des seuils. Structures par âge et âges de la vie* (Cahier des Annales de Démographie historique, Vol. 2, pp. 271–287). Paris: Société de Démographie historique.

Harpending, H. C., & Pennington, R. (1991). Age structure and sex-biased mortality among Herero pastoralists. *Human Biology, 63*(3), 329–353.

Harris, E. (1998). Dental Maturation. In S. J. Ulijaszek, F. E. Johnston, & M. A. Preece (Eds.), *Human growth and development* (pp. 45–48). Cambridge: Cambridge University Press.

Hassan, F. (1975). Determination of size, density and growth rate of hunting-gathering people. In S. Polgar (Ed.), *Population ecology and social evolution* (pp. 27–52). La Hague: Mouton.

Hassan, F. (1981). *Demographic archaeology* (Studies in archaeology). New York: Academic.

Hasselblad, V. (1966). Estimation of parameters for a mixture of normal distribution. *Technometrics, 8*(3), 431–444.

Helm, S. (1969). Secular trends in tooth eruption: A comparative study of Danish school children of 1913 and 1965. *Archives of Oral Biology, 14*, 1177–1191.

Helmuth, H. (1999). The maximum lifespan potential of Hominidae: A re-evaluation. *Homo, 50* (3), 283–296.

Henneberg, M. (1977). Proportion of dying children in palaeodemographic studies: Estimation by guess or by methodological approach. *Przegląd antropologiczny, 43*, 105–114.

Henneberg, M., & Steyn, M. (1994). Preliminary report on the palaeodemography of the K2 and Mapungubwe Populations (South Africa). *Human Biology, 66*(1), 105–120.

Henry, L. (1954). Commentaire sur l'article de L.-R. Nougier: "Essai sur le peuplement préhistorique de la France. *Population, 2*, 272–274.

Henry, L. (1957). La mortalité d'après les inscriptions funéraires. *Population, 12*(1), 149–152.

Henry, L. (1959). L'âge au décès d'après les inscriptions funéraires. *Population, 14*(2), 327–329.

Henry, L., & Blum, A. (1988). *Techniques d'analyse en démographie historique* (Manuels). Paris: INED.

Herrscher, E. (2003). Alimentation d'une population historique: analyse des données isotopiques de la nécropole Saint Laurent de Grenoble (XIIIe- XVe siècle, France). *Bulletins et mémoires de la Société d'anthropologie de Paris, 15*(3–4), 145–320.

Heuze, Y. (2004). *Chronologie et étiologie de la maturation macrostructurale des dents dé finitives*. Ph.D. thesis, Université Bordeaux I.

Hilmi, B. (2005). *Détermination du sexe à partir des mesures crâniennes*. M.Sc. thesis, Université de Nice.

Holden, M. J., & Raitt, D. F. S. (1974). *Manual of fisheries science, part 2: Methods of resource investigation and their application*. Rome: FAO. Tech. Pap, *115*, 1–214.

Holman, D. J. (2000). MLE: *A Programming Language for Building Likelihood Models*. Version 2 (Software and manual). http://faculty.washington.edu/djholman/mle/index.html

Holman, D. J., & Jones, R. E. (1998). Longitudinal analysis of deciduous tooth emergence II: Parametric survival analysis in Bangladeshi, Guatemalan, Japanese and Javanese children. *American Journal of Physical Anthropology, 105*, 209–230.

Holman, D. J., Wood, J. W., & O'Connor, K. A. (2002). Estimating age-at-death distributions from skeletal samples: A multivariate latent-trait approach. In R. D. Hoppa & J. W. Vaupel (Eds.), *Palaeodemography, age distribution from skeletal samples* (pp. 193–221). Cambridge: Cambridge University Press.

Hooton, E. A. (1930). *Indians of pecos pueblo. A study of their skeletal remains*. New Haven: Yale University Press.

Hoppa, R. D. (1996). Representativeness and bias in cemetery samples: Implications for palaeodemographic reconstructions of past populations. Unpublished Ph.D. dissertation, Department of Anthropology, McMaster University.

Hoppa, R. D. (1999). Modeling the effects of selection bias on palaeodemographic analyses. *Homo, 50*(3), 228–243.

Hoppa, R. D. (2000a). What to do with long bones: Toward a progressive palaeoauxology. *International Journal of the Science of Man, XXXVIII*(1), 23–32.

Hoppa, R. D. (2000b). Population variation in osteological aging criteria: An example from the pubic symphysis. *American Journal of Physical Anthropology, 111*, 185–191.

Hoppa, R. D. (2001). The once and future palaeodemography. In L. Sawchuck & S. Pfeiffer (Eds.), *Out of the past: The history of human osteology at the university of Toronto*. CITD Press, University of Toronto at Scarborough. http://citdpress.utsc.utoronto.ca/osteology/pfeiffer.html

Hoppa, R. D. (2002). Palaeodemography: Looking back and thinking ahead. In R. D. Hoppa & J. W. Vaupel (Eds.), *Palaeodemography: Age distributions from skeletal samples* (pp. 9–28). Cambridge: Cambridge University Press.

Hoppa, R. D., & Fitzerald, C. M. (1999). *Human growth in the past: Studies from bones and teeth* (Cambridge studies in biological and evolutionary anthropology, Vol. 25). Cambridge: Cambridge University Press.

Hoppa, R. D., & Saunders, S. R. (1998). The mad legacy: How meaningful is mean age-at-death in skeletal samples. *Human Evolution, 13*, 1–14.

Hoppa, R. D., & Vaupel, J. W. (Eds.). (2002a). *Palaeodemography. Age distribution from skeletal samples*. Cambridge: Cambridge University Press.

Hoppa, R. D., & Vaupel, J. W. (2002b). The Rostock manifesto for palaeodemography: The way from stage to age. In R. D. Hoppa & J. W. Vaupel (Eds.), *Palaeodemography: Age distributions from skeletal samples* (pp. 1–8). Cambridge: Cambridge University Press.

Horowitz, S., Armelagos, G., & Watcher, K. (1988). On generating birth rates from skeletal populations. *American Journal of Physical Anthropology, 76*, 189–196.

Houdaille, J. (1971). La mortalité des Carmélites en France de 1736 à 1792. *Population, 4*, 745–748.

Hougaard, P. (1986). A class of multivariate failure time distributions. *Biometrika, 73*, 671–678.

Howell, N. (1973). The feasability of demographic studies in "anthropological" populations. In M. Crawford & P. Workman (Eds.), *Methods and theories of anthropological genetics* (pp. 249–262). Albuquerque: University of New Mexico Press.

Howell, N. (1976). Toward a uniformitarian theory of human palaeodemography. *Journal of Human Evolution, 5*, 25–40.

Howell, N. (1979). *Demography of the Dobe !Kung*. New York: Academic.

Howell, N. (1982). Village composition implied by palaeodemographic life table: The Libben Site. *American Journal of Physical Anthropology, 59*, 263–269.

Howell, N., & Lehotay, V. (1978). Ambush: A computer program for stochastic microsimulation of small human populations. *American Anthropology, 80*, 905–922.

Hummel, S., Bramanti, B., Finke, T., & Herrmann, B. (2000). Evaluation of morphological sex determinations by molecular analyses. *Anthropologischer Anzeiger, 58*(1), 9–13.

Hunt, D. R. (1990). Sex determination in the subadult ilia: an indirect test of weaver's nonmetric sexing method. *Journal of Forensic Science, 35*(4), 881–885.

Hurme, V. O. (1948). Standards of variation in the eruption of the first six permanent teeth. *Child Development, 19*(4), 213–231.

Hurme, V. O. (1949). Ranges of normalcy in the eruption of permanent teeth. *Journal of Dentistry for Children, 16*, 11–15.

İşcan, M. Y. (1989). *Age markers in human skeletons*. Springfield: Charles C. Thomas.

İşcan, M. Y., & Kennedy, K. A. R. (Eds.). (1989). *Reconstruction of life from the skeleton*. New York: Alan R. Liss.

İşcan, M. Y., Loth, S. R., & Wright, R. K. (1984). Age estimation from the rib by phase analysis: White males. *Journal of Forensic Science, 29*, 1094–1104.

İşcan, M. Y., Loth, S. R., & Wright, R. K. (1985). Age estimation from the rib by phase analysis: White females. *Journal of Forensic Science, 30*, 853–863.

İşcan, M. Y., Loth, S. R., & Scheuerman, E. H. (1992). Age assessment from the sternal end of the rib and pubic symphysis: A systematic comparison. *Anthropology (Brno), 30*, 41–44.

Jackes, M. K. (1985). Pubic symphysis age distributions. *American Journal of Physical Anthropology, 68*, 281–299.

Jackes, M. K. (1986). The mortality of Ontario archaeological populations. In S. Pfeiffer (Ed.), *The skeletal biology of Ontario populations*. Canadian Journal of Anthropology/Revue canadienne d'anthropologie physique, 5, 33–48.

Jackes, M. K. (1988). Demographic change at Mesolithic-Neolithic transition: Evidence from Portugal. *Rivista di Antropologia,* suppl. del, *LXVI*, 141–158.
Jackes, M. K. (1992). Palaeodemography: Problems and techniques. In S. R. Saunders & M. A. Katzenberg (Eds.), *Skeletal biology of past people: Research methods* (pp. 189–224). New York: Wiley-Liss.
Jackes, M. K. (1993). On paradox and osteology. *Current Anthropology, 34*(4), 434–439.
Jackes, M. K. (2000). Building the basis for palaeodemographic analysis: Adult age determination. In M. A. Katzenberg & S. R. Saunders (Eds.), *Biological anthropology of the human skeleton* (pp. 407–456). New York: Wiley-Liss.
Jackes, M. K., & Meiklejohn, C. (2008). The palaeodemography of Central Portugal and the Mesolithic- Neolithic transition. In J.-P. Bocquet-Appel (Ed.), *Recent advances in palaeodemography: Data, techniques, patterns* (pp. 209–258). New York: Springer Verlag.
Jackes, M. K., Lubell, D., & Meiklejohn, C. (1997). Healthy but mortal: Human biology and the first farmers of Western Europe. *Antiquity, 71*, 639–685.
Johanson, G. (1971). Age determination from human teeth. *Odontologisk Revy, 22*(suppl. 2), 21.
Johansson, S. R., & Horowitz, S. (1986). Estimating mortality in skeletal populations: Influence of the growth rate on the interpretation of levels and trends during the transition to agriculture. *American Journal of Physical Anthropology, 71*, 233–250.
Johnston, F. (1962). Growth of the long bones of infants and young children at Indian Knoll. *American Journal of Physical Anthropology, 20*, 225–264.
Johnston, F. E., & Zimmer, L. O. (1989). Assessment of growth and age in the immature skeleton. In M. Y. İşcan & K. A. R. Kennedy (Eds.), *Reconstruction of life from the skeleton* (pp. 11–21). New York: Alan R. Liss.
Jousset, N., Franco, A., Gard, C., Penneau, M., & Rouge-Maillart, C. (2006). Détermination de l'âge des adultes en post mortem: intérêt de l'utilisation des critères de Gustafson. *Antropo, 11*, 271–277.
Kagerer, P., & Grupe, G. (2001a). On the validity of individual age-at-death diagnosis by incremental line counts in human dental cementum. Technical considerations. *Anthropologischer Anzeiger, 59*, 331–342.
Kagerer, P., & Grupe, G. (2001b). Age-at-death diagnosis and determination of life history parameters by incremental lines in human dental cementum as an identification aid. *Forensic Science International, 118*, 75–82.
Kannisto, V. (2001). Mode and dispersion of the length of life. *Population, an English Selection, 13*(1), 159–171.
Katz, D., & Suchey, J. M. (1986). Age determination of the male pubis symphyseal aging patterns in the male. *American Journal of Physical Anthropology, 69*, 427–435.
Kemkes-Grottenthaler, A. (2002). Aging through the ages: Historical perspectives on age indicator methods. In R. D. Hoppa & J. W. Vaupel (Eds.), *Palaeodemography: Age distributions from skeletal samples* (pp. 48–72). Cambridge: Cambridge University Press.
Kerley, E. R. (1965). The microscopic determination of age in human bone. *American Journal of Physical Anthropology, 23*, 149–164.
Kerley, E. R. (1969). Age determination of bone fragments. *Journal of Forensic Sciences, 14*(1), 59–67.
Kerley, E. R. (1970). Estimation of skeletal age: After about age 30. In T. D. Stewart (Ed.), *Personal identification* (pp. 57–70). Washington: Smithsonian Institution.
Kerley, E. R., & Ubelaker, D. H. (1978). Revisions in the microscopic method of estimating age at death in human cortical bone. *American Journal of Physical Anthropology, 49*(4), 545–546.
Keyser-Tracqui, C., Ricaut, F., Crubézy, E., & Ludes, B. (2002). Populations anciennes et ADN ancien: état actuel de la question. *Antropo, 2*, 1–8.
Kimura, D. K. (1977). Statistical assessment of the age-length key. *Journal of the Fisheries Research Board of Canada, 34*, 317–324.
Kimura, D. K., & Chikuni, S. (1987). Mixture of empirical distributions: An iterative application of the age-length key. *Biometrics, 43*, 23–35.
Klein, J. P., Pelz, C., & Zhang, M. (1999). Modeling random effects for censored data by a multivariate normal regression model. *Biometrics, 55*, 497–506.

Knudson, K. J., & Stojanowski, C. M. (2008). New directions in bioarchaeology: Recent contributions to the study of human social identities. *Journal of Archaeological Research, 16*(4), 397–432.

Komlos, J., & Nefedov, S. (2002). A compact macromodel of pre-industrial population growth. *Historical Methods, 35*(2), 92–94.

Konigsberg, L. W., & Frankenberg, S. R. (1992). Estimation of age structure in anthropological demography. *American Journal of Physical Anthropology, 89*, 235–256.

Konigsberg, L. W., & Frankenberg, S. R. (1994). Palaeodemography: 'Not quite dead'. *Evolutionary Anthropology, 3*, 95–105.

Konigsberg, L. W., & Frankenberg, S. R. (2002). Deconstructing death in palaeodemography. *American Journal of Physical Anthropology, 117*, 297–309.

Konigsberg, L. W., & Frankenberg, S. R. (2003). Reply to comments by Jackes: Inter-observer error and goodness-of-fit tests in palaeodemography. *American Journal of Physical Anthropology, 121*, 387–388.

Konigsberg, L. W., & Herrmann, N. P. (2002). Markov Chain Monte Carlo estimation of hazard model parameters in palaeodemography. In R. D. Hoppa & J. W. Vaupel (Eds.), *Palaeodemography. Age distribution from skeletal samples* (pp. 222–242). Cambridge: Cambridge University Press.

Konigsberg, L. W., & Holman, D. J. (1999). Estimation of age at death from dental emergence and implications for studies of prehistoric somatic growth. In R. D. Hoppa & C. M. Fitzgerald (Eds.), *Human growth in the past: Studies from bones and teeth* (pp. 264–289). Cambridge: Cambridge University Press.

Konigsberg, L. W., Buikstra, J. E., & Bullington, J. (1989). Palaeodemographic correlates of fertility: A reply to Corruccini, Brandon, Handler and Holland. *American Antiquity, 54*(3), 626–636.

Konigsberg, L. W., Frankenberg, S. R., & Walker, R. B. (1997). Regress what on what? Palaeodemographic age estimation as a calibration problem. In R. Paine (Ed.), *Integrating archaeological demography: Multidisciplinary approaches to prehistoric population* (pp. 64–88). Carbondale: Southern Illinois University Press (Center for Archaeological Investigations).

Krogman, W. M. (1962). *The human skeleton in forensic medicine* (pp. 244–276). Springfield: Charles C. Thomas.

Kruithof, R. (1937). Telefoonverkeersrekening. *De Ingenieur, 52*, E15–E25.

Kunos, C. A., Simpson, S. W., Russel, K. F., & Hershkovitz, I. (1999). First rib metamorphosis: Its possible utility for human age-at-death estimation. *American Journal of Physical Anthropology, 110*, 303–323.

Kvaal, S., & Solheim, T. (1995). Incremental lines in human dental cementum in relation to age. *Journal of Oral Science, 103*, 223–230.

La Rome impériale. (1997). *La Rome impériale: démographie et logistique (Actes de la table ronde, Rome, mars 1994)* (de l'École française de Rome, Vol. 230). Paris: De Boccard.

Lalys, L. (2002). *Biométrie radiologique de l'avant-bras et de la main, appliquée à l'estimation de l'âge des individus immatures*. Ph.D. thesis, Université de la Méditerranée-Aix-Marseille II, Faculté de médecine.

Lalys, L., Adalian, P., Chaumoître, K., Signoli, M., & Leonetti, G. (2006). Biométrie radiologique des os de la main. Application à l'estimation de l'âge des individus immatures. In L. Buchet, C. Dauphin, & I. Séguy (Eds.), *La paléodémographie. Mémoire d'os, mémoire d'hommes* (pp. 263–279). Antibes: APDCA.

Lambert, A. (1973). *Un logit standard pour l'Amérique latine*, Louvain-la-Neuve, Université catholique de Louvain, Département de démographie.

Lamendin, H. (1988). Appréciation d'âge par la méthode de Gustafson simplifiée. *Le Chirugien-dentiste de France, 427*, 43–47.

Lamendin, H., Humbert, J.-F., Tavernier, J.-C., Brunel, G., & Nossintchouk, R.-M. (1990). Estimation of age by a method of 2 dental criteria. *Le Chirugien-dentiste de France, 539*, 93–96.

Lamendin, H., Baccino, E., Humbert, J.-F., Tavernier, J.-C., Nossintchouk, R.-M., & Zerilli, A. (1992). A simple technique for age estimation in adult corpses: The two-criteria dental method. *Journal of Forensic Sciences, 37*(5), 1373–1379.

Landry, A. (1934). (repub. 1982). *La révolution démographique. Études et essais sur les problèmes de la population, présentation d'Alain Girard* (Les classiques de l'économie et de la population). Paris: INED.

Langaney, A. (1988). *Les hommes: passé, présent, conditionnel.* Paris: Armand Colin.

Langenscheidt, F. (1985). *Methodenkritische Untersuchungen zur Paläodemographie am Beispiel zweier fränkischer Gräberfelder. Materialen für Bevölkerungswissenschaft,* 2 Numéro spécial, Wiesbaden.

Langlois, J.-Y. (1983). Abbaye de Maubuisson (Saint-Ouen-l'Aumône, Val-d'Oise): rapport de fouille 1983. SDAVO, Pontoise, unpublished.

Langlois, J.-Y. (1991). *L'abbaye cistercienne de Maubuisson (Saint-Ouen-l'Aumône, Val-d'Oise): la salle du chapitre: architecture et pratiques funéraires.* M.Sc. dissertation, Université de Paris I Panthéon-Sorbonne.

Langlois, J.-Y., & Wabont, M. (1996). Différentes composantes d'une communauté cistercienne à l'Époque moderne, à travers l'histoire, l'archéologie et l'anthropologie. In L. Buchet (Ed.), *L'identité des populations archéologiques* (pp. 287–302). Antibes: APDCA.

Lanphear, K. M. (1989). Testing the value of skeletal samples in demographic research: A comparison with vital registration samples. *International Journal of Anthropology, 4*(3), 185–193.

Larsen, C. S. (1997). *Bioarchaeology. Interpreting behavior from the human skeleton.* Cambridge: Cambridge University Press.

Larsen, C. S. (2006). The agricultural revolution as environmental catastrophe: Implications for health and lifestyle in the Holocene. *Quaternary International, 150*, 12–20.

Lassen, C., Hummel, S., & Herrman, B. (1996). PCR based sex identification of ancient human bones by amplification of X- and Y- chromosomal sequences: A comparison. *Ancient Biomolecules (Netherlands), 1*(1), 25–33.

Lassen, C., Hummel, S., & Herrmann, B. (2000). Molecular sex identification of stillborn and neonate individuals ("Traufkinder") from the burial site Aegerten. *Anthropologischer Anzeiger, 58*, 1–8.

Lazorthes, G., & Lhez, A. (1939). Les caractères sexuels de la grande échancrure sciatique. *Bulletins et mémoires de la Société d'anthropologie de Paris, t. 10, VIII*(4–6), 180–185.

Ledermann, S. (1969). *Nouvelles tables-types de mortalité* (Travaux et documents, Vol. 53). Paris: INED.

Ledermann, S., & Bréas, J. (1959). Les dimensions de la mortalité. *Population, 11*(4), 637–682.

Legoux, P. (1966). *Détermination de l'âge dentaire des fossiles de la lignée humaine.* Paris: Maloine.

Legoux, R., Buchet, L., Calligaro, T., Dhénin, D., Liéger, A., Poirot, J.-P., & Rodet-Belarbi, I. (2005). La nécropole mérovingienne de Cutry (Meurthe-et-Moselle). *Revue archéologique de l'Est, 54.* http://rae.revues.org/index1139.html. Saint-Germain-en-Laye, Mémoires publiés par l'Association Française d'Archéologie Mérovingienne, XIV.

Leonetti, G. (1998). *Apport de nouveaux paramètres anthropologiques à l'identification en mé decine légale.* Ph.D. thesis, université de la Méditerranée, Aix-Marseille II.

Leontief, W. W. (1949). *The structure of the American economy 1919–1939. An empirical application of equilibrium analysis.* Oxford/New York: Oxford University Press.

Leridon, H., & Toulemon, L. (1997). *Démographie: approche statistique et dynamique des populations.* Paris: Economica.

Lett, D. (1995). Faire le deuil d'un enfant mort sans baptême au Moyen Âge: la naissance du limbe pour enfants aux XIIe-XIIIe s. *Devenir, 7*(1), 101–112.

Lett, D. (1997). *L'enfant des miracles. Enfance et société au Moyen Âge (xiie-xiiie siècles).* Paris: Aubier.

Leveau, P., Sillières, P., & Vallat, J.-P. (1993). *Campagnes de la Méditerranée romaine: Occident*. Paris: Hachette.

Leveau, P., Trément, F., Walsh, K., & Barker, G. (Eds.). (1999). *Environmental reconstruction in Mediterranean landscape archaeology (the archaeology of Mediterranean landscapes)*. Oxford: Oxbow Books.

Lewis, A. B., & Garn, S. M. (1960). The relationship between tooth formation and other maturational factors. *The Angle Orthodontist, 30*, 70–77.

Lipsinic, F. E., Paunovich, E., Houston, D. G., & Robison, S. F. (1986). Correlation of age and incremental lines in the cementum of human teeth. *Journal of Forensic Sciences, 31*, 982–989.

Lipták, P. (1980). *Embertan és emberszármazástan*. Budapest: Tankönyvkiadó.

Lo Cascio, E. (1997). Le procedure di recensus della tarda Repubblica al tardoantico e il calcolo della popolazione di Roma. In *La Rome impériale: démographie et logistique, Actes de la table ronde (Rome, 25 mars 1994)* (collection de l'École française de Rome, Vol. 230, pp. 3–76). Paris-Rome, Ecole française de Rome.

Lo Cascio, E. (1999). Popolazione e risorse agricole nell'Italia del II secolo a. C. In D. Vera (Ed.), *Demografia, sistemi agrari, regimi alimentari nel mondo antico. Atti del Convegno Internazionale di Studi (Parma, ottobre 1977)* (pp. 21–246). Bari: Edipuglia.

Logan, W. H., & Kronfield, R. (1933). Development of the human jaws and surrounding structures from birth to the age of fifteen years. *The Journal of the American Dental Association, 20*, 379–427.

Long, J. S. (1997). *Regression models for categorical and limited dependent variables*. Thousand Oaks: Sage.

Longacre, W. A. (1976). Population dynamics at Grasshopper Pueblo, Arizona, 169–184. In E. B. Zubrow (Ed.), *Demographic anthropology. Quantitative approaches*. Albuquerque: University of New Mexico Press.

Lotka, A. (1934). *Théorie analytique des associations biologiques. Première partie: principes* (Vol. 187). Paris: Hermann & Cie.

Lotka, A. (1939). *Théorie analytique des associations biologiques. Deuxième partie: analyse dé mographique avec application particulière à l'espèce humaine* (Vol. 780). Paris: Hermann & Cie.

Love, B., & Muller, H. G. (2002). A solution to the problem of obtaining a mortality schedule for palaeodemographic data. In R. D. Hoppa & J. W. Vaupel (Eds.), *Palaeodemography. Age distribution from skeletal samples* (pp. 181–192). Cambridge: Cambridge University Press.

Lovejoy, C. O. (1971). Methods for detection of census error in palaeodemography. *American Anthropologist, 73*(1), 101–109.

Lovejoy, C. O., Meindl, R. S., Pryzbeck, T. R., et al. (1977). Palaeodemography of the Libben site. *Science, 198*, 291–293.

Lovejoy, C. O., Meindl, R. S., Mensforth, R. P., & Barton, T. J. (1985a). Multifactorial determination of skeletal age at death: A method and blind tests of its accuracy. *American Journal of Physical Anthropology, 68*, 1–14.

Lovejoy, C. O., Meindl, R. S., Pryzbeck, T. R., & Mensforth, R. P. (1985b). Chronological metamorphosis of the auricular surface of the ilium: A new method for the determination of adult skeletal age at death. *American Journal of Physical Anthropology, 68*(1), 15–28.

Lucy, D. (1997). *Human age estimation from skeletal and dental evidence*. Ph.D. thesis, University of Bradford.

Lucy, D., Pollard, A. M., & Roberts, C. A. (1995). A comparison of three dental techniques for estimating age at death in humans. *Journal of Archaeological Science, 22*, 417–428.

Lucy, D., Aykroyd, R. G., Pollard, A. M., & Solheim, T. (1996). A Bayesian approach to adult human age estimation from dental observations by Johanson's age changes. *Journal of Forensic Sciences, 41*, 189–194.

Luy, M. A., & Wittwer-Backofen, U. (2005). Das Halley-Band für paläodemographische Mortalitätsanalysen. *Zeitschrift für Bevölkerungswissenschaft, 30*(2–3), 219–244.

Luy, M. A., & Wittwer-Backofen, U. (2008). The Halley band for palaeodemographic mortality analysis. In J.-P. Bocquet-Appel (Ed.), *Recent advances in palaeodemography: Data, techniques, patterns* (pp. 119–141). Dordrecht/London: Springer Verlag.

Maber, M., Liversidge, H. M., & Hector, M. P. (2006). Accuracy of age estimation of radiographic methods using developing teeth. *Forensic Science International, 159*(suppl. 1), S68–S73.

Macchiarelli, R., & Salvadei, L. (1994). Palaeodemography and selective funerary practices at Latium Vetus, Middle-Tyrrhenian Italy. *Anthropologischer Anzeiger, 52*, 37–52.

Majó, T. (1992). Ontogénèse de l'os coxal et détermination sexuelle: l'importance de l'ilium. *Bulletins et mémoires de la Société d'anthropologie de Paris, 1–2*, 61–74.

Majó, T. (1997). Diagnostic auxologique de l'os coxal. In L. Buchet (Ed.), *L'enfant, son corps, son histoire. (Actes des 7es Journées anthropologiques de Valbonne, juin 1994)* (pp. 165–175). Antibes: APDCA.

Majó, T., Tillier, A.-M., & Bruzek, J. (1993). Test des fonctions discriminantes de Schutkowski impliquant l'ilium pour la détermination du sexe dans des séries d'enfants de sexe et d'âge connus. *Bulletins et mémoires de la Société d'anthropologie de Paris, 1–2*, 61–68.

Makeman, W. M. (1860). On the law of mortality. *Journal of Institute of Actuaries, 13*, 325–358.

Malthus, T. R. (1798). *An essay on the principle of population as it affects the future improvement of society, with remarks on the speculations of Mr. Godwin, M. Condorcet, and other writers.* J. Johnson, London.

Mannucci, A., Sullivan, K. M., Ivanov, P. L., & Gill, P. (1994). Forensic application of a rapid and quantitative DNA sex test by amplification of the X-Y homologous gene amelogenin. *International Journal of Legal Medicine, 106*, 190–193.

Maples, W. R., & Rice, P. M. (1979). Some difficulties in the Gustafson dental age estimations. *Journal of Forensic Sciences, 24*(1), 168–172.

Margerison, B. J., & Knusel, C. J. (2002). Palaeodemographic comparison of a catastrophic and an attritional death assemblage. *American Journal of Physical Anthropology, 119*, 134–143.

Márquez Morfín, L., & Hernández, E. (2001). *Principios básicos, teóricos y metodológicos de la palaeodemografía*. Mexico: INAH, ENAH.

Martin, R. (1928) (2nd ed.). *Lehrbuch der Anthropologie,* Jena, G. Fischer., 3 vol.; 1957, 3rd ed. revised and corrected by K. Saller, *Lehrbuch der Anthropologie in systematischer Darstellung,* Stuttgart, G. Fischer.

Masset, C. (1971). Erreurs systématiques dans la détermination de l'age par les sutures crâniennes. *Bulletins et mémoires de la Société d'anthropologie de Paris, 7*, 85–105.

Masset, C. (1973a). Influence du sexe et de l'âge sur la conservation des os humains. In *L'Homme d'hier et d'aujourd'hui, recueil d'études en hommage à André Leroi-Gourhan* (pp. 333–343). Paris: Cujas.

Masset, C. (1973b). La démographie des populations inhumées: essai de paléodémographie. *L'Homme, XIII*(4), 95–131.

Masset, C. (1974). *Problèmes de démographie préhistorique.* Thesis in prehistory, Université de Paris I Panthéon-Sorbonne.

Masset, C. (1976a). Sur la mortalité chez les anciens Indiens de l'Illinois. *Current Anthropology, 17*(1), 128–132.

Masset, C. (1976b). Sur quelques fâcheuses méthodes de détermination de l'âge des squelettes. *Bulletins et mémoires de la Société d'anthropologie de Paris, 7*, 329–336.

Masset, C. (1982). *Estimation de l'âge au décès par les sutures crâniennes.* Ph.D. thesis, Université Paris Diderot Paris 7.

Masset, C. (1987). Le recrutement d'un ensemble funéraire. In H. Duday, Cl. Masset (Eds.) Anthropologie Physique et Archéologie. Méthodes d'étude des sépultures. Paris: Editions du CNRS, pp. 111–134.

Masset, C. (1989). Age estimation on the basis of the cranial sutures. In M. Y. İşcan (Ed.), *Age markers in human skeletons* (pp. 71–103). Springfield: Charles C. Thomas.

Masset, C. (1993). Encore l'âge des adultes. *Bulletins et mémoires de la Société d'anthropologie de Paris, 5*(ns), 217–224

Masset, C. (1994). La paléodémographie. *Histoire et mesure, IX*(3–4), 381–394.

Masset, C. (1995). Palaéodémographie: problèmes méthodologiques. *Cahiers d'anthropologie et biométrie humaine, XIII*(1–2), 27–38.

Masset, C. (1997). Le peuplement préhistorique de l'Europe: l'apport de l'anthropologie. In J.-P. Bardet & J. Dupâquier (Eds.), *Histoire des populations de l'Europe. Tome I: Des origines aux prémices de la révolution démographique* (pp. 69–80). Paris: Fayard.

Masset, C. (1999). Problems and prospects in palaeodemography. In K. Bintliff & J. Sbonias (Eds.), *The archaeology of the Mediterranean landscape, vol. 1: Reconstruction past population trends in Mediterranean Europe (3000 BC-AD 1800)* (pp. 241–247). Oxford: Oxbow Books.

Masset, C., & Parzysz, B. (1985). Démographie des cimetières? Incertitude statistique des estimateurs en paléodémographie. *L'Homme, XXV*(2), 147–154.

Masset, C., & Sellier, P. (1990). Les anthropologues, les morts et les vivants. *Les Nouvelles de l'archéologie, 40*, 5–8.

Masset, C., & Castro e Almeida, M. E. (1990). *Âge et sutures crâniennes, Catania, Atti della Accademia Mediterranea delle Scienze.*

McCaa, R. (1998). Calibrating palaeodemography: The uniformitarian challenge turned. Paper presented at the American association of physical anthropology annual meeting (Salt Lake City, April 1998). www.hist.umn.edu/~rmccaa/palaeo98

McCaa, R. (2005). Palaeodemography of the Americas from ancient times to colonialism and beyond. In R. H. Steckel & J. C. Rose (Eds.), *The backbone of history: Health and nutrition in the western hemisphere* (pp. 94–124). New York: Cambridge University Press.

McCullagh, P. (1980). Regression models for ordinal data. *Journal of the Royal Statistical Society, 42*, 109–142.

McKelvey, R. D., & Zavoina, W. (1975). A statistical model for the analysis of ordinal level dependent variables. *Journal of Mathematical Sociology, 4*, 103–120.

McKern, T. W., & Stewart, T. D. (1957). *Skeletal age changing in young American males, analyzed from the standpoint of identification*, Nattick, US Army quartermaster research and development command, technical report, EP-45.

Megyesi, M. S., Ubelaker, D. H., & Sauer, N. J. (2006). Test of the Lamendin aging method on two historic skeletal samples. *American Journal of Physical Anthropology, 131*(3), 363–367.

Meiklejohn, C., & Zvelebil, M. (1991). Health status of European populations at the agricultural transition and the implications for the adoption of farming. In H. Bush & M. Zvelebil (Eds.), *Health in past populations* (British Archaeological Reports International Series, Vol. 567, pp. 129–145). Oxford, British Archaeological Reports.

Meindl, R. S., & Lovejoy, C. O. (1985). Ectocranial suture closure: A revised method for the determination of skeletal age at death based on the lateral-anterior sutures. *American Journal of Physical Anthropology, 68*, 7–66.

Meindl, R. S., & Lovejoy, C. O. (1989). Age changes in the pelvis: Implications for palaeodemography. In M. Y. İşcan (Ed.), *Age markers in the human skeleton* (pp. 137–168). Springfield: Charles C. Thomas.

Meindl, R. S., & Russell, K. F. (1998). Recent advances in method and theory in palaeodemography. *Annual Review of Anthropology, 27*(1), 375–399.

Meindl, R. S., Lovejoy, C. O., Mensforth, R. P., & Don, C. L. (1985a). Accuracy and direction of error in the sexing of the skeleton: Implications for palaeodemography. *American Journal of Physical Anthropology, 68*, 79–85.

Meindl, R. S., Lovejoy, C. O., Mensforth, R. P., & Walker, R. A. (1985b). A revised method of age determination using the os pubis, with a review and tests of accuracy of other current methods of pubic symphyseal aging. *American Journal of Physical Anthropology, 68*, 29–45.

Meindl, R. S., Bedford, M. E., Russel, K. E., Lovejoy, C. O., & Simpson, S. W. (1995). Testing the test of the multifactorial aging method: A reply to Fairgrieve and Oost. *American Journal of Physical Anthropology, 97*(1), 85–87.

Meindl, R. S., Mensforth, R. P., & Lovejoy, C. O. (2008). The Libben site: A hunting, fishing and gathering village from the Eastern Lake Woodlands of North America. Analysis and implications for palaeodemography and humans origins. In J.-P. Bocquet-Appel (Ed.), *Recent Advances in palaeodemography. Data, techniques, patterns* (pp. 259–276). Dordrecht: Springer Verlag.

Menken, J., Trussell, J., Stempel, D., & Babakol, O. (1981). Proportional hazards life table models: Illustrative analysis of socio-demographic influences on marital dissolution in the US. *Demography, 18*, 181–200.

Menken, J., Coale, A. J., & Heuveline, P. (1995). Modèles démographiques: les cinquante premières années de Population. *Population, 6*(Special issue), 1545–1564.

Mensforth, R. P. (1990). Palaeodemography of the Carlston Annis (Bt-5) Late Archaic skeletal population. *American Journal of Physical Anthropology, 82*, 81–99.

Meslé, F., & Vallin, J. (2001). *Tables de mortalité françaises pour les xixe et xxe siècles et projections pour le xxie siècle* (Données statistiques, Vol. 4). Paris: INED.

Milner, G. R., Wood, J. W., & Boldsen, J. L. (2000). Palaeodemography. In S. R. Saunders & M. A. Katzenberg (Eds.), *Biological anthropology of the human skeleton* (pp. 467–497). New York: Wiley-Liss.

Milner, G. R., Wood, J. W., & Boldsen, J. L. (2008) (2nd ed.). Advances in palaeodemography. In M. A. Katzenberg & S. R. Saunders (Eds.), *Biological anthropology of the human skeleton* (pp. 561–600). New York: Wiley.

Mitra, S. (1983). A simple model for linking life tables by survival-mortality ratios. *Demography, 20*(2), 227–234.

Mittler, D. M., & Sheridan, S. G. (1992). Sex determination in subadults using auricular surface morphology: A forensic science perspective. *Journal of Forensic Sciences, 37*, 1068–1075.

Molleson, T. (1991). Demographic implications of the age structure of early English cemetery samples. In L. Buchet (Ed.), *Ville et campagne en Europe occidentale (v[e]- xiiie siècles) (Actes des 5es Journées anthropologiques de Valbonne, mai 1990)* (Dossiers de documentation archéologique, Vol. 14, pp. 113–122). Paris: CNRS Éditions.

Molleson, T. (1993). Demographic structure of the population. In D. E. Farwell & T. Molleson (Eds.), *Poundbury, vol. 2, the cemeteries* (Dorset natural history & archaeological society, monograph series, Vol. 11, pp. 207–214). Dorchester: Dorset Natural History & Archaeological Society.

Molleson, T. (1995). Rates of ageing in the eighteenth century. In S. R. Saunders & A. Herring (Eds.), *Grave reflections, portraying the past through cemetery studies* (pp. 99–222). Toronto: Canadian Scholars' Press.

Molleson, T. (1997). Patterns of growth. In L. Buchet (Ed.), *L'enfant, son corps, son histoire (Actes des 7es Journées anthropologiques de Valbonne, juin 1994)* (pp. 201–210). Antibes: APDCA.

Molleson, T., & Cox, M. (1993). *The spitalfields project, vol. 2, the anthropology – the middling sort* (Research Report, Vol. 86). York: Council of British Archaeology.

Moore, J. A., Swedlung, A. C., & Armelagos, G. L. (1976). The use of life tables in palaeodemography. In A. C. Swedlung (Ed.), *Population studies in archaeology and biological anthropology: A symposium* (Memoirs of the society for American archaeology, Vol. 30, pp. 55–70). Washington.

Moorrees, C. F. A., Fanning, E. A., & Hunt, E. E., Jr. (1963a). Formation and resorption of three deciduous teeth in children. *American Journal of Physical Anthropology, 21*, 205–213.

Moorrees, C. F. A., Fanning, E. A., & Hunt, E. E., Jr. (1963b). Age variation of formation stages for ten permanent teeth. *Journal of Dental Research, 42*(6), 1490–1502.

Müller, H.-G., Love, B., & Hoppa, R. D. (2001). Semiparametric method for estimating palaeodemographic profiles from age indicator data. *American Journal of Physical Anthropology, 117*, 1–14.

Müller, H.-G., Love, B., & Hoppa, R. D. (2002). Semiparametric methods for estimating palaeodemographic profiles from age indicator data. *American Journal of Physical Anthropology, 117*, 1–14.

Muller-Bolla, M., Lupi-Pégurier, L., Quatrehomme, G., Velly, A.-M., & Bolla, M. (2003). Age estimation from teeth in children and adolescents. *Journal of Forensic Sciences, 48*(1), 140–148.

Murail, P. (1996). *Biologie et pratiques funéraires des populations d'époque historique: une dé marche méthodologique appliquée à la nécropole gallo-romaine de Chantambre (Essonne, France)*. Ph.D. thesis, Université de Bordeaux I.

Murail, P. (1997). Pratiques funéraires et paléodémographie: les sujets immatures de la nécropole gallo-romaine de Chantambre (Essonne). In L. Buchet (Ed.), *L'enfant, son corps, son histoire (Actes des 7es Journées anthropologiques de Valbonne, juin 1994)* (pp. 227–237). Antibes: APDCA.

Murail, P., & Sellier, P. (1995). Farewell to estimators. Paper presented at the XVIes Rencontres internationales d'Archéologie et d'Histoire d'Antibes, October 1995, Paper abstracts.

Murail, P., Bruzek, J., & Braga, J. (1999). A new approach to sexual diagnosis in past populations. Practical adjustments from van Vark's procedure. *International Journal of Osteoarchaeology, 9*(1), 39–53.

Murail, P., Bruzek, J., Houët, F., & Cunha, E. (2005). DSP: A probabilistic sex diagnosis tool using worldwide variability in hip-bone measurements. *Bulletins et mémoires de la Société d'anthropologie de Paris, 17*(3–4 ns), 167–176.

Murray, C. J. L., Ferguson, B. D., Lopez, A. D., Guillot, M., Salomon, J. A., & Ahmad, O. (2003). Modified logit life table system: Principles, empirical validation, and application. *Population Studies, 57*(2), 165–182.

Nagaoka, T., & Hirata, K. (2007). Reconstruction of palaeodemographic characteristics from skeletal age at death distributions: Perspectives from Hitotsubashi, Japan. *American Journal of Physical Anthropology, 134*(3), 301–311.

Nagaoka, T., Hirata, K., Yokota, E., & Matsu'ura, S. (2006). Palaeodemography of a medieval population in Japan: Analysis of human skeletal remains from the Yuigahama-minami site. *American Journal of Physical Anthropology, 131*(1), 1–14.

Nelson, W. (1982). *Applied life data analysis*. New York: Wiley.

Nemeskéri, J. (1971). Les conditions archéologiques et anthropologiques des recherches paléodémographiques. In *Congrès international de la population (Londres 1969)* (Vol. IV). Liège: UIESP.

Nemeskéri, J., & Harsányi L. (1958). La méthode de détermination de l'âge sur le squelette et son application (in Hungarian). *Magyar Tudományos Akadémia Biológiai Csoportjának Közlemé nyei*, 1–2, 115–164.

Nemeskéri, J., Harsányi, L., & Acsádi, G. (1960). Methoden zur Diagnose des Lebensalters von Skelettfunden. *Anthropologischer Anzeiger, 24*, 70–95.

Notestein, F. W. (1945). Population: The long view. In T. W. Schultz (Ed.), *Food of the world* (pp. 36–57). Chicago: Chicago University Press.

Notestein, F., et al. (1944). *La population future de l'Europe et de l'Union soviétique. Perspectives démographiques 1940–1970*. Geneva: UN.

Novotný, V. (1975). Diskriminanzanalyse der Geschlechtsmerkmale auf dem os coxae beim Menschen, XIII. Czechosl. Anthrop. Congr. Brno, Sept. 1–4.

O'Connor, K. A. (1995). *The age pattern of mortality: A micro-analysis of Tipu and a meta-analysis of twenty-nine palaeodemographic samples*. Unpublished Ph.D. dissertation, University at Albany-SUNY.

O'Connor, K. A., Holman, D. J., Boldse, J., Wood, J. W., & Gage, T. B. (1997). Competing and mixed hazard models of mortality: Osteological applications (abstract). *American Journal of Physical Anthropology, 24* (suppl.), 180.

OECD. (1980). *Mortality in developing countries*. III. Vol. V–VI: *New model life tables for use in developing countries* (Development centre studies). Paris: OECD.

Olivier, G. (1960). *Pratique anthropologique*. Paris: Vigot.

Omran, A. R. (1971). The epidemiologic transition: A theory of the epidemiology of population change. *The Milbank Memorial Fund Quarterly, 49*(4), 509–538. part 1.

Orban-Segebarth, R. (1984). Procédé métrique pour la diagnose du sexe de l'os coxal. *Bulletins et mémoires de la Société d'anthropologie de Paris, 1*, series XIV, 5–12.

Orchard, T., & Woodbury, M. A. (1972). A missing information principle: theory and applications. In *Proceedings of the VIth Berkeley symposium on mathematical statistical probability* (Vol. 1, pp. 697–715).

Orlando, L. (2005). *L'anti-Jurassic Park. Faire parler l'ADN fossile. (Collection Regards sur la science)*. Paris: Editions Belin- Pour la Science.

Ortega Muñoz, A. (2004). La palaeodemografía: ¿un instrumento para simular el comportamiento demográfico del pasado? Análisis comparativo con la demografía histórica en la Ciudad de México del siglo XIX. *Estudios demográficos y urbanos, 19*(1), 181–215.

Ouzoulias, P., Pellecuer, C., & Raynaud, C. et al. (2001). *Les campagnes de la Gaule à la fin de l'Antiquité. Actes du colloque de Montpellier (mars 1998)*. Antibes: APDCA.

Paillard, D. (1994). Lisieux de la fin de l'Antiquité au haut Moyen Âge: les éléments du IVe s., la nécropole Michelet. In *Lisieux avant l'an Mil, essai de reconstitution*, exhibition catalogue, Alençon, Musées de la Ville de Lisieux, pp. 62–99.

Paillard, D. (1999). Trois nouveaux sites archéologiques à Lisieux. *Bulletin de la Société historique de Lisieux, 43*, 19–23.

Paillard, D., Buchet, L., & Alduc-Le Bagousse, A. (2006). Nombre d'inhumés, nombre d'habitants. Estimations archéologiques et anthropologiques: Lisieux (Calvados), IVe s de notre ère. In L. Buchet, C. Dauphin, & I. Séguy (Eds.), *La paléodémographie. Mémoire d'os, mémoire d'hommes. Actes des 8es Journées anthropologiques de Valbonne, juin 2003* (pp. 209–223). Antibes: APDCA.

Paine, R. R. (1989). Model life table fitting by maximum likelihood estimation: A procedure to reconstruct palaeodemographic characteristics from skeletal age distributions. *American Journal of Physical Anthropology, 79*, 51–62.

Paine, R. R. (1992). *Population dynamics at Copán, Honduras, A. D. 400–1250: A study in archaeological demography*. Ph.D. dissertation, Department of Anthropology, Ann Arbor, Pensylvania State University, University Microfilms.

Paine, R. R. (Ed.). (1997a). *Integrating archaeological demography: Multidisciplinary approaches to prehistoric population* (Occasional papers, Vol. 24, cf p. 313). Carbondale: Center for Archaeological Investigations.

Paine, R. R. (1997b). Uniformitarian models in osteological palaeodemography. In R. R. Paine (Ed.), *Integrating archaeological demography: Multidisciplinary approaches to prehistoric population* (Occasional papers, Vol. 24, pp. 191–204). Carbondale: Center for Archaeological Investigations.

Paine, R. R. (2000). If a population crashes in prehistory, and there is no palaeodemographer there to hear it, does it make a sound? *American Journal of Physical Anthropology, 112*, 181–190.

Paine, R. R., & Boldsen, J. L. (1997). Long-term trends in mortality patterns in pre-industrial Europe. *American Journal of Physical Anthropology*, suppl. 24, 183.

Paine, R. R., & Boldsen, J. L. (2002). Linking age at death distribution and ancient population dynamics. In R. D. Hoppa & J. W. Vaupel (Eds.), *Palaeodemography. Age distribution from skeletal samples* (pp. 169–180). Cambridge: Cambridge University Press.

Paine, R. R., & Harpending, H. C. (1996). The reliability of palaeodemographic fertility estimators. *American Journal of Physical Anthropology, 101*, 151–160.

Paine, R. R., & Harpending, H. C. (1998). Effect of sample bias on palaeodemographic fertility estimates. *American Journal of Physical Anthropology, 105*, 231–240.

Paine, R. R., Freter, A. C., & Webster, D. L. (1996). A mathematical projection of population growth in the Copán Valley, Honduras, AD 400-800. *Latin American Antiquity, 7*, 51–60.

Paquin, J. D., Rest, M., Marie, J. P., Mort, J. S., Pidoux, I., Poole, A. R., & Roughley, P. J. (1983). Biochemical and morphologic studies of cartilage from the adult human sacro-iliac joint. *Arthritis and Rheumatism, 26*, 887–895.

Parner, E. T., Heidmann, J. M., Væth, M., & Poulsen, S. (2001). A longitudinal study of time trends in the eruption of permanent teeth in Danish children. *Archives of Oral Biology, 46*(5), 425–431.

Parsons, F. G., & Keene, L. (1920). Sexual differences in the skull. *Journal of Anatomy, 54*, 58–65.

Pedinielli, A., Bartoli, C., Foti, B., & Leonetti, G. (2004). Translucidité dentinaire et estimation de l'âge. *Journal de médecine légale, droit médical, 47*(8), 363–368.

Pennington, R. (1996). Causes of early human growth. *American Journal of Physical Anthropology, 99*, 259–274.

Perez-Guevara, S., Bartoli, C., Giustiniani, J., Adalian, P., Piercecchi-Marti, M.-D., Panuel, M., Foti, B., & Leonetti, G. (2006). Validation of biometric study of fetal mandibles by 3D tomodensitometric images. *Bulletin du Groupement international pour la recherche scientifique en stomatologie et odontologie, 47*(1), 11–17.

Perréard-Lopreno, G. (2007). *Adaptation structurelle des os du membre supérieur et de la clavicule à l'activité: analyse de l'asymétrie des propriétés géométriques de sections transverses et de mesures linéaires dans une population identifiée (collection Simon)*. Ph.D. thesis, Université de Genève, Département d'anthropologie et d'écologie.

Perréard-Lopreno, G., & Eades, S. (2003). Une démarche actualiste en paléoanthropologie: la collection de squelettes de référence. In M. Besse, L.-I. Stahl Gretsch, & P. Curdy (Eds.), *ConstellaSion. Hommage à Alain Gallay*. (Cahiers d'archéologie romande, Vol. 95, pp. 463–472) Lausanne: Cahiers d'archéologie romande.

Perrenoud, A. (1975). L'inégalité sociale devant la mort à Genève au XVIIe siècle. *Population*, N° spécial "Démographie historique", *30*, 221–243.

Perrin, C.-E. (1963). Note sur la population de Villeneuve-Saint-Georges au IXe siècle. *Le Moyen Âge, LXIX*, 75–86.

Peterson, W. (1975). A demographer's view of prehistoric demography. *Current Anthropology, 16*, 227–245.

Petrioli, L. (1982). *Nouvelles tables-types de mortalité. Applications et populations stables*. Siena: University of Siena.

Piasecki, E. (1985). Life tables in demography. Truth or fiction. *Studia demograficzne, 80*(2), 31–41 (abstract in English).

Pickles, A. (1985). *An introduction to likelihood analysis*. Norwich: Geo Books.

Pilet, C. (1980). *La nécropole de Frénouville. Étude d'une population de la fin du IIIe à la fin du VIIe siècle*. Oxford, British Archaeological Reports International Series, 83, 3 vol.

Pilet, C., Alduc-Le Bagousse, A., Blondiaux, J., Buchet, L., Grévin, G., & Pilet-Lemière, J. (1990). Les nécropoles de Giberville (Calvados), fin du ve siècle -fin du viie siècle ap. J.-C. *Archéologie Médiévale, XX*, 3–140.

Pilet, C., Alduc-Le Bagousse, A., Blondiaux, J., Buchet, L., & Pilet-Lemière, J. (1992). Le village de Sannerville, Lirose. Fin de la période gauloise au viie siècle ap. J-.C. *Archéologie Médié vale, XXII*, 1–189.

Pilet, C., Alduc-Le Bagousse, A., & Buchet, L. et al. (1994). La nécropole de Saint-Martin-de-Fontenay (Calvados). Recherches sur le peuplement de la plaine de Caen du ve siècle avant J.-C. au VIIe siècle après J.-C. *Gallia* (suppl. 54). Paris: CNRS edition.

Piontek, J., & Weber, A. (1990). Controversy on palaeodemography. *International Journal of Anthropology, 5*(1), 71–83.

Piontek, J., Wiercińska, A., & Wierciński, A. (1996). Mortality structure in Mesolithic, Neolithic and Early Bronze Age populations of Central Europe and Ukraine: A new methodological approach. *Anthropologie (Brno), 34*, 307–313.

Piontek, J., Jerszyńska, B., & Segeda, S. (2001). Long bones growth variation among prehistoric agricultural and pastoral populations from Ukraine (Bronze Era to Iron Age). *Variability and Evolution, 9*, 61–73.

Pison, G. (1989). *Dynamique d'une population traditionnelle. Démographie, apparentement, et mariage dans une population d'effectif limité: les Peul Bandé (Sénégal oriental)* (Travaux et documents, Vol. 99). Paris: INED.

Pison, G. (2007). The population of the world (2007). *Population and Societies, 436*.

Post, J. B. (1973). Ages at menarche and menopause: Some medieval authorities. *Population Studies, 25*(1), 83–87.

Poulain, M. (1981). *Contribution à l'analyse spatiale d'une matrice de migrations internes*. Louvain-la-Neuve: Cabay.
Pressat, R. (1979). *Dictionnaire de démographie* (1st ed.). Paris: Puf.
Pressat, R. (1983) (4th ed., 1st éd. 1961). *L'analyse démographique: concepts, méthodes, résultats*. Paris: Puf.
Pressat, R. (1995). *Éléments de démographie mathématique*. Paris: Aidelf.
Preston, S. (1976). *Mortality patterns in national populations: With special reference to recorded causes of death* (Studies in population). New York/San Francisco/London: Academic.
Preston, S., & Coale, A. J. (1982). Age structure, growth, attrition, and accession: A new synthesis. *Population Index, 48*(2), 217–259.
R Development Core Team. (2008). *R: A language and environment for statistical computing*, R foundation for statistical computing. Vienna, Austria. URL http://www.R-project.org
Raoult, D., & Drancourt, M. (2002). Cause of black death. *The Lancet Infectious Diseases, 2*, 459.
Raoult, D., Aboudharam, G., Crubézy, É., Larrouy, G., Ludes, B., & Drancourt, M. (2000). Molecular identification by 'suicide PCR' of *Yersinia pestis* as the agent of the medieval black death. *Proceedings of the National Academy of Sciences of the USA, 97*, 12800–128003.
Riché, P. (1973). L'enfant dans le Haut Moyen Age. *Annales de démographie historique*, 95–98. Paris : La Haye, Mouton & Cie.
Rigeade, C., & Signoli, M. (2005). Les sépultures de catastrophe: une gestion originale des cadavres en temps de crises démographiques. In N. Gualde, J.-P. Goubert, & S. Barry (Eds.), *Actes du colloque Épidémies et sociétés dans le monde occidental xiv*^{e}*-xxi*^{e} *siècle* (Revue sociologie et santé, pp. 117–131).
Rigeade, C., Willot, J.-M., Demolon, P., Rabino Massa, E., & Signoli, M. (2006). Approche anthropologique de sépultures de catastrophe du XVIIIe siècle (rue Martin du Nord, Douai, France). *Comptes rendus Palevol, 5*(7), 901–907.
Robert, P. (2006). *Le choix bayésien. Principes et pratique* (Statistique et probabilités appliquées). Paris: Springer-Verlag-France.
Robert-Lamblin, J. (1983). *Ammassalik (Groenland oriental): fin ou persistance d'un isolat? Étude anthropo-démographique du changement*. Ph.D. thesis, Université Paris Descartes.
Robert-Lamblin, J. (1986). *Les Ammassalimint au xx*^{e} *siècle: analyse du changement social au Groenland oriental* (Mémoires des Cahiers ethnologiques, Vol. 1). Bordeaux: Centre d'études et de recherches ethnologiques.
Robling, A. G., & Stout, S. D. (2000a). Methods of determining age at death using bone microstructure. In M. A. Katzenberg & S. R. Saunders (Eds.), *Biological anthropology of the human skeleton* (pp. 187–205). New York: Wiley-Liss.
Robling, A. G., & Stout, S. D. (2000b). Histomorphometry of human cortical bone: Applications to age estimation. In M. A. Katzenberg & S. R. Saunders (Eds.), *Biological anthropology of the human skeleton* (pp. 187–205). New York: Wiley-Liss.
Roche, A. F. (Ed.). (1979). *Secular trends in human growth, maturation and development*. Chicago: University of Chicago Press.
Rogers, T. (1990). *A test of the auricular surface method of estimating age-at-death and a discussion of its usefulness in the construction of palaeodemographic life tables*. Presented at the 18th annual meeting of the Canadian association of physical anthropology, Banff, Alberta.
Rogers, T., & Saunders, S. (1994). Accuracy of sex determination using morphological traits of the human pelvis. *Journal of Forensic Sciences, 39*, 1047–1056.
Rohrbasser, J.-M., & Théré, C. (2004). Antoine deparcieux (1703–1768) and demographic data collection. *The History of the Family, 9*(1), 115–122.
Roth, E. A. (1992). Applications of demographic models to palaeodemography. In S. R. Saunders & M. A. Katzenberg (Eds.), *Skeletal biology of past peoples: Research methods* (pp. 175–188). New York: Wiley-Liss.
Rousset, M. M., Boualam, N., Delfosse, C., & Roberts, W. E. (2003). Emergence of permanent teeth: Secular trends and variance in a modern sample. *Journal of Dentistry for Children (Chicago), 70*(3), 208–214.

Rovillé-Sausse, F. (1997). L'enfant et son milieu: variations des rythmes de croissance dans une population d'immigrés de la première génération. In L. Buchet (Ed.), *L'enfant, son corps, son histoire (Actes des 7es Journées anthropologiques de Valbonne, juin 1994)* (pp. 177–185). Antibes: APDCA.

Rozoy, J.-G. (1988). Le Magdalénien en Europe: démographie, groupes régionaux. *Bulletin de la Société préhistorique du Luxembourg, 10*, 139–158.

Sansilbano-Collilieux, M. (1993). Note sur la discordance entre âge dentaire, âge osseux et âge statural: l'exemple des sujets immatures de la nécropole médiévale de Saint-Martin de Cognac. *Bulletins et mémoires de la Société d'anthropologie de Paris, 1–2*, 103–108.

Sansilbano-Collilieux, M. (1994). Biologie et espace funéraire au Moyen Âge. Les nécropoles de l'ancien évêché de Poitiers (fin IVe siècle) et Saint-Martin-de-Cognac (VIIe- XVe siècles), thèse de doctorat, université de Bordeaux I, 2 vol.

Sansilbano-Collilieux, M., Bougault, D., Darlas, Y., & Sabatier, J.-P. (1994). Incidence du sexe et de l'âge sur le contenu minéral osseux. In L. Buchet (Ed.), *La femme pendant le Moyen Âge et l'Époque moderne, (Actes des 6es Journées anthropologiques de Valbonne, juin 1992)* (Dossiers de documentation archéologique, Vol. 17, pp. 159–171). Paris: CNRS Éditions.

Sardi, M. L., & Ramírez Rozzi, F. (2007). Developmental connections between cranial components and the emergence of the first permanent molar in humans. *Journal of Anatomy, 210*(4), 406–417.

Sattenfield, L., & Harpending, H. C. (1983). Stable populations and skeletal age. *American Antiquity, 48*, 489–498.

Saunders, S. R., & Herring, A. (1995). *Grave reflections: Portraying the past through cemetery studies*. Toronto: Canadian Scholars' Press.

Saunders, S. R., & Hoppa, R. D. (1993). Growth deficit in survivors and non-survivors: Biological mortality bias in nonadult skeletal samples. *Yearbook of American Journal of Physical Anthropology, 36*, 127–151.

Saunders, S. R., & Katzenberg, M. A. (1992). *Skeletal biology of past peoples: Research methods*. New York: Wiley-Liss.

Saunders, S. R., & Katzenberg, M. A. (Eds.). (2000). *Biological anthropology of the human skeleton*. New York: Wiley-Liss.

Saunders, S. R., Fitzgerald, C., Rogers, T., Dudar, C., & McKillop, H. (1992). A test of several methods of skeletal age estimation using a documented archaeological sample. *Canadian Society of Forensic Science Journal, 25*, 97–118.

Saunders, S. R., de Vito, C., & Herring, A. (1993). Accuracy tests of tooth formation age estimations for human skeletal remains. *American Journal of Physical Anthropology, 92*, 113–188.

Saunders, S. R., Herring, A., Sawchuk, L. A., & Boyce, G. (1995). The nineteenth-century cemetery at St. Thomas's Anglican Church Belleville: Skeletal remains, parish records and censuses. In S. R. Saunders & A. Herring (Eds.), *Grave reflections: Portraying the past through cemetery studies* (pp. 93–118). Toronto: Canadian Scholars' Press.

Saunders, S. R., Hoppa, R. D., Macchiarelli, M., & Bondioli, L. (2000). Investigating variability in human dental development in the past. *Anthropologie, 38*(1), 101–107.

Sauter, M.-R., & Privat, F. (1954). Sur un nouveau procédé métrique de détermination sexuelle du bassin osseux. *Bulletin der Schweizerischen Gesellschaft für Anthropologie und Ethnologie, 31*, 60–84.

Sauter, M.-R., & Simon, C. (1980). La paléoanthropologie et la paléodémographie: leurs buts, leurs méthodes et leurs limites. *Archives suisses d'anthropologie générale, 44*(1), 35–44.

Scheuer, J. L., & Black, S. (2000). *Developmental juvenile osteology*. London: Academic.

Scheuer, J. L., & Bowman, J. E. (1995). Correlation of documentary and skeletal evidence in the St. Bride's Crypt population. In S. R. Saunders & A. Herring (Eds.), *Grave reflections: Portraying the past through cemetery studies* (pp. 49–70). Toronto: Canadian Scholars' Press.

Schmitt, A. (2001). *Variabilité de la sénescence du squelette humain. Réflexions sur les indicateurs de l'âge au décès: à la recherche d'un outil performant*. Ph.D. thesis, Université de Bordeaux 1.

Schmitt, A. (2002). Estimation de l'âge au décès des sujets adultes à partir du squelette: des raisons d'espérer. *Bulletins et mémoires de la Société d'anthropologie de Paris, 14*(1–2), 51–73.
Schmitt, A. (2004). Age-at-death assessment using the os pubis and the auricular surface of the ilium: A test on an identified Asian sample. *International Journal of Osteoarchaeology, 14*, 1–6.
Schmitt, A. (2005). Une nouvelle méthode pour estimer l'âge au décès des adultes à partir de la surface sacro-pelvienne iliaque. *Bulletins et mémoires de la Société d'anthropologie de Paris, 17*(1–2), 89–102.
Schmitt, A., & Broqua, C. (2000). Approche probabiliste pour estimer l'âge au décès à partir de la surface auriculaire de l'ilium. *Bulletins et mémoires de la Société d'anthropologie de Paris, 12* (3–4), 279–301.
Schour, I., & Massler, M. (1940). Studies in tooth development: The growth pattern of human teeth. *The Journal of the American Dental Association, 27*, 1778–1793 and 1918–1931.
Schour, I., & Massler, M. (1941). The development of the human dentition. *The Journal of the American Dental Association, 28*, 1153–1160.
Schutkowski, H. (1986). Geschlechtsdifferente Merkmale an kindlichen Skeletten. Kenntnisstand und diagnostische Bedeutung. *Zeitschrift für Morphologie und Anthropologie, 76*, 149–168.
Schutkowski, H. (1987). Sex determination of fetal and neonate skeletons by means of discriminant analysis. *International Journal of Anthropology, 2*, 347–352.
Schutkowski, H. (1989). Age and sex diagnosis of the skeleton of immature individuals [in German]. *Anthropologischer Anzeiger, 47*(1), 1–9.
Schutkowski, H. (1993). Sex determination of infants and juvenile skeletons: I. Morphognostic features. *American Journal of Physical Anthropology, 90*, 199–205.
Scott, E. C. (1979). Dental wear scoring technique. *American Journal of Physical Anthropology, 51*, 213–218.
Séguy, I. (1996). Structures des populations et modes de renouvellement. In L. Buchet & C. Pilet (Eds.), *L'identité des populations archéologiques (Actes des XVIes Rencontres internationales d'Archéologie et d'Histoire d'Antibes, octobre 1995)* (pp. 389–402). Antibes: APDCA.
Séguy, I. (1997). Les aspects religieux et profanes dans le traitement funéraire réservé au nouveau-né au Moyen Âge et à l'Époque moderne. In L. Buchet (Ed.), *L'enfant son corps et son histoire (Actes des 7es Journées anthropologiques de Valbonne, juin 1994)* (pp. 97–113). Antibes: APDCA.
Séguy, I. (2010). La muerte de los párvulos: entre dogma y creencias populares. Francia, de finales de la Antigüedad a la época Moderna. *Traces* (Mexico), N° 58, 29–39.
Séguy, I., & Buchet, L. (2006). Âge biologique, âge démographique, âge sociologique. Estimation de l'âge des enfants inhumés et analyse paléodémographique. In *Enfants d'aujourd'hui. Diversité des contextes. Pluralité des parcours (Actes du Colloque de l'Aidelf, Dakar, 2002)* (pp. 76–86). Paris: Aidelf.
Séguy, I., & Signoli, M. (2007). Caractérisation des crises démographiques en Basse-Provence (1650–1730). In M. Signoli, D. Chevé, P. Adalian, G. Boëtsch, & O. Dutour (Eds.), *Peste: entre épidémies et sociétés/Plague: Epidemics and Societies* (pp. 197–204). Florence: Firenze University Press. Atti 28.
Séguy, I., & Signoli, M. (2008). Quand la naissance côtoie la mort: pratiques funéraires et religion populaire en France au Moyen Âge et à l'Époque moderne. In Francesc Gusi, Susanna Muriel, Carme Olària (coords.), *Nasciturus, Infans, Puerulus Vobis Mater Terra. La muerte en la infancia/La mort dans l'enfance/La mort a l'infancia/The death in the childhood,* Castelló de la Plana, Servicio de Investigaciones Arqueológicas y Prehistóricas, Colección de Prehistoria y Arqueología, pp. 497–512.
Séguy, I., & Wabont, M. (1999). La méthode des vecteurs de probabilité à l'épreuve des faits: les moniales de Maubuisson (Val-d'Oise, XVIIe- XVIIIe siècles). Paper presented to the Société d'Anthropologie de Paris, January 1999.
Séguy, I., Couvert, N., Belaigues-Rossard, M., & Bringé, A. (2006a). Des tables-types de mortalité pour les populations pré-industrielles. In L. Buchet, C. Dauphin, & I. Séguy (Eds.), *La paléodé mographie. Mémoire d'os, mémoire d'hommes (Actes des 8es Journées anthropologiques de Valbonne, juin 2003)* (pp. 303–321). Antibes: APDCA.

Séguy, I., Pennec, S., Tzortzis, S., & Signoli, M. (2006b). Modélisation de l'impact démographique de la peste, à travers l'exemple de Martigues (Bouches-du-Rhône). In L. Buchet, C. Dauphin, & I. Séguy (Eds.), *La paléodémographie. Mémoire d'os, mémoire d'hommes (Actes des 8es Journé es anthropologiques de Valbonne, juin 2003)* (pp. 323–330). Antibes: APDCA.

Séguy, I., Buchet, L., & Bringé, A., et al. (2008). Model life tables for pre-industrial populations. First applications in palaeodemography. In J.-P. Bocquet-Appel (Ed.), *Recent advances in palaeodemography. Data, techniques, patterns* (pp. 109–141). Dordrecht: Springer Verlag.

Séguy, I., Caussinus, H., Courgeau, D., & Buchet, L. (2013). Estimating the age structure of a buried adult population: A new statistical approach applied to archaeological digs in France. *American Journal of Physical Anthropology, 150*(2), 170–183.

Sellier, P. (1995). Palaéodémographie et archéologie funéraire: les cimetières de Mehrgarh, Pakistan. *Palaéorient, 21*(2), 123–143.

Sellier, P. (1996). La mise en évidence d'anomalies démographiques et leur interprétation: population, recrutement et pratiques funéraires du tumulus de Courtesoult. In J.-F. Piningre (Ed.), *Nécropoles et société au premier âge du fer: le tumulus de Courtesoult (Haute-Saône)* (Documents d'Archéologie Française, Vol. 54, pp. 136–142 et pp. 188–200). Paris: Éditions de la Maison des Sciences de l'Homme.

Sellier, P. (1997). Le potentiel des études anthropologiques pour l'archéologie: l'exemple de la paléodémographie. *Les Nouvelles de l'archéologie, 66*, 49–52.

Signoli, M. (1998). *Étude anthropologique de crises démographiques en contexte épidémique. Aspects paléo- et biodémographiques de la Peste en Provence*. Ph.D. thesis, Université de la Méditerranée-Aix-Marseille II, faculté de Médecine.

Signoli, M. (2006). *Études anthropologiques des crises démographiques en contexte épidémique: aspects paléo- et biodémographique de la peste en Provence* (British archaeological reports international series, Vol. 1515). Oxford: Archaeopress.

Signoli, M., Chausserie-Laprée, J., & Dutour, O. (1995). Étude anthropologique d'un charnier de la peste de 1720–1721 à Martigues. *Préhistoire et anthropologie méditerranéennes, 4*, 173–189.

Signoli, M., Séguy, I., Biraben, J.-N., & Dutour, O. (2002). Palaeodemography and historical demography in the context of an epidemic: Plague in Provence in the eighteenth century. *Population, English Edition, 57*(6), 821–847.

Signoli, M., Séguy, I., & Buchet, L. (2005). Les populations historiques. Le recrutement funéraire à travers trois échantillons paléodémographiques. In O. Dutour, J.-J. Hublin, & B. Vandermeersch (Eds.), *Origine et évolution des populations humaines* (pp. 241–257). Paris: Éditions du Comité des Travaux historiques et scientifiques.

Siler, N. (1979). A competing-risk model for animal mortality. *Ecology, 60*, 750–757.

Siler, N. (1983). Parameters of mortality in human populations with widely varying life spans. *Statistics in Medicine, 2*, 373–380.

Simon, C. (1982). Nécropole de Sézegnin (Avusy, Genève), nécropole de Thoiry (Ain, France). Étude anthropologique et paléodémographique. *Archives suisses d'anthropologie générale, 46* (1), 77–174.

Simon, C. (1983). Tentative de comparaison de structures de mortalité à différentes époques. *Archives suisses d'anthropologie générale, 47*(1), 81.

Simon, C. (1986). La surmortalité féminine, mythe ou réalité? *Bulletin de la Société d'anthropologie du Sud-Ouest, XXI*(2), 71–76.

Simon, C. (1987). Évolution de la synostose des sutures crâniennes sur quelques populations anciennes. In H. Duday & C. Masset (Eds.), *Anthropologie physique et archéologie. Méthodes d'étude des sépultures (Actes du colloque de Toulouse, 1982)* (pp. 239–244). Paris: CNRS Éditions.

Simon, C., & Leemans, É. (1991). Problème posé par une population moyenâgeuse de la campagne genevoise (Satigny-Genève). In L. Buchet (Ed.), *Ville et campagne en Europe*

occidentale (v^{e}- xiiie siècles) (Dossiers de documentation archéologique, Vol. 14, pp. 135–151). Paris: CNRS Éditions.
Simon, C., Gerbore, R., Passard, F., et al. (1996). La population de la Grande Oye à Doubs (Doubs): analyse anthropologique et données culturelles: concordances et contrastes dans l'étude du peuplement romano-burgonde de la seconde moitié du vi^{e} s. à la fin du viie s. ap. J.-C. In L. Buchet (Ed.), *L'Identité des populations archéologiques (Actes des rencontres arché ologiques d'Antibes, octobre 1995)* (pp. 351–366). Antibes: APDCA.
Siven, C. H. (1991a). On estimating mortalities from osteological age data. *International Journal of Anthropology, 6*(2), 97–110.
Siven, C. H. (1991b). On reconstructing the (once) living population from osteological age data. *International Journal of Anthropology, 6*(2), 111–118.
Skytthe, A., & Boldsen, J. L. (1993). A method for construction of standards for determination of skeletal age at death. *American Journal of Physical Anthropology, 16*(suppl), 182.
Smith, B. H. (1991). Standards of human tooth formation and dental age assessment. In M. A. Kelley & C. S. Larsen (Eds.), *Advances in dental anthropology* (pp. 143–168). New York: Wiley-Liss.
Smith, D. W. (1993). *Human longevity*. New York: University Oxford Press.
Smith, B. H., Crummett, T. L., & Brandt, K. L. (1994). Ages of eruption of primate teeth: A compendium for aging individuals and comparing life histories. *Yearbook of Physical Anthropology, 37*, 177–231.
Snow, C. C. (1983). Equations for estimating age at death from the pubic symphysis: A modification of the McKern-Stewart method. *Journal of Forensic Sciences, 28*, 864–870.
Solheim, T. (1990). Cementum apposition as an indicator of age. *Scandinavian Journal of Dental Research, 98*, 510–519.
Solheim, T. (1993). A new method for dental age estimation in adults. *Forensic Science International, 59*, 37–147.
Solheim, T., & Kvaal, S. (1993). Dental root surface structure as an indicator of age. *Journal for Odonto-Stomatology, 11*, 9–21.
Soulier, D., & Perrot, D. (1993). Une nouvelle contribution à la détermination individuelle de l'âge d'après l'état de synostose du segment S 3 de la suture sagittale. *Palaeobios, 9*(1–2), 59.
Soulier, P., Ruffier, O., Toupet, C., & Trompetta, P.-J. (1978). Abbaye de Maubuisson (Saint-Ouen-l'Aumône, Val-d'Oise): rapport de fouille 1978. Pontoise, SDAVO (Service départemental d'archéologie du Val-d'Oise).
Soulier, P., Ruffier, O., & Toupet, C. (1979). Fouilles archéologiques à l'abbaye de Maubuisson (Saint-Ouen-l'Aumône, Val-d'Oise): rapport de fouille 1979. Pontoise, SDAVO (Service départemental d'archéologie du Val-d'Oise).
Statistics Sweden. (1955). *Historisk statistik för Sverige. Del. 1, Befolkning 1720–1967*. Stockholm: The Swedish Institute.
Steckel, R. H. (2003). What can be learned from skeletons that might interest economists, historians, and other social scientists? *The American Economic Review, 93*(2), 213–220.
Steckel, R. H. (2005). Young adult mortality following severe physiological stress in childhood: Skeletal evidence. *Economics and Human Biology, 3*, 314–328.
Steckel, R. H., & Rose, J. (Eds.). (2000). *The backbone history: Long term trends in health and nutrition in the Americas*. Cambridge: Cambridge University Press.
Steele, D. G. (1976). The estimation of sex on the basis of the talus and calcaneus. *American Journal of Physical Anthropology, 45*, 581–588.
Stephan, F. F. (1942). An iterative method of adjusting sample frequency tables when expected marginal totals are known. *The Annals of Mathematical Statistics, 13*(2), 166–178.
Stewart, T. D. (1970). Identification of the scars of parturition in the skeletals remains of females. In T. D. Stewart (Ed.), *Personal identification* (pp. 127–135). Washington: Smithsonian Institution.
Stewart, T. D. (1979). *Essentials of forensic anthropology, especially as developed in the United States*. Springfield: Thomas Charles C.

Stloukal, M., & Hanáková, H. (1978). Die Länge der Längsknochen altslawischer Bevölkerungen unter besonderer Berücksichtigung von Wachstumfragen. *Homo, 29*, 53–69.

Stone, R., Bates, J., & Bacharach, M. (1963). *A programme for growth, vol. 3 Input–output relationships 1954–1956,* London/Cambridge: Chapman and Hall/University of Cambridge, Department of Applied Economics.

Stone, A. C., Milner, G. R., Pääbo, S., & Stoneking, M. (1996). Sex determination of ancient human skeletons using DNA. *American Journal of Physical Anthropology, 99*(2), 231–238.

Storey, G. R. (1997). Estimating the population of ancient Roman cities. In R. R. Paine (Ed.), *Integrating archaeological demography: Multidisciplinary approaches to prehistoric population* (Occasional papers, Vol. 24, pp. 101–130, cf p. 319). Carbondale: Center for Archaeological Investigations.

Storey, R., & Hirth, K. (1997). Archaeological and palaeodemographic analyses of the El Cajon skeletal population. In R. Paine (Ed.), *Integrating archaeological demography: Multidisciplinary approaches to prehistoric population* (Occasional papers, Vol. 24, pp. 131–149, cf p. 319). Carbondale: Center for Archaeological Investigations.

Stout, S. D. (1989a). The use of cortical bone histology to estimate age at death. In M. Y. İşcan (Ed.), *Age markers in the human skeleton* (pp. 195–207). Springfield: Charles C. Thomas.

Stout, S. D. (1989b). Histomorphometric analysis of human skeletal remains. In M. Y. İşcan & K. A. Kennedy (Eds.), *Reconstruction of life from the skeleton* (pp. 41–52). New York: Wiley and Liss.

Stout, S. D., & Gehlert, S. J. (1980). The relative accuracy and reliability of histological aging methods. *Forensic Science International, 15*(3), 181–190.

Suder, W. (1988). *Census populi: Bibliographie de la démographie de l'antiquité romaine*. Bonn: Habelt.

Sullivan, K. M., Mannucci, A., Kimpton, C. P., & Gill, P. (1993). A rapid and quantitative DNA sex test: Fluorescence-based PCR and analysis of X-Y homologous gene amelogenin. *BioTechniques, 15*(4), 636–641.

Sundick, R. I. (1978). Human skeletal growth and age determination. *Homo, 29*(4), 228–249.

Suzanne, C., Rebato, E., & Chiarelli, B. (2003). *Anthropologie biologique: évolution et biologie humaine* (Anthropologie). Brussels: De Boeck.

Swedlung, A. C. (1975). *Population studies in archaeology and biological anthropology: A symposium* (Memoirs of the society for American archaeology, Vol. 30). Washington: Society for American Archaeology.

Swedlung, A. C., & Armelagos, G. J. (1976). *Demographic anthropology*. Dubuque: WC Brown Publisher.

Tanner, J. M. (1962). *Growth of adolescents* (2nd ed.). Oxford: Blackwell Scientific.

Theureau, C. (1996). Proposition pour améliorer l'évaluation de l'âge au décès des adultes à partir de plusieurs indicateurs pris sur le squelette. *Bulletins et mémoires de la Société d'anthropologie de Paris, 8*(3–4), 441–450.

Theureau, C. (1998). La population archéologique de Tours (IVe- XVIIe siècles). Étude anthropologique. *Revue archéologique du Centre de la France* (Recherches sur Tours, 7), suppl. 14.

Theureau, C. (2006a). Synthèse anthropologique. In É. Lorans (Ed.), *Saint-Mexme de Chinon, v^e^-xxe siècles* (Archéologie et histoire de l'art, Vol. 22, pp. 247–266). Paris: CTHS.

Theureau, C. (2006–2007). Étude anthropologique d'un cimetière de paroisse rurale: les sépultures (8e–19e s.) de Rigny (Rigny-Ussé, Indre-et-Loire). *Revue archéologique du Centre de la France*, Tome 45–46 (online).

Thillaud, P. L. (1996). *Palaéopathologie humaine*. Sceaux: Kronos B. Y Editions.

Thionet, P. (1963). Sur certaines variantes des projections du tableau d'échanges inter-industriels. *Bulletin de l'Institut international de statistique,* XL (1re livraison), *105*, 431–446.

Thionet, P. (1964). Note sur le remplissage d'un tableau à double entrée. *Journal de la Société de statistique de Paris n° 10-11-12*, 228–247.

Thuillier, G. (2004). Pour une histoire des tables de mortalité aux XVIIIe et XIXe siècles. *Bulletin d'histoire de la sécurité sociale, 50*, 413–426.
Todd, T. W. (1920). Age changes in the pubic bone: I. The white male pubis. *American Journal of Physical Anthropology, 3*, 285–339.
Todd, T. W. (1927). Skeletal records of mortality. *The Scientific Monthly, 24*(6), 481–496.
Todd, T. W., & Lyon, D. (1924). Cranial Suture Closure. Its progress and age relationship. I. Endocranial suture closure. Adult males and white stock. *American Journal of Physical Anthropology, 7*, 325–384.
Todd, T. W., & Lyon, D. (1925). Cranial suture closure. Its progress and age relationship. II. Endocranial Suture Closure in Adult Males of White Stock; III. Endocranial Closure in Adult Males of Negro Stock; Ectocranial Closure in Adult Males of Negro Stock. *American Journal of Physical Anthropology, 8*, 23–71 and 149–168.
Toupet, C., Wabont-Lemoine, M., & Soulier, P. (1981). Abbaye cistercienne de Maubuisson, Saint-Ouen-l'Aumône, Val-d'Oise: lavabo monumental du cloître: rapport d'étude 1980. Pontoise, SDAVO (Service départemental d'archéologie du Val-d'Oise).
Townsend, N., & Hammel, E. A. (1990). Age estimation from the number of teeth erupted in young children: An aid to demographic surveys. *Demography, 27*(1), 165–174.
Treffort, C. (1996). *L'église carolingienne et la mort: christianisme, rites funéraires et pratiques commémoratives*. Lyon, Presses universitaires de Lyon (Centre interuniversitaire d'histoire et d'archéologie médiévales).
Treffort, C. (1997). La vie et le corps de l'enfant au VIe siècle. Perception, signification et utilisation du thème de l'enfance dans l'œuvre de Grégoire de Tours. In L. Buchet (Ed.), *L'Enfant, son corps, son histoire (Actes des 7es Journées anthropologiques de Valbonne, juin 1994)* (pp. 115–127). Antibes: APDCA.
Trément, F. (1999a). *Archéologie d'un paysage: les étangs de Saint Blaise (Bouches du Rhône)* (Documents d'archéologie française, Vol. 74). Paris: Maison des Sciences de l'Homme.
Trément, F. (1999b). Prospection archéologique et démographie en Provence. In K. Bintliff & J. Sbonias (Eds.), *Reconstructing past population trends in Mediterranean Europe (3000 BC-AD 1800)* (The archaeology of the Mediterranean landscape, Vol. 1, pp. 93–113). Oxford: Oxbow Books.
Trussell, J., & Hammerslough, C. (1983). A hazard-model analysis of the covariates of infant and child mortality in Sri Lanka. *Demography, 20*, 1–26.
Trussell, J., & Menken, J. (1984). Estimating levels, trends and determinants of child mortality in countries with poor statistics. *Population and Development Review, 10*(suppl), 325–346.
Tugault, Y. (1970). Méthode d'analyse d'un tableau origine-destination de migrations. *Population, 25*(1), 59–68.
Tzortzis, S. (2009a). Documenter l'épidémie de peste de 1720–1721 à Martigues (Bouches-du-Rhône, France). In L. Buchet, C. Rigeade, I. Séguy, & M. Signoli (Eds.), *Vers une anthropologie des catastrophes (Actes des 9es Journées Anthropologiques de Valbonne, mai 2007)* (pp. 517–544). Paris/Antibes: INED/APDCA.
Tzortzis, S. (2009b). Archives biologiques et archives historiques. Une approche anthropologique de l'épidémie de peste de 1720–1721 à Martigues (Bouches-du-Rhône, France), thèse de doctorat, université de la Méditerranée-Aix-Marseille II, Faculté de médecine, 2 vol.
Tzortzis, S., & Séguy, I. (2008). Pratiques funéraires en lien avec les décès de nouveau-nés. À propos d'un cas dauphinois durant l'Époque moderne: la chapelle Saint-Jean à l'Argentière-la-Bessée (Hautes-Alpes). *Socio-anthropologie, 20*, 75–92.
Ubelaker, D. H. (1978) (2nd revised ed. 1989). *Human skeletal remains. Excavation, analysis, interpretation* (Manuals on archeology, Vol. 2). Chicago/Washington, DC: Aldine/Taraxacum.
Ubelaker, D. H. (1987). Estimating age at death from immature human skeletons: An overview. *Journal of Forensic Sciences, 32*, 1254–1263.
Ubelaker, D. H. (1989). The estimation of age at death from immature human bone. In M. Y. İşcan (Ed.), *Age markers in the human skeleton* (pp. 55–70). Springfield: Charles C. Thomas.

Ullrich, H. (1975). Estimation of fertility by means of pregnancy and childbirth alteration at the pubis, the ilium and the sacrum. *Ossa, 2*, 23–39.

United Nations. (1955). *Age and sex patterns of mortality: Model life-tables for under-developed countries* (Population studies, no. 22). New York: United Nations.

United Nations. (1956). *Schémas de variation de la mortalité selon l'âge et le sexe. Tables-types pour les pays en voie de développement* (Études démographiques, Vol. 22). New York: United Nations.

United Nations. (1982). *Model life tables for developing countries.* New York. http://www.un.org/esa/population/publications/Model_Life_Tables/Model_Life_Tables.htm

Usher, B. M. (2002). Reference samples: The first step linking biology and age in the human skeleton. In R. D. Hoppa & J. W. Vaupel (Eds.), *Palaeodemography: Age distributions from skeletal samples* (pp. 29–47). Cambridge: Cambridge University Press.

Valkovics, E. (1982). The conditions of the use of population models on basis of the findings of palaeodemographic cemetery excavations. *Demografie, 5*(2–3), 295–315.

Vallois, H.-V. (1937). La durée de la vie chez l'homme fossile. *L'Anthropologie, 47*, 499–532.

Vallois, H.-V. (1948). Technique anthropométrique. *La Semaine des hôpitaux de Paris, 24*, 374–383.

Van der Leeuw, S., Favory, F., & Fiches, J.-L. (Eds.). (2003). *Archéologie et systèmes socio-environnementaux: études multiscalaires sur la vallée du Rhône dans le programme ARCHAEOMEDES* (CRA- Monographies, Vol. 27), Paris: CNRS Éditions.

Van Gerven, D. P., & Armelagos, G. J. (1983). 'Farewell to palaeodemography?' Rumors of its death have been greatly exaggerated. *Journal of Human Evolution, 12*, 353–360.

Van Ossel, P. (1992). Établissements ruraux de l'antiquité tardive dans le Nord de la Gaule, Paris. *Gallia* (suppl. 51). Paris: CNRS edition.

Vassin, S. (1994). Epidemiological transition of mortality curves in terms of the Brass logit system. *European Journal of Population, 10*, 43–68.

Vaupel, J. W., Manton, K. G., & Stallard, E. (1979). The impact of heterogeneity in individual frailty on the dynamics of mortality. *Demography, 16*, 439–454.

Verdin, F. (1999). Occupation du sol et pression démographique indigène autour de Marseille au second âge du Fer. In J.-N. Corvisier & M. Bellancourt-Valdher (Eds.), *La démographie historique* (Cahiers scientifiques de l'Université d'Artois, Vol. 11, pp. 51–68). Arras: Artois Presses Université.

Véron, J. (1994). Les dimensions du temps en démographie. In E. Vilquin (Ed.), *Le Temps et la dé mographie. Chaire Quételet 1993* (pp. 371–384). Louvain-la-Neuve: Academia/L'Harmattan.

Vidal, P. (1999). *Antibes-Îlôt Terminus. Document final de synthèse de fouille préventive,* Association pour les Fouilles archéologiques nationales-antenne Méditerranée, Service régional de l'Archéologie- Provence-Alpes-Côte d'Azur.

Vlček, E., & Mrklas, L. (1975). Modifications of the Gustafson method of determination of age according to teeth on prehistorical and historical osteological material. *Scripta Medica, 48*, 203–208.

Vollset, S. E. (1993). Confidence intervals for a binomial proposition. *Statistics in Medicine, 12*, 809–824.

von Lenhossék, M. (1917). Über Nahtverknocherung im Kindesalter. *Archiv für Anthropologie, Neue Folge, 15*, 164–180.

Waldron, T. (1994). *Counting the dead: The epidemiology of skeletal populations.* New York: Wiley.

Walker, P. L. (1995). Problems of preservation and sexism in sexing: Some lessons from historical collections for palaeodemographers. In S. R. Saunders & A. Herring (Eds.), *Grave reflections. Portraying the past through cemetery studies* (pp. 31–48). Toronto: Canadian Scholars' Press.

Walker, P. L. (2008). Sexing skulls using discriminant function analysis of visually assessed traits. *American Journal of Physical Anthropology*, 39–50.

Walker, R. A., & Lovejoy, C. O. (1985). Radiographic changes in the clavicle and proximal femur and their use in the determination of skeletal age at death. *American Journal of Physical Anthropology, 68*, 67–78.
Walker, P. L., Johnson, J. R., & Lambert, M. (1988). Age and sex biases in the preservation of human skeletal remains. *American Journal of Physical Anthropology, 76*, 183–188.
Walker, P. L., Dean, G., & Shapiro, P. (1991). Estimating age from tooth wear in archaeological populations. In S. Hillson (Ed.), *Dental anthropology* (pp. 169–178). Cambridge: Cambridge University Press.
Walker, R. A., Lovejoy, C. O., & Meindl, R. S. (1994). Histomorphological and geometric properties of human femoral cortex in individuals over 50: Implications for histomorphological determination of age-at-death. *American Journal of Human Biology, 6*(5), 659–667.
Ward, R., & Weiss, K. (1976). The demographic evolution of human populations. In R. H. Ward & K. M. Weiss (Eds.), *A perspective on demographic methods and effects in social history* (pp. 1–23). London: Academic.
WEA (Workshop of European Anthropologists). (1980). Recommendations for age and sex diagnoses of skeletons. *Journal of Human Evolution, 9*, 517–549.
Weaver, D. S. (1980). Sex differences in the ilia of a known sex and age sample of fetal and infant skeletons. *American Journal of Physical Anthropology, 52*, 191–195.
Weibull, W. (1951). A statistical distribution function of wide applicability. *Journal of Applied Mechanics, 18*(3), 293–297.
Weiss, K. M. (1973). *Demographic models for archaeology* (Memoirs of the society for American archaeology, Vol. 27). Washington: Society for American Archaeology.
Weiss, K. M. (1990). The biodemography of variation of human frailty. *Demography, 27*, 185–206.
Welker, H. (1866). Kraniologische Mitteilungen. *Archiv für Anthropologie, 1*, 89–162.
White, T. D., & Folkens, P. A. (1991). *Human osteology*. San Diego: Academic.
White, T. D., Black, M. T., & Folkens, P. A. (2012). *Human osteology* (3rd ed.). Elsevier Academic Press.
Whittington, S. L. (1991). Detection of significant demographic differences between subpopulations of prehistoric Maya from Copán, Honduras, by survival analysis. *American Journal of Physical Anthropology, 85*(2), 167–184.
Wiechmann, I., & Grupe, G. (2005). Detection of *Yersinia pestis* DNA in two early medieval skeletal finds from Aschheim (Upper Bavaria, 6th century AD). *American Journal of Physical Anthropology, 126*(1), 48–55.
Willekens, F. (1977). The recovery of detailed migration patterns from aggregate data: an entropy maximizing approach. Laxenburg, IIASA (International Institute for Applied Systems Analysis), Research Memorandum RM-77-58.
Willekens, F., Por, A., & Raquillet, R. (1981). Entropy, multiproportional and quadratic techniques for inferring patterns of migration from aggregate data. Mathematical theories, algorithms, applications, and computer programs. Laxenburg: IIASA (International Institute for Applied Systems Analysis), Working Paper WP-79-88, pp. 83–124.
Wittwer-Backofen, U. (1987). Überblick über den aktuellen Stand paläodemographischer Forschung. *Homo, 38*, 151–160.
Wittwer-Backofen, U. (1991). Nekropole und Siedlung: Möglichkeiten und Grenzen der Rekonstruktion prähistorischer Bevölkerungsstrukturen. *Mitteilungen der Berliner Gesellschaft für Anthropologie, Ethnologie und Urgeschichte, 12*, 31–37.
Wittwer-Backofen, U., & Buba, H. (2002). Age estimation by tooth cementum annulation: Perspectives of a new validation study. In R. Hoppa & J. W. Vaupel (Eds.), *Palaeodemography. Age distributions from skeletal samples* (Cambridge studies in biological and evolutionary anthropology, Vol. 31, pp. 107–128). Cambridge: Cambridge University Press.
Wittwer-Backofen, U., Gampe, J., & Vaupel, J. W. (2004). Tooth cemetum annulation for age estimation: Result from a large known-age validation study. *American Journal of Physical Anthropology, 123*(2), 119–129.

Wood, J. W. (1998). A theory of preindustrial population dynamics: Demography, economy, and well-being in Malthusian systems. *Current Anthropology, 39*(1), 99–135.

Wood, J. W., & Dewitte-Aviña, S. (2003). Was the black death yersinial plague? *The Lancet Infectious Diseases, 3*, 327–328.

Wood, J. W., & Milner, G. R. (1994). Reply to Cohen's the osteological paradox reconsidered. *Current Anthropology, 35*, 31–637.

Wood, J. W., Holman, D. J., Weiss, K. M., Buchanan, A. V., & Lefor, B. (1992a). Hazards models for human population biology. *American Journal of Physical Anthropology,* Yearbook 35, *15* (suppl.), 43–87.

Wood, J. W., Milner, G. R., Harpending, H. C., & Weiss, K. M. (1992b). The osteological paradox: problems of inferring prehistoric health from skeletal samples. *Current Anthropology, 33*(4), 343–370.

Wood, J. W., Weeks, S. C., Bentley, G. R., & Weiss, K. M. (1994). Human population biology and the evolution of aging. In D. E. Crews & R. M. Garruto (Eds.), *Biological anthropology and aging: Perspectives on human variation over life span* (pp. 19–75). Oxford: Oxford University Press.

Wood, J. W., Holman, D. J., O'Connor, K. A., & Ferrell, R. J. (2001). *Mortality models for palaeodemography.* Seattle: University of Washington, Seattle population research center, Center for studies in demography and ecology, Working papers series n°2001–2002, 53).

Wood, J. W., Holman, D. J., O'Connor, K. A., & Ferrell, R. J. (2002). Mortality models for palaeodemography. In R. D. Hoppa & J. W. Vaupel (Eds.), *Palaeodemography. Age distributions from skeletal samples* (Cambridge studies in biological and evolutionary anthropology, Vol. 31, pp. 129–168). Cambridge: Cambridge University Press.

Woods, R. (1993). On the historical relationship between infant and adult mortality. *Population Studies, 47*, 193–219.

Wright, L. E., & Yoder, C. J. (2003). Recent progress in bioarchaeology: Approaches to the osteological paradox. *Journal of Archaeological Research, 11*(1), 43–70.

Wrigley, E. A., Davies, R. S., Oeppen, J. E., & Schofield, R. S. (1997). *English population history from family reconstitution, 1580–1837*. Cambridge: Cambridge University Press.

Yilmaz, U. (2003). Marked by life. *The influence of social and biological context on osteological sex traits.* Doctoral thesis, University of Tromsø, Archaeology.

Zaba, B. (1979). The four-parameter logit life table system. *Population Studies, 33*(1), 79–150.

Zerner, M. (1979). La population de Villeneuve-Saint-Georges et de Nogent-sur-Marne au IXe siècle, d'après le polyptique de Saint-Germain-des-Prés. *Annales de la Faculté de lettres de Nice, 37*, 17–24.

Zerner-Chardavoine, M. (1981). Enfants et jeunes au IXe siècle. La démographie du polyptyque de Marseille (813–814). *La Provence historique, XXXI*(126), 355–384.

Zvelebil, M. (1981). *From forager to farmer in the Boreal zone. Reconstructing economic patterns through catchment analysis in prehistoric Finland* (British archaeological reports international series, 115, 2 vol.). Oxford: British Archaeological Reports

Index

I. Séguy and L. Buchet, *Handbook of Palaeodemography*, INED Population Studies 2, DOI 10.1007/978-3-319-01553-8,

MIX
Papier aus verantwortungsvollen Quellen
Paper from responsible sources
FSC® C105338

If you have any concerns about our products,
you can contact us on
ProductSafety@springernature.com

In case Publisher is established outside the EU,
the EU authorized representative is:
Springer Nature Customer Service Center GmbH
Europaplatz 3, 69115 Heidelberg, Germany

Printed by Libri Plureos GmbH
in Hamburg, Germany